Industrial and Laboratory Measuring Systems

Sensors, Distributed, Modular and Wireless Systems

RIVER PUBLISHERS SERIES IN ENERGY MANAGEMENT

Series Editors

MICHELE ALBANO
Aalborg University,
Denmark

The "River Publishers Series in Energy Management" is a series of comprehensive academic and professional books focussing on management theory and applications for energy related industries and facilities. Books published in the series serve to provide discussion and exchange information on management strategies, techniques, methodologies and applications, with a focus on the energy industry.

Topics include management systems, handbooks for facility management, safety, security, industrial strategies, maintenance and financing, impacting organizational communications, processes and work practices. Content is also featured for energy resilient and high-performance buildings.

The main aim of this series is to serve as a useful reference for academics, researchers, managers, engineers, and other professionals in related matters with energy management practices.

Books published in the series include research monographs, edited volumes, handbooks and textbooks. The books provide professionals, researchers, educators, and advanced students in the field with an invaluable insight into the latest research and developments.

Topics covered in the series include, but are not limited to:

- Facility management;
- Safety and security;
- Management systems and solutions;
- Industrial energy strategies;
- Financing and costs;
- Energy resilient buildings;
- Green buildings management.

For a list of other books in this series, visit www.riverpublishers.com

Industrial and Laboratory Measuring Systems

Sensors, Distributed, Modular and Wireless Systems

Josef Vedral

Czech Technical University in Prague,

Czech Republic

River Publishers

Routledge
Taylor & Francis Group
NEW YORK AND LONDON

Published 2024 by River Publishers
River Publishers
Alsbjergvej 10, 9260 Gistrup, Denmark
www.riverpublishers.com

Distributed exclusively by Routledge
605 Third Avenue, New York, NY 10017, USA
4 Park Square, Milton Park, Abingdon, Oxon OX14 4RN

Industrial and Laboratory Measuring Systems: Sensors, Distributed, Modular and Wireless Systems / Josef Vedral.

Routledge is an imprint of the Taylor & Francis Group, an informa business

ISBN 978-87-7004-150-8 (hardback)
ISBN 978-87-7004-244-4 (paperback)
ISBN 978-87-7004-246-8 (online)
ISBN 978-8-770-04245-1 (ebook master)

Contents

Annotation

The book describes the types and properties of computer-controlled laboratory and industrial measuring systems for collecting and processing signals of typical physical quantity sensors. An overview of the properties of these sensors used in laboratory and industrial environments is provided. The following chapters describe the properties of computers used in measurement systems, including plug-in cards and the IEEE-488 protocol. The following chapters describe the types and properties of distributed, modular, data networks, and wireless measuring systems. The final chapters describe methods of digitization, reconstruction, and signal processing in measurement systems in time and frequency domains.

The book is intended for users and designers of digital measuring systems, working in industry, scientific, and research workplaces and students of university relevant technical focus.

Preface

The book was created on the basis of my pedagogical and research and development activities at the department of measurement of the *Faculty of Electrical Engineering of the Czech Technical University in Prague* and in the design and development laboratories of the companies *Laboratory Instruments in Prague* and the *Research Institute of Nuclear Technology Instruments*, where the author devoted himself to the development of measurement systems for data collection and their processing, especially in areas of gas and liquid chromatography, and nuclear spectroscopy systems.

In connection with the publication of the book, I thank assoc. prof. Ing. Jiří Novak, Ph.D., assoc. prof. Ing. Jaroslav Roztočil, CSs, and Ing. Martin Šimůnek from the *Department of Measurement of the Czech Technical University in Prague* for useful consultations. I also thank prof. Ing. Aleš Prochazka, CSc, member of IEEE Society and teacher from the *Department of Mathematics, Informatics and Cybernetics of the University of Chemical Engineering* in Prague for a careful review of the book.

Finally, I would like to thank my family, especially my wife Helena, for moral and practical help in writing and editing the book.

In Prague 2024
Josef Vedral

List of Figures

List of Tables

1

Specifics of Measuring Systems

Computer-controlled industrial measurement systems are designed to collect, process, and transfer measured data from sensors of physical quantities to a computer. The transmission speed of all these systems is determined by the usable bandwidth (BW) of their transmission channel and its signal-to-noise ratio (SNR),

$$CT = BW.\log_2(1 + SNR) = BW.\log_2\left(1 + \frac{P}{N}\right), \qquad (1.1)$$

which is determined by the ratio of the power of the useful component of the transmitted signal P to the power of its noise component N, which has a power of $2^{-n} + 2/\sqrt{12}$ with n-bit coding.

The transmission rate is given either in *bits per* 1s (b/s) or *bytes* per 1s (B/s), where 1 *byte* represents one 8-bit word. For example, transferring an 8-bit word at 1 MB/s corresponds to a transfer rate of 8 Mb/s.

Measuring systems date their existence to the beginning of the seventies of the twentieth century. Since then, in addition to independent digital measuring devices (stand-alone), measuring systems consist of several measuring devices, whose cooperation was controlled via the serial–parallel bus IEE-488, or its equivalent IEC 625. This bus, originally referred to as HP-IB, was designed in 1975 by the Hewlett Packard company and was gradually adopted by other measuring equipment manufacturers. In 1987, the IEEE-488.1 standard was published, followed by the new IEEE-488.2 standard, defining their functional and operational characteristics. This 8-bit bus enables the connection of up to 14 functional units and achieves a data transfer rate of up to 1 Mb/s with a maximum length of 2 m connecting cables.

The CAMAC (*computer automated measurement and control*) system is considered to be one of the oldest centralized measurement systems, which has found application especially in nuclear technology in the collection and processing of data from nuclear radiation sensors. This system was standardized by the EUR 4100 standard in 1972 and by the IEEE 583 standard in

1975. The CAMAC system has a 24-bit data bus, allowing data transfer at speeds up to 1 Mb/s and up to 10 Mb/s when using block transfers. In the following years, VME (*Versa Modul Europe bus*) modular measurement systems began to be used with 16- or 32-bit bus, enabling the connection of up to 20 functional modules with a transfer speed of up to 40 Mb/s. Currently, the most powerful centralized systems are the VXI and PXI systems with 32- or even with 64-bit buses, enabling the connection of up to 12 functional modules with a transfer speed of up to 264 Mb/s.

In addition to these centralized systems, decentralized measuring systems with serial data transmission are used in test laboratories and industrial environments. The original systems with an RS-232 bus with a transmission speed of up to 20 kB/s and a maximum length of connecting cable of hundreds of meters were replaced in the late 1980 by the industrial standard RS-485, enabling the connection of up to 31 functional units with a transmission speed of up to 10 Mb/s and maximum connection length of 1.2 km. To effectively suppress interfering signals, the RS-485 interface is galvanically isolated by impulse transformers or optically coupled isolators.

In 1996, the USB (*universal serial bus*) serial interface was launched, which is supported by the USB-IF implementation forum. Its original version USB 1.0 (1996) with four wires and one-way data transfer (simplex transfer) had a transfer rate of 1.5 Mb/s (187.5 kB/s), and the USB 1.1 version (1996) had a transfer rate of up to 12 Mbps (1.5 Mb/s). The faster version USB 2.0 (2001) allows data to be transferred at speeds of up to 480 Mb/s (60 Mb/s). Version USB 3.0 (Superspeed USB) with two-sided data transfer (Duplex Transfer) (2011) allows data to be transferred at a speed of up to 5 Gb/s (625 Mb/s), version USB 3.1 Gen. 2 has a transfer rate of up to 10 Gb/s, the USB 3.2 version with double-sided transfer DT (*double duplex*) has a transfer rate of up to 20 Gb/s, and the fastest USB 4.0 version (2019) reaches transfer rates of up to 40 Gb/s.

A special group of measuring systems consists of systems with plug-in cards that are inserted into the internal buses of personal computers. Older computers with ISA buses (PC/104), or PC Card (PCMCIA) with 16-bit buses, enabled data transfer with a speed of 8 or 20 Mb/s. Modern PCI bus (IPIC), or Compact PCI buses with 32- to 64-bit data buses allow data transfer with a transfer rate of up to 264 Mb/s.

Wireless radio modem networks most often use GSM transmission networks (*global system mobile*) for data transmission. Either infrared rays or high-frequency connections such as Bluetooth, Wifi, ZigBee, and LoRAN are used to transmit data over short distances.

An important criterion in the design and operation of measurement systems is the organization of the flow of information in measurement systems.

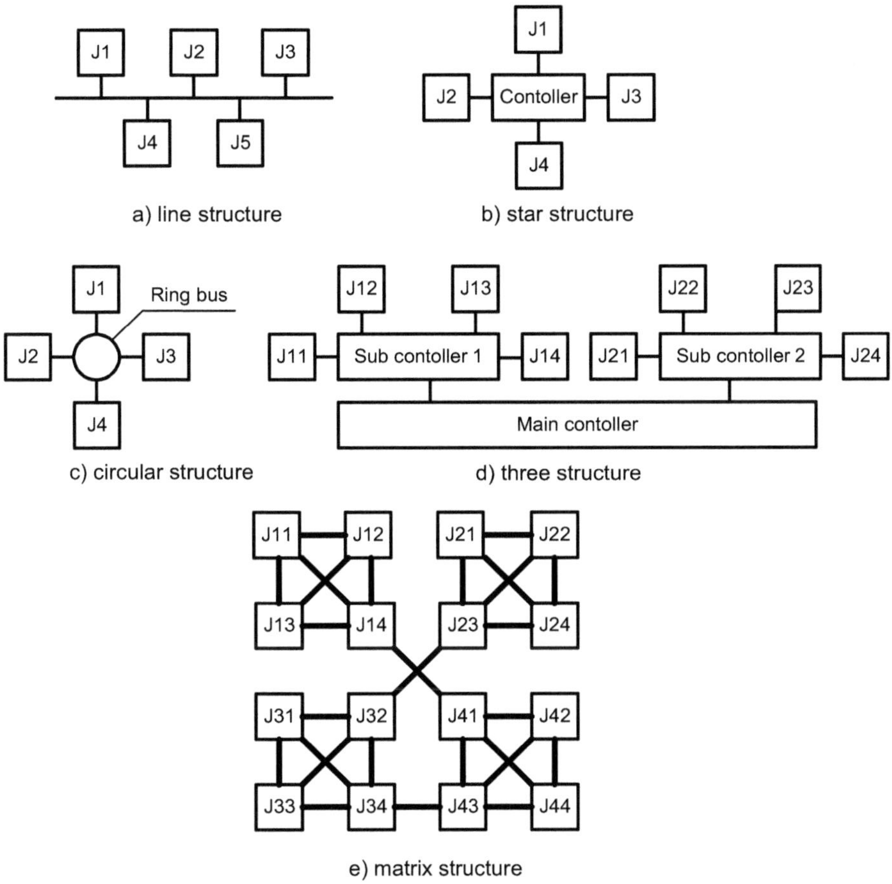

Figure 1.1 Distribiuting of measuring units.

We recognize serial and parallel data transfer methods. In the serial method of data transmission, data is transmitted by a series of two-valued signals (bits). In the parallel method of data transfer, data is transferred in several bit words (bytes), referred to as byte. Therefore, for the serial method of transmission, the transmission speed is given in bit/s, and for the parallel method of data transmission, also in units of byte/s.

From the point of view of the location of the *measuring units* in the measuring system, we recognize the bus, star, ring, and matrix structures of their arrangement (Figure 1.1).

The *line structure*, which is most often used in simple measurement systems, has one common data and control bus and one controller. This arrangement is not suitable for systems with a larger number of units, since only one unit is always the transmitter and the other units are the receivers. In addition, this structure does not allow high-speed data transmission.

The *star structure* allows the units to communicate only through the central unit (controller). The advantage of the structure is its greater reliability, as the controller excludes a faulty unit from the communication of other units.

A *ring structure* uses a two-point connection of adjacent units, each of which has one input and one output. The disadvantage of this structure is that in the event of a failure of one of the units, the communication of the other units is disabled.

The *tree structure* is an extended version of the star and ring structure. Communication between secondary units is provided by secondary buses with lower transmission speed requirements, and communication between main controllers is ensured by the main bus with high transmission speed. The matrix structure, used in large-scale distributed measurement systems, allows units to be connected by several independent paths, which increases the resistance of these systems to simultaneous failure even in several buses.

More complex measurement systems are constructed hierarchically. At the lowest level, measurement subsystems are arranged, which are located in the immediate vicinity of the measurement points. Data from these subsystems is sent to the main system controller. It not only receives the initially processed measured data but can also send commands related to the execution of the measurement procedure. The master controller of the measurement system can also use memory resources, display data, and data storage devices.

Figure 1.2 shows the hierarchical structure of the measuring system, where one of the subsystems is formed by an industrial RS-485 interface and the other uses a USB interface for data transfer. Both of these subsystems are connected to the superior *main controller* of the measuring system via a LAN (*local area network*).

Each of these interfaces provides pairing and synchronization functions, data transfer functions, and their management. The *pairing function* consists in assigning individual measuring system units with the same interface. The *synchronization function* ensures time synchronization during data transfer. In asynchronous data transmission, the data is generated by a frequency of the transmitter, which is different from the frequency of the receiver. In synchronous data transmission, both of these frequencies are the same and time-synchronized. Synchronous data transfer therefore enables faster and more reliable data transfer.

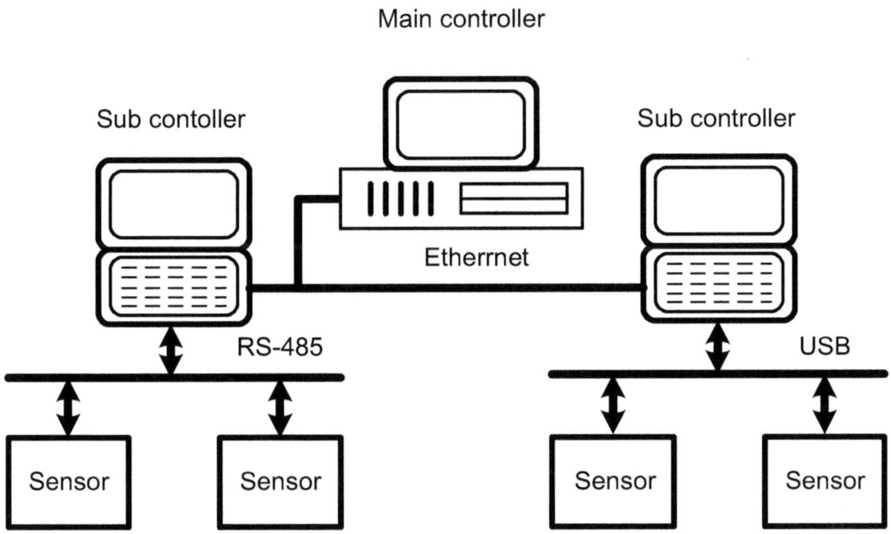

Figure 1.2 Hierarchical structure of the measurement system.

An important feature of data transmission in any transmission system is data transmission reliability and security against unauthorized interception. These requirements are ensured in classic modular and distributed measurement systems by using check parities, by confirming received messages with the ARG (*automatic repeat request*).

In modern transmission measurement systems, *cyclic redundant codes* (CRC) are used to increase the reliability of data transmission, allowing to reduce the error rate to 10^{-9}. A very effective means of illegal interception of data is the use of quadrature phase pulse modulations QPSK (*quadrature phase shift keying*) and methods with a spread frequency spectrum DSS (*direct sequence system*), using for wireless data transmission frequency modulation of carrier signals pseudo-randomly generated by binary signals PRBFM (*pseudo-random binary frequency modulation*).

Sensors of physical quantities are an integral part of most measuring systems, which are designed to convert their values into electrical signals such as voltage and current. Each measuring system is thus made up of sensors, measuring circuits and interface circuits, enabling the transfer of data to the DAQ (*data acquisition unit*).

The sensors themselves are made up of sensors that convert measured physical quantities into measuring quantities, such as voltage, current, charge, electrical resistance, capacity and inductance, and frequency. These quantities are transformed into voltages or currents in measuring circuits and

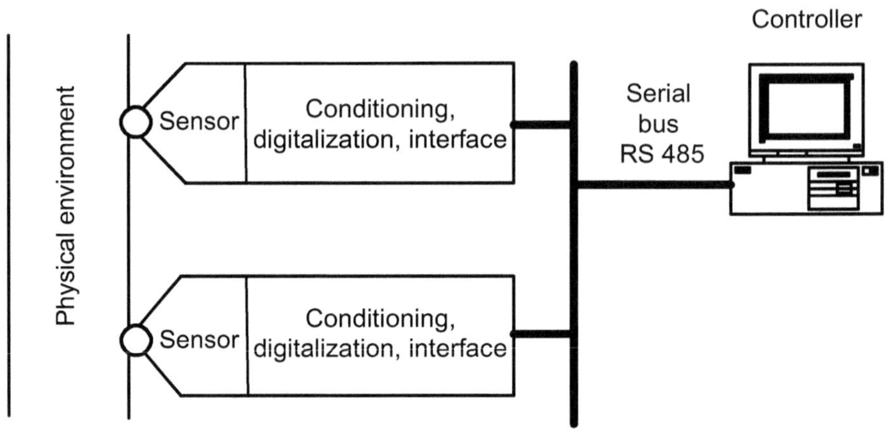

Figure 1.3 Block diagram of the measuring chain with data transfer via RS 485 serial line.

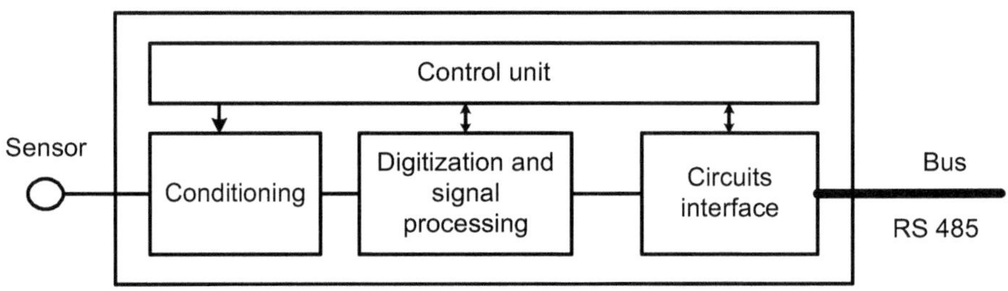

Figure 1.4 Connection of intelligent sensors with RS-485 interface.

subsequently digitized by built-in A/D converters, whose data are transmitted through interface circuits via local or centralized bus measuring systems to collection units, which are usually industrial or personal computers (see Figure 1.3).

Smart sensors are sensors that contain in one compact unit (housing) not only sensors, measuring circuits, but also processor circuits, enabling not only digitization and processing of signals but also their transmission via data channels, e.g., RS-485 and IEEE-1451, to a superior computer controller (server) (Figure 1.4).

Signal processing in these sensors usually involves correction of errors and non-linearities of their transfer characteristics. Sophisticated intelligent sensors also enable self-diagnosis and recording of measured data in the internal data memory.

In addition, the software of the measuring systems enables the use of advanced methods of processing measured data and their visualization in the superior controllers. These methods include, for example, statistical evaluation of measured data, their time and frequency analysis using discrete Fourier transformation, and adaptive filtering. Some software equipment of measuring systems also allow the use of advanced diagnostic methods, using, for example, elements of *artificial intelligence*.

References

[1] Ebelin, J. *Measurement system application and design*. Mc. Graw Hill Publishing Company, 1990. ISBN 0-07017-338-9.
[2] Garrett, P. H. *Advanced Instrumentation and Computer I/O Design*. Real-Time System Computer Interface Engineering, John Viley, 2000. ISBN 978-0-780-36013-6.
[3] Boyes, W., (ed.), *Instrumentation Reference Book*. 3rd ed., Boston, MA: Butterworth-Heinemann, 2003.
[4] Park, J., Mackey, S. *Practical Data Acquisition for Instrumentation and Control System*. Amsterdam: Elsevier, 2003. ISBN 978-0750-65796-9.
[5] Nawrocki, W. *Measurement systems and sensors*. Second edition, Artec House, 2016. ISBN: 1-58053-945-9.

2

Typical Sensors of Physical Quantities

Sensors of physical quantities are an integral part of laboratory and industrial measuring systems. Sensors for position, displacement and rotation, mechanical stress, force, vibrometric quantities, height of liquid levels, bulk materials, pressure and flow of liquids, temperature and thermal energy, and electrical and magnetic quantities are most often used in these environments. In addition to these sensors, chemical sensors for the acidity of solutions, gas analyzers, and radioactive radiation sensors are also used in laboratories and industrial environments.

2.1 Position Sensors

Position sensors – potentiometric, inductive, capacitive, magnetostrictive, ultrasonic, and optoelectronic sensors – are used for combined position sensing.

2.1.1 Potentiometric position sensors

These sensors use sensing potentiometers with a linear dependence of resistance on the position of their slider to sense longitudinal or circular movement. The resistance path of these sensors is formed either by a winding with a resistance wire with a diameter of several tenths of an mm, or by a conductive resistance path, steamed on a fiberglass substrate. The advantages of potentiometric sensors are their simplicity, low noise, and long service life.

Potentiometric sensors of longitudinal movement are produced in lengths ranging from a few cm to tens of cm and in the range of nominal resistances from units to hundreds of kΩ. These sensors achieve a resolution of up to 10^{-6} and their non-linearity is less than 10^{-5}. Circular motion sensors are produced in single or multi-turn design also in the range of nominal resistances from units to hundreds of kΩ. Their resolution reaches up to 10^{-6} angular degrees and their non-linearity is less than 10^{-5} angular degrees.

a) line potenciometric sensor b) circular potentiometric sensors

Figure 2.1 Potentiometric position sensors.

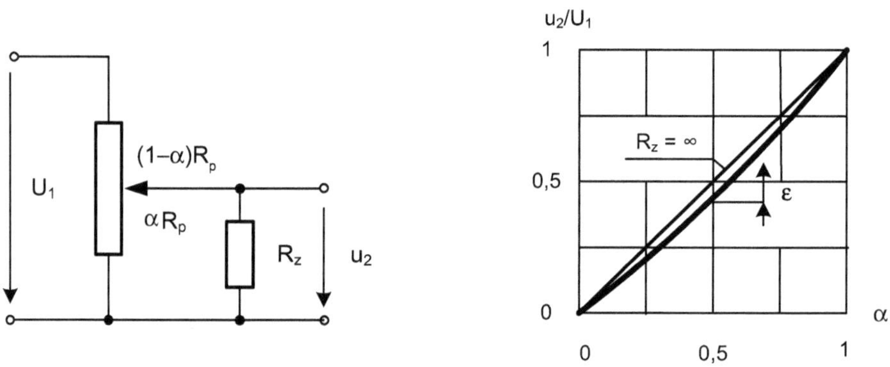

a) loaded potentiometer b) conversion characteristic of the potentiometer

Figure 2.2 Replacement diagram of a loaded potentiometer and its conversion characteristic.

Figure 2.1 shows typical potentiometric longitudinal and circular motion sensors.

If the sensing potentiometer with nominal resistance R_p is supplied with voltage U_1, then when its output is loaded with resistance R_z, there is a voltage on the slider of the potentiometer (Figure 2.2)

$$u_2 = \frac{\alpha}{1 + \alpha \frac{1 - \alpha}{k}} U_1, \tag{2.1}$$

where α determines the position of the potentiometer slider and $k = R_z / R_p$.

The maximum proportional deviation of the transfer characteristic of the sensor from its linear course occurs at the position of the slider $\alpha = 0.5$ and is

$$\varepsilon_m \left(\alpha = 0.5\right) = \frac{0.25}{k + 0.25}. \tag{2.2}$$

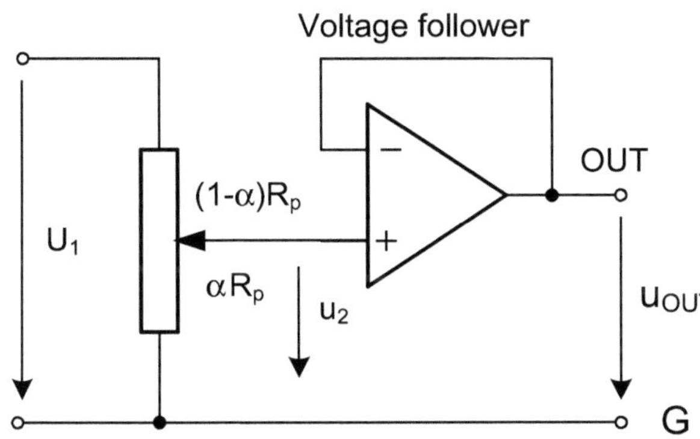

Figure 2.3 Connecting the sensing potentiometer to the voltage follower.

Isolation amplifiers with high input impedance are therefore used to sense the voltage on the potentiometer slider. For impedance separation, in the simplest case, a connection with an operational amplifier in the function of a *voltage follower* with a nominal gain of +1 is used (Figure 2.3).

When using BiFET operational amplifiers with JFET input transistors whose input resistance ranges from tens to hundreds of GΩ, non-linearity of the order of 10^{-5} can be achieved with nominal resistances of potentiometers in the range of tens to hundreds of kΩ.

The input residual voltage U_0 and input current I_{BP} of the operational amplifier at its non-inverting input cause the proportional additive error of the sensor transfer characteristic

$$\varepsilon_0 = \frac{U_0 + \alpha \cdot I_{BP} \cdot R_p(1-\alpha)}{U_1}, \tag{2.3}$$

which has a maximum value of $e_{0m} = (U_0 + 0.25 \cdot I_{BP} \cdot R_p)/U_1$ at the slider position $\alpha = 0.5$.

The *differential amplification* A_D and *common mode rejection ratio*, CMRR, of the op-amp causes the proportional *multiplicative error* of the transfer characteristic of the sensor

$$\varepsilon_m = \frac{1}{A_D} + \frac{1}{\mathrm{CMRR}}. \tag{2.4}$$

If, for example, the voltage tracker has an input resistance of 1 GΩ, then this resistance causes a maximum proportional non-linearity of the sensor conversion characteristic of $2.5 \cdot 10^{-5}$ at the nominal resistance of the potentiometer of 100 kΩ. If this potentiometer is connected to a voltage source of 5 V, the input residual voltage of the operational amplifier is 100 μV, and its input

Table 2.1 Parameters of low-offset operational amplifiers.

Type	A_D	CMRR	U_0	dU_0/dq	I_0	u_n (Df = 1 Hz)
MAX4238	100 dB	120 dB	100 nV	10 nV/°C	20 pA	25 nV
LTC2050	140	145	0.5 μV	10 nV/°C	12 pA	22 nV
AD8628	140	130	1 μV	2 nV/°C	15 pA	22 nV
LT1028	130 dB	120 dB	10 μV	200 nV/°C	12 pA	1 nV
TLE2027	130 dB	140 dB	10 μV	200 nV/°C	15 pA	2.5 nV
OPA277	120 dB	135 dB	20 μV	100 nV/°C	35 pA	8 nV
AD797	140 dB	145 dB	25 μV	200 nV/°C	100 nA	0.9 nV
MAX4475	135 dB	135 dB	70 μV	300 nV/°C	65 nA	4.5 nV

current is 50 pA, then the relative additive error of the transfer characteristic of the sensor is $3 \cdot 10^{-5}$.

For example, the *differential amplification* of the operational amplifier $A_D = 10^5$ and its *CMRR* = 100 dB cause a proportional multiplicative error of the sensor conversion characteristic of $2 \cdot 10^{-5}$. The parameters of typical operational amplifiers with high differential gain A_D, high *CMRR* suppression factor, low input residual voltage U_0, and input residual current I_0, including their spectral noise voltages u_n, are shown in Table 2.1.

If the sensing potentiometer is far away from the amplifier, not only the resistances of its leads but also DC- and AC-induced interference signals can be applied. In this case, the sensing potentiometer is connected to a bridge formed by the sensing potentiometer and identical resistors (Figure 2.4).

Bridge differential voltage

$$u_D = u_2 - u_1 \tag{2.5}$$

is amplified by the *differential gain* of the instrumentation amplifier G_D.

The *common voltage*, given by the mean value of both input voltages

$$u_C = \frac{u_1 + u_2}{2}, \tag{2.6}$$

is suppressed by the factor CMRR of the instrument amplifier, which is determined by the ratio of its *differential gain* $G_D = u_{\mathrm{OUT}}/u_D$ to the *common gain* $G_C = u_{\mathrm{OUT}}/u_C$

$$\mathrm{CMRR} = \frac{G_D}{G_C}. \tag{2.7}$$

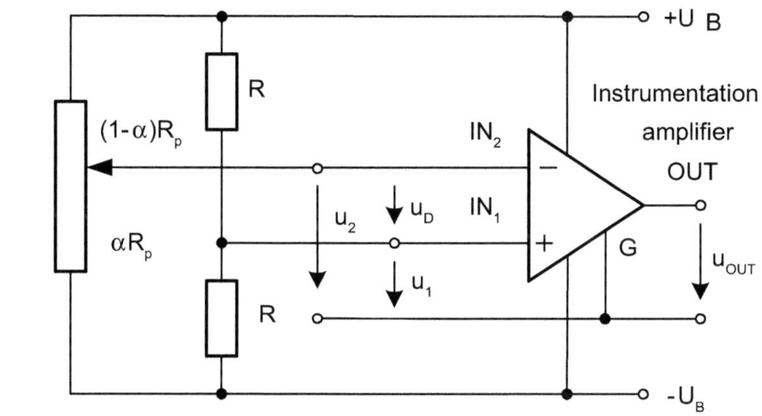

Figure 2.4 Connecting the sensing potentiometer to the instrumentation amplifier.

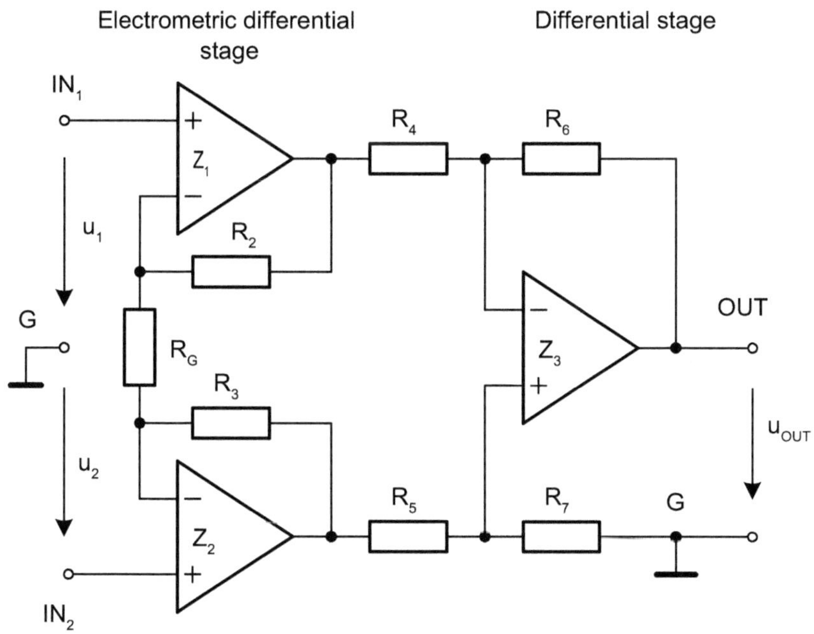

Figure 2.5 Symmetrical instrumentation amplifier.

In the simplest case, the *instrument amplifier* consists of a symmetrical differential electrometric stage with high input resistances and a differential amplifier with paired resistors $R_2 = R_3$, $R_4 = R_5$, and $R_6 = R_7$ (Figure 2.5).

The *differential gain* of this symmetrical instrumentation amplifier is determined by the product of the differential gains of both parts of the instrumentation amplifier

Table 2.2 Parameters of typical instrument amplifiers.

Type	G_D	**CMRR**	U_0	I_0	dU_0/dq
INA115	$1–10^3$	86 dB	2 mV	25 pA	5 μV/°C
INA111	$1–10^3$	100 dB	1 mV	20 pA	5 μV/°C
LTC1100	10–100	90 dB	10 μV	65 pA	100 nV/°C
AD8221	$1–10^3$	90 dB	25 μV	400 pA	0.3 μV/°C
AD8225	$1–10^3$	86 dB	150 μV	1 nA	2 μV/°C
LTC2053	1	105 dB	10 μV	10 nA	5 μV/°C

$$G_D = \left(1 + 2\frac{R_2}{R_G}\right) \cdot \frac{R_6}{R_4}. \tag{2.8}$$

The factor CMRR of the *instrumentation amplifier* is determined by the pair mismatch of the resistances of the resistors and the suppression factor of the operational amplifiers themselves. The differential gains of G_D instrumentation amplifiers are usually adjustable with internal resistors in binary or decimal ranges of $1–10^3$, i.e., 0–60 dB.

Table 2.2 shows the parameters of typical monolithic instrument amplifiers.

2.1.2 Inductance sensors

Inductive sensors with an open and closed magnetic circuit are used to sense the position.

Inductance sensors with an open magnetic circuit use the dependence of the inductance of the coil on its distance from the electrically conductive object whose movement is being sensed. The eddy currents caused by the high-frequency magnetic field in this material create a secondary magnetic field in it, the intensity of which opposes the intensity of the magnetic field of the excitation coil, causing a decrease in the impedance of the sensing coil.

If the excitation coil is also the coil of the oscillating circuit, amplitude modulation of the oscillating voltage occurs during the oscillating movement of the measured object. By demodulating this course, we obtain a voltage that corresponds to the time course of the measured object's deflection. In order to eliminate external interference, the sensing coil including the demodulator is placed in an electromagnetically shielded cover (Figure 2.6).

Inductive sensors with an open magnetic circuit are used not only as *inductive proximity sensor* but also for non-contact measurement of their distance l from a moving object. In both cases, the measurement range is limited by the *Skin effect*, in which, during alternating magnetization of the

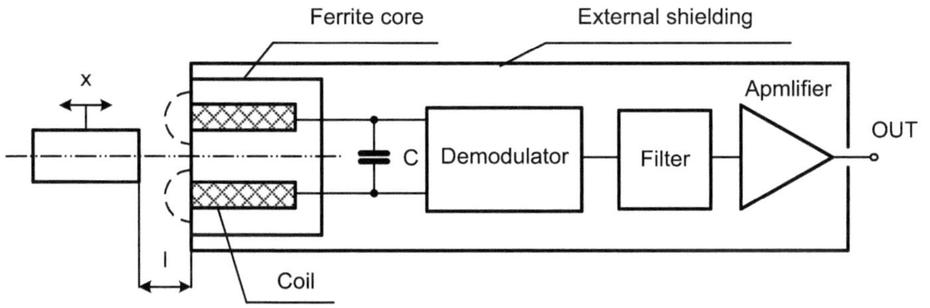

Figure 2.6 Inductance sensor with an open magnetic circuit.

Table 2.3 Penetration depths of typical metals.

f [kHz]	δ_{Fe} [mm]	δ_{Al} [mm]	δ_{Cu} [mm]
0.05	0.88	11.2	8.9
0.1	0.68	8.5	6.6
1	0.2	2.7	2.1
10	0.08	0.84	0.66
100	0.02	0.3	0.2

conductor, the intensity of the magnetic field decreases exponentially from the surface of the conductor to its center. In connection with this phenomenon, the *penetration depth* is defined

$$\delta = \sqrt{\frac{2 \cdot \rho}{2 \cdot \pi \cdot f \cdot \mu}}, \tag{2.9}$$

at which the intensity of the alternating electromagnetic field drops to a value of $1/e$, i.e., to 37% of the intensity of this field on the surface of the material. The penetration depth depends on the frequency f, the relative conductivity ρ, and the permeability μ of the conductor. When using these inductive sensors, the frequency of the excitation signal must therefore be chosen so that the sensor data does not depend on the thickness of the measured object d, i.e., so that $d >> \delta$ applies. Table 2.3 shows the penetration depths of the most common metals that are used in non-contact measurement of movement.

In terms of eliminating the influence of the stray alternating magnetic field on the function of induction sensors, these sensors are produced in two different versions. In the coupling version, the sensor coil is embedded in an electromagnetically shielded cylindrical cover, through which the sensor can also be fixed to an electrically conductive mounting panel. If the support panel is made of electrically non-conductive material, a non-coupled version of the sensor is used, in which the sensing coil is not shielded (Figure 2.7).

a) sensor in binding design b) sensor in non-binding design

Figure 2.7 Examples of coupling and non-coupling design of inductance sensors.

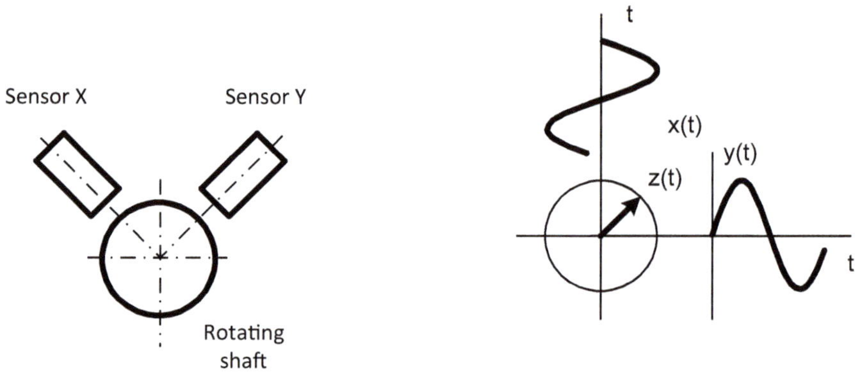

a) location of ferromagnetic probes b) orbit of the rotating shaft

Figure 2.8 Measuring the kinematic path of a rotating shaft.

If, for example, we place two of these sensors to the ferromagnetic shaft of a rotating machine so that their axes intersect at 90°, it is possible to evaluate the kinematic path of the machine shaft, its orbit, based on the display of their signals (Figure 2.8).

From the course of the kinematic path of the shaft, it is possible to determine the instantaneous values of deflections x, y in mutually perpendicular directions and the instantaneous kinematic deflection of the shaft, given by the vector sum of both deflections

$$z(t) = \sqrt{x^2(t) + y^2(t)}. \tag{2.10}$$

Inductance sensors with a closed magnetic circuit use the change in inductance of coils wound on a ferromagnetic core, whose magnetic flux is interrupted by an air gap of variable length, or flat (Figure 2.9).

Assuming that the stray magnetic flux of the magnetic circuit is negligible and its yoke moves in the x-direction, i.e., at $d_2 = 0$, it applies to the magnetic reluctance of the magnetic circuit

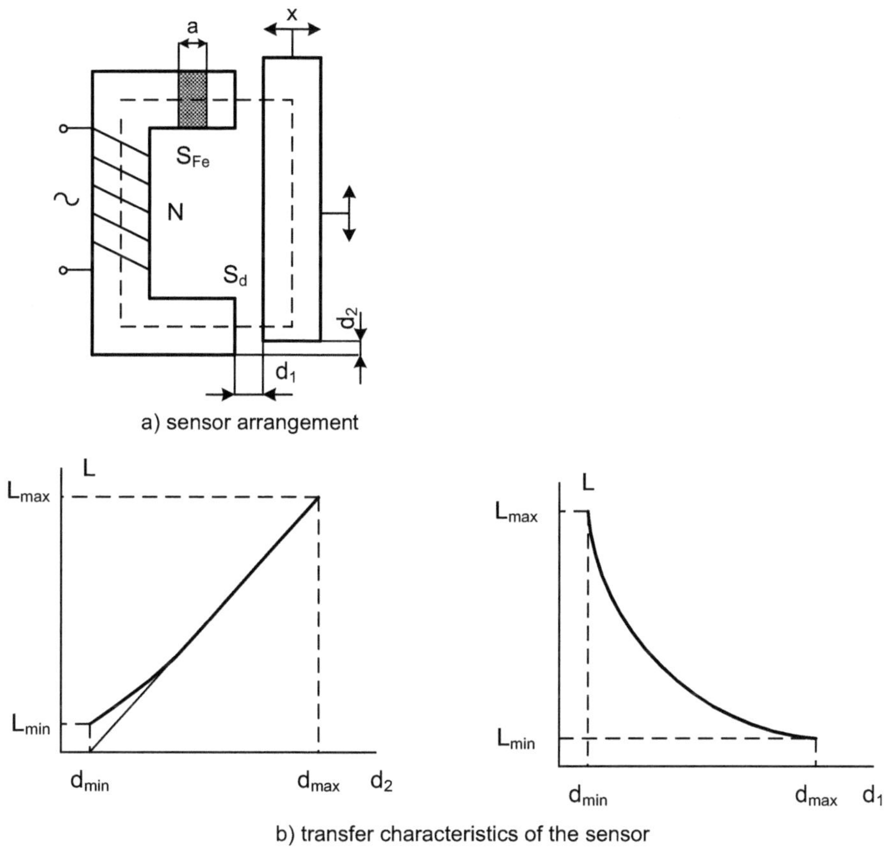

Figure 2.9 Inductance sensor with a ferromagnetic core.

$$R_m = \frac{l_{\mathrm{Fe}}}{\mu_0 \cdot \mu_{\mathrm{Fe}} \cdot S_{\mathrm{Fe}}} + 2\frac{d_1}{\mu_0 \cdot S_d}. \tag{2.11}$$

where l_{Fe} is the mean length of the magnetic field lines of the magnetic circuit, $\mu_0 = 4 \cdot \pi \cdot 10^{-7}$, H/m is the vacuum permeability, μ_{Fe} is the relative permeability of the ferromagnetic core, S_{Fe} is the cross section of the ferromagnetic core, and S_d is the cross section of the air gap. Since the relative permeability of the core $\mu_{\mathrm{Fe}} >> \mu_0$, the inductance of a coil with N turns, wound on a ferromagnetic core, is

$$L \approx \frac{N^2}{R_m} \approx \frac{N^2}{2 \cdot d_1}\mu_0 \cdot S_d. \tag{2.12}$$

The hyperbolic dependence of the inductance of the coil in the range of changes $d_1/\Delta d_1 < 0.1$ can be approximated by its linear dependence with a root mean square deviation of 1%.

If the displacement of the yoke in the *y*-direction is used to measure the displacement, then at a constant distance d_1, the inductance of the coil is directly proportional to the displacement d_2

$$L \approx \frac{N^2}{R_m} \approx \frac{N^2}{2 \cdot d_1} \mu_0 \cdot a \cdot d_2, \qquad (2.13)$$

where *a* is the width of the pole extensions of the ferromagnetic circuit.

Differential inductance sensors with a ferromagnetic core form a symmetrical ferromagnetic core, between whose poles a ferromagnetic yoke moves (Figure 2.10).

Under the same assumptions as in the previous case, it applies to the inductance of anti-series connected coils

$$L \approx \frac{N^2}{d_0 + \Delta d} \mu_0 \cdot S_d \approx N^2 \cdot \mu_0 \cdot S_d \cdot (d_0 - \Delta d). \qquad (2.14)$$

Compared to the previous sensor, these differential sensors have twice the sensitivity $\Delta L/\Delta d$ and are therefore suitable for sensing even small changes in movement in the range of up to a few millimeters.

2.1.3 Transformer position sensors

Linear variable differential transformer (LVDT) linear motion transformer sensors are made up of a primary coil with N_1 turns and two symmetrically placed secondary coils with the same number of turns $N_2 = N_3$, whose outputs are connected opposite to each other. A ferromagnetic core moves between these systems of coils (Figure 2.11).

When alternating current *I* passes through the primary coil of the transformer sensor, alternating voltages are induced in its secondary coils

$$U_1 = \omega \cdot M_{12} \cdot I, \qquad U_2 = \omega \cdot M_{13} \cdot I, \qquad (2.15\text{ a, b})$$

the difference of which is directly proportional to the difference in mutual inductances M_{12} and M_{13} between the primary and secondary coils of the transformer

$$\Delta U = U_1 - U_2 = \omega \cdot (M_{12} - M_{13}) \cdot I. \qquad (2.16)$$

The dependence of the differential voltage is, to a certain extent, directly proportional to the position of the ferromagnetic core, while when the position of the core changes around the zero value ($\Delta x = 0$), there is a change in the

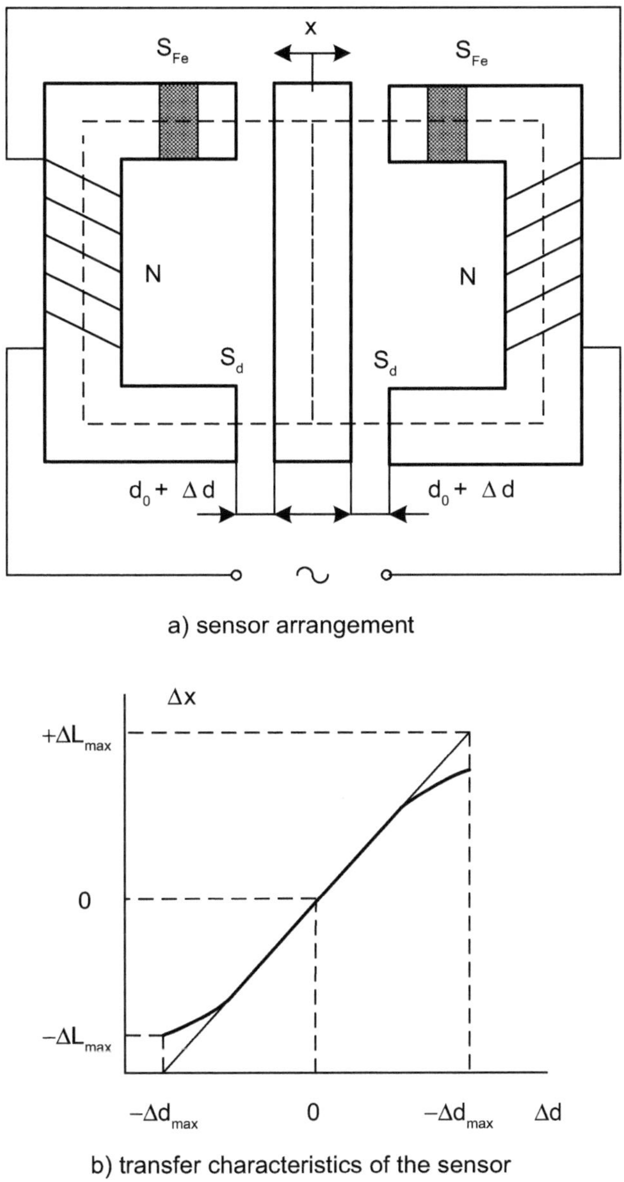

Figure 2.10 Differential inductance sensor with a ferromagnetic core.

phase shift between the differential voltage and the excitation current of the primary coil of the transformer.

Synchronous detectors are used to rectify the differential alternating voltage of the sensing secondary coils, which rectify only that component of the alternating voltage whose frequency is equal to the frequency of the

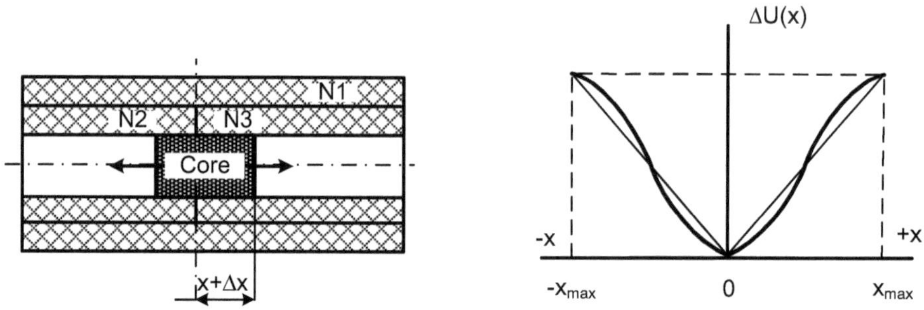

a) LVDT sensor b) transfer characteristic of LVDT sensor

Figure 2.11 Linear differential transformer position sensor.

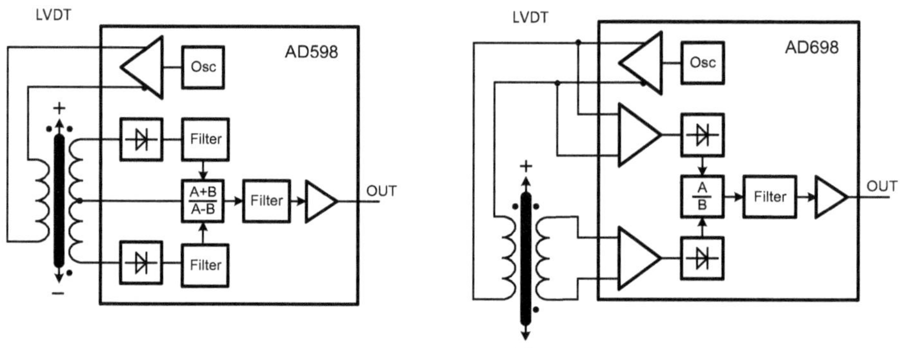

a) five-wire connection b) four-wire connection

Figure 2.12 Methods of connecting LVDT sensors.

excitation current and suppress all other higher harmonic components of the voltage of the sensing coils, arising from the non-linear course of the magnetization curve of the ferromagnetic core. Examples of connecting LVDT sensors with AD598 and AD698 circuits are shown in Figure 2.12.

Both circuits contain a sine oscillator, the frequency of which can be adjusted according to the type of LVDT sensor in the range of several units to tens of kHz. In the AD598 circuit, the output voltages of the LVDT secondary windings are first rectified, then added and subtracted, and then this sum is divided by their difference. In the AD698 circuit, the voltages of the primary and secondary coils of the LVDT sensor are first rectified and then divided. Both of these parametric operations increase the linearity of the conversion characteristics of transformer sensors, which reaches up to several tenths of a percent in the range of measured lengths of several tens of mm to tens of cm.

a) LVDT sensor b) RVDT sensor

Figure 2.13 Transformer sensors.

In addition to these transformer sensors of linear motion, transformer sensors of circular motion *rotary variable differential transformer* (RVDT) are also produced, whose angular accuracy reaches up to several tenths of an angular degree (Figure 2.13).

Inductosyns are transformer inductance sensors that are made up of a fixed coil in the shape of a meander with a pitch of turns p and two moving coils also in the shape of meanders. These coils are made by depositing conductive layers on glass, ceramic, or fiberglass insulating rulers. When the sensing coils move over the fixed coil, due to their variable mutual inductances, alternating voltages are generated, the phase shift of which always changes by one-fourth of their pitch (Figure 2.14).

To determine the displacement of moving coils with the same number of meanders relative to the fixed coil, current excitation of the fixed coil is used, which induces alternating voltages in the moving coils

$$u_1(t) = U_1 \cdot \sin(\omega t) \cdot \sin\varphi, \qquad u_2(t) = U_1 \cdot \sin(\omega t) \cdot \cos\varphi, \qquad (2.17\text{ a, b})$$

phase shifted by the phase angle

$$\phi = \frac{T_\varphi}{T_\mathrm{p}} = 2 \cdot \pi \cdot \frac{d}{p}, \qquad (2.18)$$

where T_φ is the time shift of both voltages and T_p is the time of their periods.

Inductosyn position sensors are produced in lengths of several tens of cm to units of m, their resolution reaches tens of μm, and the maximum speed of movement reaches units of m/s. Inductors are manufactured to detect circular motion, the meander winding of which is applied to non-magnetic circular discs (Figure 2.15).

The achievable resolution of circular inductosyns reaches up to several thousandths of angular degrees at a maximum number of revolutions up to

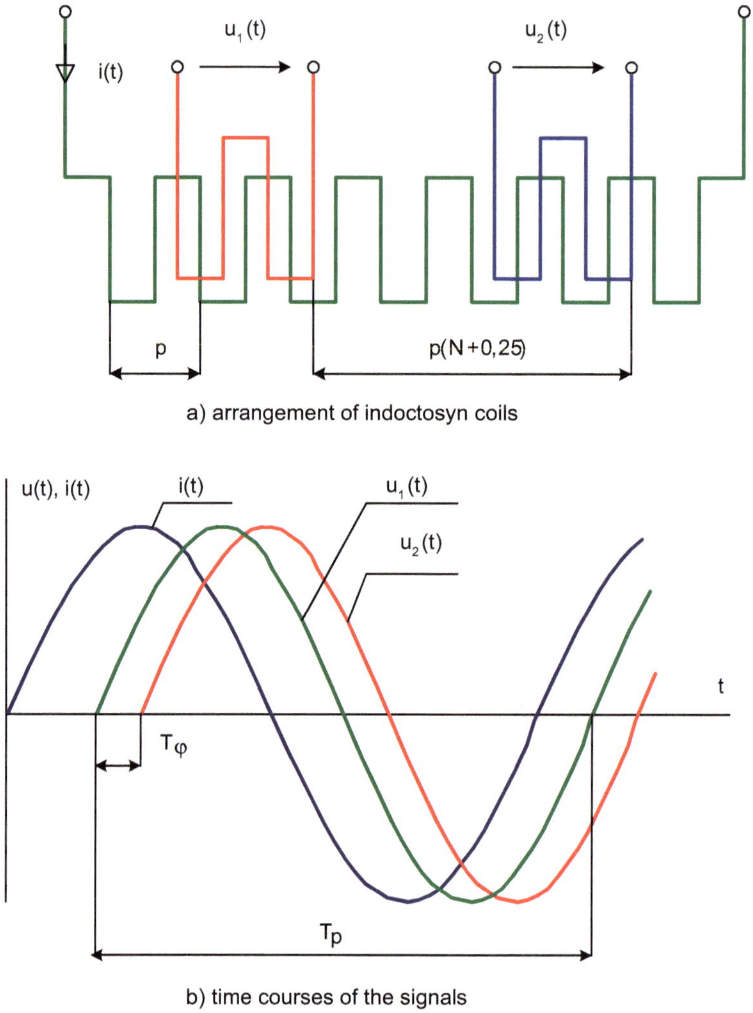

Figure 2.14 Induktosyn.

tens per second. Therefore, these sensors are used to measure the position in numerically controlled machine tools. Their advantage is principled simplicity and resistance to external interference.

Selsyns are inductive sensors of circular motion, formed by a system of three fixed stator coils, whose axes are geometrically shifted by 120°. A rotor coil rotates between these coils (Figure 2.16).

If we excite the rotor coil of *selsyn* with an alternating current of circular frequency ω, then voltages are induced in its rotor coils, phase shifted by 120°

Figure 2.15 Discs of circular inductosyns.

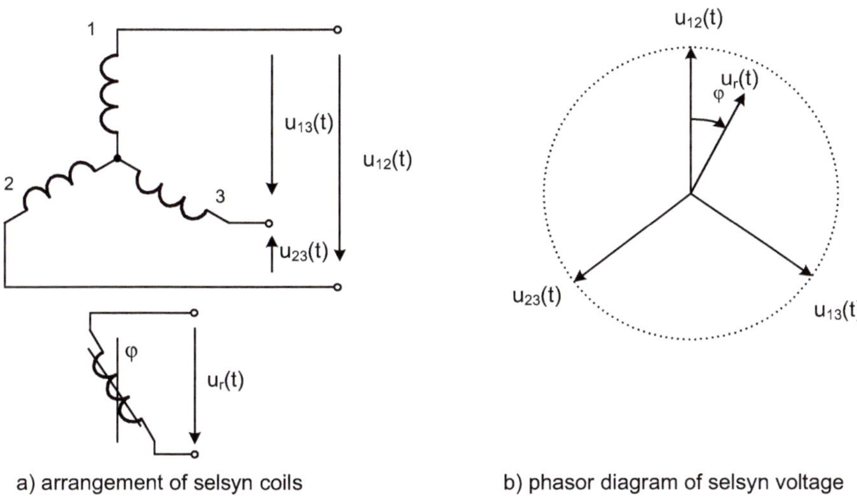

Figure 2.16 Principle of activity of selsyn.

$$u_{12}(t) = U \cdot \sin(\omega t + \phi), \qquad u_{23}(t) = U \cdot \sin(\omega t + 120^\circ + \phi), \quad (2.19\text{ a, b})$$

$$u_{23}(t) = U \cdot \sin(\omega t + 240^\circ + \phi), \quad (2.20)$$

where φ is the rotation angle of the rotor coil with respect to the stator coils of selsyn.

To convert these phase-shifted three-phase voltage waveforms into two phase-shifted voltages, a *Scott transformer* is used, consisting of two independent transformers whose conversion ratios are ½ and √3/2 (Figure 2.17).

If we excite the stator coils of the selsyn with a three-phase symmetrical current, then a voltage is induced in its rotor coil

$$u_r(t) = U_r \cdot \sin(\omega t + \phi), \quad (2.21)$$

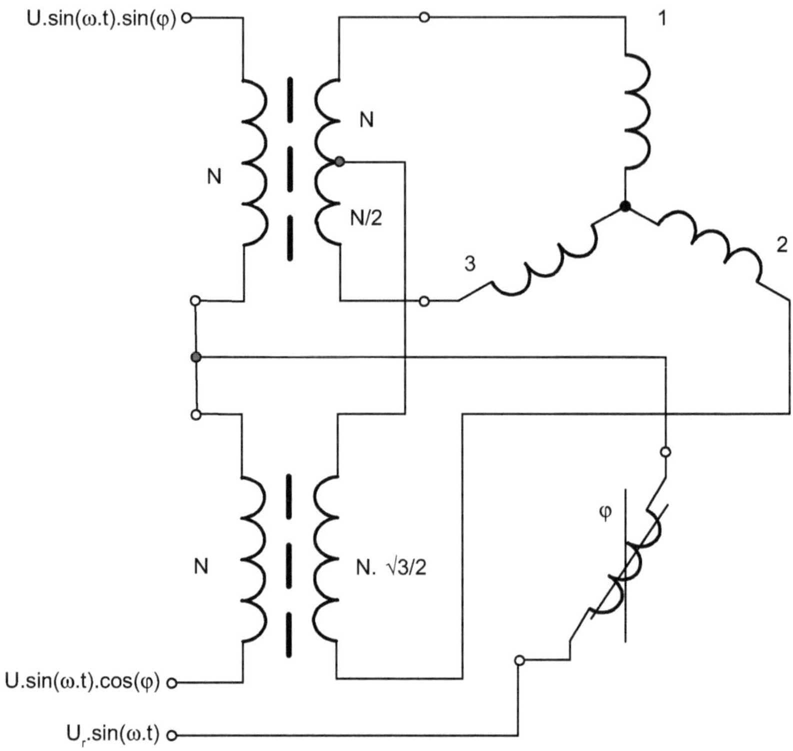

Figure 2.17 Scott transformer.

where φ is the angle between the axis of the rotor coil and the axis system of the stator coils.

The disadvantage of selsyns is the need to excite them with a three-phase symmetrical system of signals and the need to transform their output signals into a more easily applicable two-phase Cartesian system.

Resolvers are induction sensors of circular motion, formed by a system of two mutually perpendicular stator coils and two mutually perpendicular rotor coils (Figure 2.18).

When exciting one of the stator coils with identical alternating voltages

$$u_{1s}(t) = U_s \cdot \sin(\omega t), \tag{2.22}$$

voltages are induced in the rotor coils of the resolver

$$u_{r1}(t) = U_r \cdot \sin(\omega t) \cdot \sin\varphi, \qquad u_{r2}(t) = U_r \cdot \sin(\omega t) \cdot \cos\varphi, \tag{2.23 a, b}$$

where φ is the angle between the rotation of the rotor coils of the revolver relative to the axis of the stator coils.

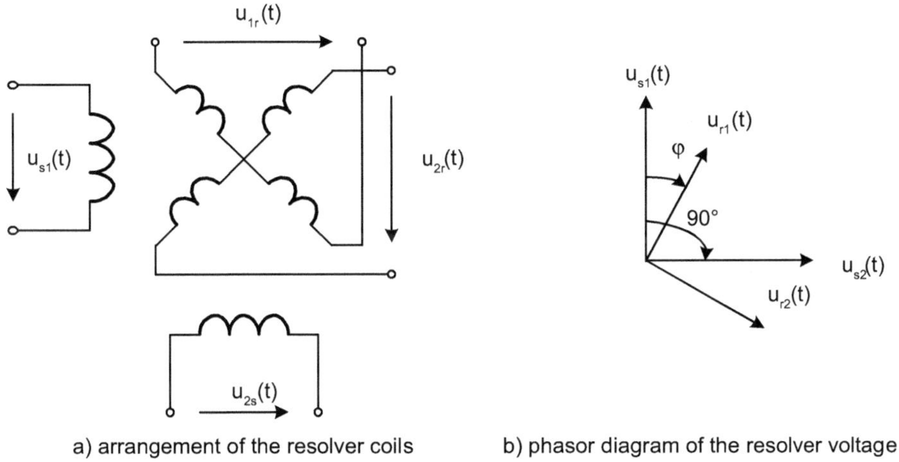

Figure 2.18 Principle of resolver operation.

Similar voltage time courses can be achieved by exciting one of the rotor coils, and alternating voltages phase-shifted by 90° can be sensed from the stator coils.

To convert these signals into a digital equivalent, SD converters (synchro to digital convertors) are used, which contain two *multiplying D/A converters* MDAC1 and MDAC2, which multiply the voltage of the stator coils with the output digital signal of the return counter. The counting direction of the reversible counter is controlled by the output two-valued signal of the phase detector, which compares the time sequence of the difference in the output voltages of the D/A converters (Figure 2.19).

In steady state, the state of the return counter is determined by the digital equivalent of the angle of rotation of the resolver rotor coils. Typical representatives of these converters are, for example, circuits of the ADS1210 series from Analog Devices with a binary resolution of 12–16 bits, which corresponds to an angular resolution in the range of 0.08°–0.0055°.

2.1.4 Position sensors with Hall probe

These sensors are used not only for two-value position determination but also for the incremental method of displacement measurement and measurement of rotations of rotating permanent magnets.

If the *Hall probe*, consisting of a heavily doped silicon substrate, is exposed to a magnetic flux Φ acting on the surface S, whose orientation is

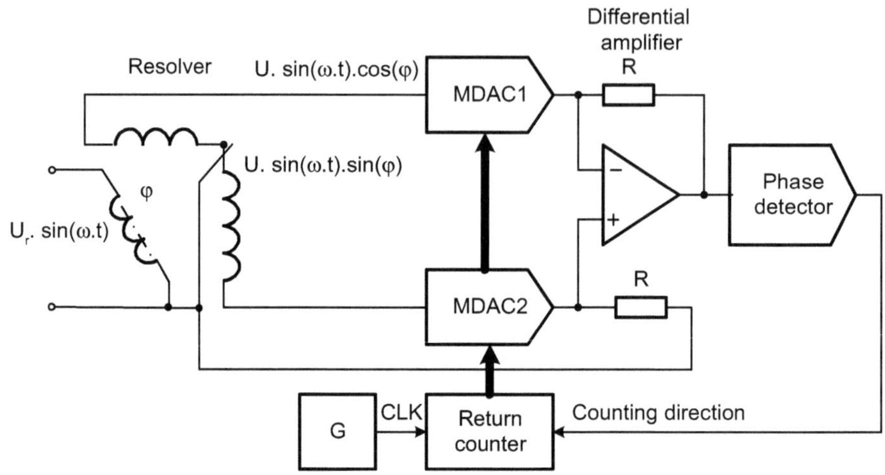

Figure 2.19 Connecting the resolver to the SD converter.

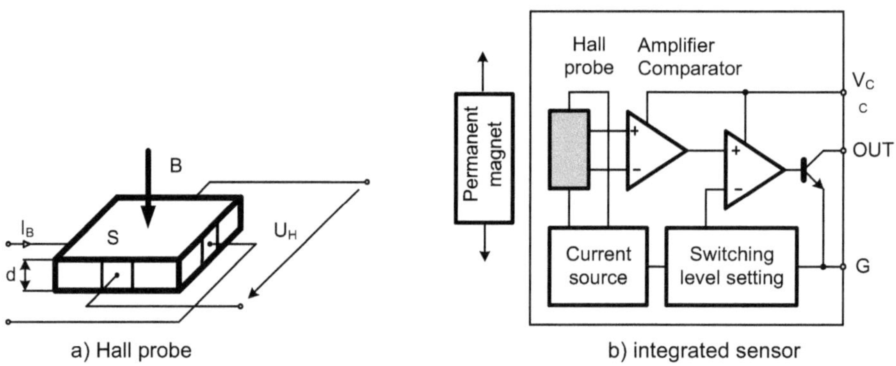

Figure 2.20 Hall probe sensor.

perpendicular to the direction of the excitation current I_B, a voltage is generated on its perpendicularly located sensing electrodes

$$U_H = k_H \frac{\Phi \cdot I_B}{S \cdot d} \tag{2.24}$$

where k_H is the *Hall constant* and d is the thickness of the Hall probe.

Position sensors with a Hall probe are formed by their own *Hall probe*, the voltage of which, after amplification, is compared by a voltage comparator with an adjustable switching level. If the output voltage of the amplifier is greater than the set voltage, the transistor switch is closed. In the opposite case, when the output voltage of the amplifier is less than the set voltage level, the transistor switch is opened (Figure 2.20).

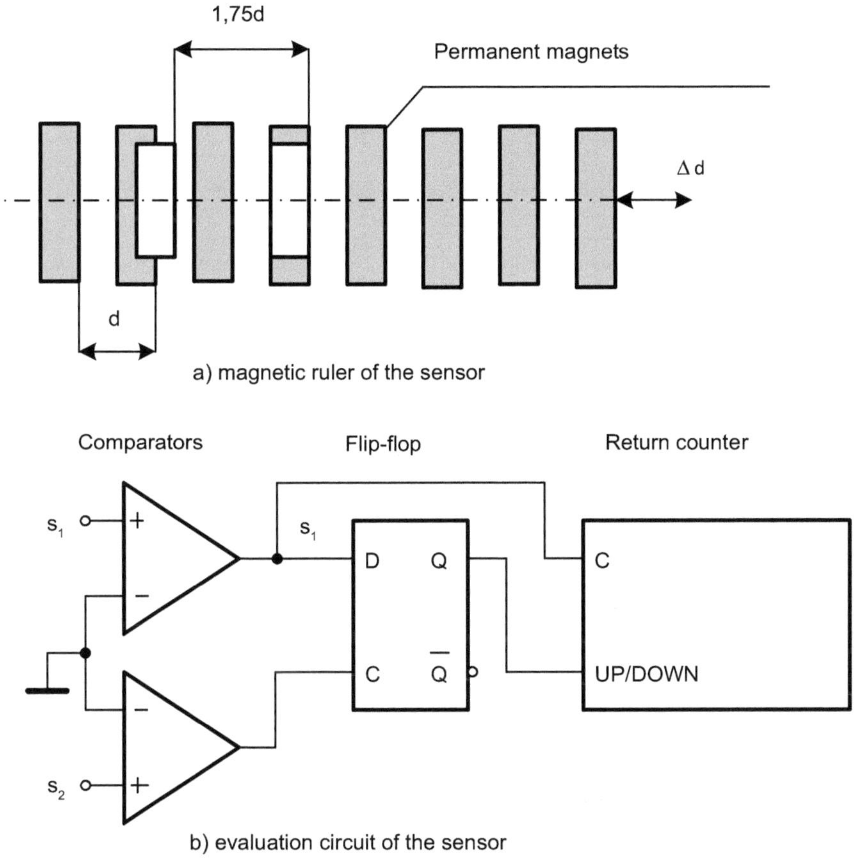

Figure 2.21 Incremental displacement sensor with Hall probes.

Incremental motion sensors with a Hall probe are made up of equidistantly distributed permanent magnets placed on a moving object, while to determine the direction of movement of the object with respect to two sensing Hall probes, these probes are placed above the object with a ¼ pitch of the placement of the permanent magnets (Figure 2.21).

As the object moves, pulse signals are generated in the *Hall probes*, the number of which corresponds to the number of pitches of the permanent magnets. The size of the object's movement is then determined by the state of the *reversible counter*, which counts the number of pulses of these signals, while the direction of the scale's movement is determined by the state of the outputs of the D-type flip-flop circuit.

If the rising edge of the signal s_1 of the first sensor precedes the rising edge of s_2 of the second sensor, then the output of the D-type flip-flop is at the H level and the return counter reads forward the number of pulses given by

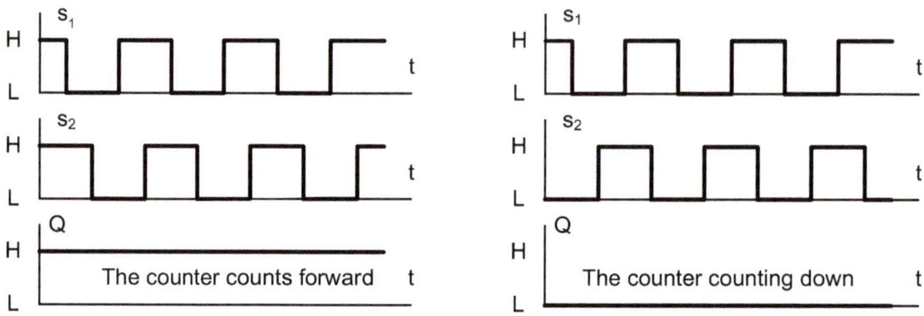

Figure 2.22 Time courses of signals of sensors with Hall probes.

a) with magnetic ruler

b) with magnetic tape

Figure 2.23 Magnetic incremental displacement sensors.

the amount of displacement and the number of permanent magnets per unit length. If the rising edge of the signal s_1 of the first sensor is delayed before the rising edge s_2 of the second sensor, then the output of the D-type flip-flop is at the L level, and the reverse counter reads backward, i.e., it subtracts its state from the previous state in which it read forward (Figure 2.22).

For easier application of linear incremental magnetic sensors, instead of firmly glued permanent magnets, magnetic self-adhesive *magnet tape* is used, on which permanent magnets are placed with a spacing of 2–5 mm with an accuracy of up to ±20 mm/m.

Figure 2.23 shows an incremental magnetic sensor of the MSH series with a magnetic tape and an MLC series sensor with a magnetic ruler from MEGATRON. Both types of sensors have a resolution of 5 μm and allow displacements to be measured at speeds of up to 3 m/s.

2.1.5 Magnetostrictive position sensors

These sensors use magnetostriction of ferromagnetic materials made from iron, nickel, and cobalt alloys. If a long rod made of ferromagnetic material is magnetized in its axis by a direct magnetic field and if current pulses pass through the rod, then torsional stress of this rod occurs.

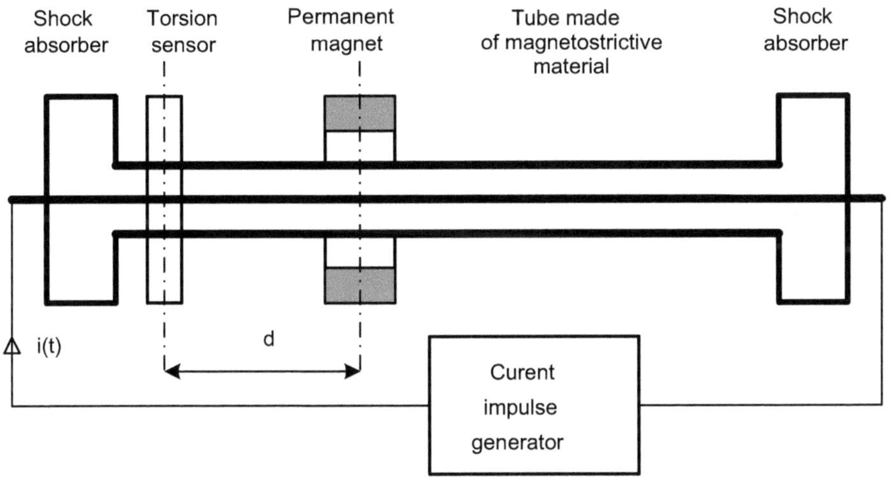

Figure 2.24 Magnetostrictive position sensors.

Magnetostrictive position sensors are formed by a ferromagnetic tube, on the axis of which there is a conductor through which current pulses flow. An annular permanent magnet strung on a pipe creates an axial direct current magnetic field, which, with the pulsed circular magnetic field of the conductor, created by the passage of current pulses in the rod, generates pulsed torques in it, which are sensed at the end of the pipe by magnetic or piezoelectric torsion sensors. To prevent reflections of these torque pulses at the ends of the rod, the ends of the rod are fixed in shock absorbers (Figure 2.24).

From the propagation speed of the mechanical disturbance in the pipe material, reaching units of 3 km/s, and the time between the occurrence of the excitation current and the corresponding voltage pulse of the induction sensor Δt, the position d of the annular permanent magnet can be determined in the range of lengths from 100 mm to 5 m, with a resolution of tenths of mm and accuracy of up to 0.1%

$$d = v \cdot \Delta t. \tag{2.25}$$

To measure these time intervals in the range of 30 ms to 1.7 ms, counters with a resolution of up to 10^{-5} are used, and their maximum sampling frequency reaches hundreds of MHz.

In practice, magnetostrictive length sensors with one or two movable permanent magnets are used, allowing measuring the differences in distances between them (Figure 2.25).

a) sensor with one magnet b) sensor with two magnets

Figure 2.25 Magnetostrictive longitudinal motion sensors.

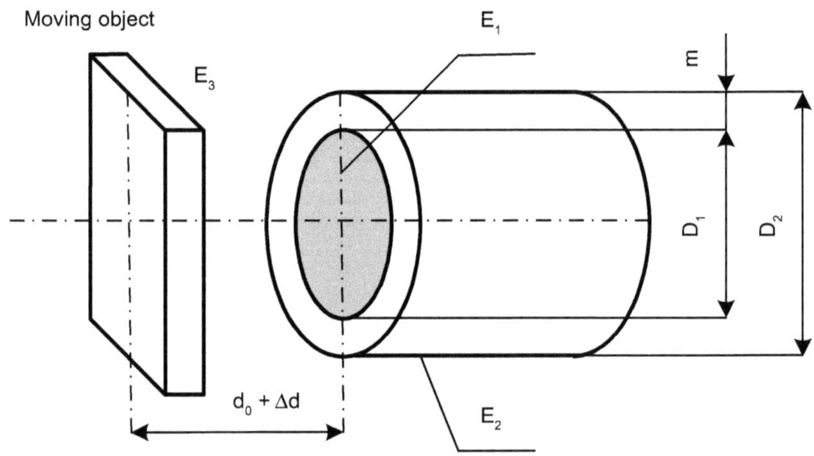

Figure 2.26 Capacitive proximity sensor.

2.1.6 Capacitive position sensors

Capacitive positive sensors use changes in capacitor capacities, caused by changes in the geometric dimensions of their electrodes and changes in the dielectric permittivity of insulating materials that are placed between these electrodes to determine the location of objects.

Capacitive proximity sensors are made up of a sensing ring electrode E_1, which is coaxially located inside the shielding cylindrical electrode E_2 and near which there is an aperture E_3, connected to the object whose position is being sensed (Figure 2.26).

Assuming that the distance d of the sensing electrode E_1 is at least five times smaller than its diameter D_1, the intensity of the electromagnetic field between the electrode E_1 and the non-conducting aperture E_3 can be considered constant, and for the change in the sensor capacity depending on the change in the distance between the electrode E_1 and the aperture E_3 applies

$$\Delta C = \varepsilon_v \cdot \frac{S}{d_0 + \Delta d} \approx \varepsilon_v \cdot \pi \frac{D_1^2}{4}\left(d_0 - \Delta d\right), \tag{2.26}$$

where S is the area of the sensing electrode and e_{iz} is the permittivity of the insulator from which the diaphragm is made.

If the aperture is made of a *conductive material* and is *not grounded,* then the capacitance of the sensor is given by the sum of the capacitances between its electrodes E_1 and E_2 and the capacitance between electrodes E_1 and E_3,

$$\Delta C = \varepsilon_v \frac{2 \cdot S}{2(d_0 + \Delta d)} \approx \varepsilon_v \cdot \pi \frac{D_1^2}{2}\left(d_0 - \Delta d\right), \tag{2.27}$$

where ε_v is the permittivity of air, which is approximately equal to the permittivity of vacuum $\varepsilon_0 = 1256 \cdot 10^{-6}$ H/m.

If the diaphragm is made of a *conductive material* and if it is additionally *grounded,* then the resulting capacitance of the sensor is also given by the sum of the capacitance between its electrodes E_1 and E_2 and the capacitance between electrodes E_1 and E_3, while between both pairs of electrodes there is air with permittivity ε_v. It then applies to the resulting capacity of this arrangement

$$\Delta C = \varepsilon_v \frac{2 \cdot S}{(d_0 + \Delta d)} \approx \varepsilon_v \cdot \pi \frac{D_1^2}{4}\left(d_0 - \Delta d\right). \tag{2.28}$$

The disadvantage of all these types of sensors is the significant non-linearity of the dependence of their capacities on the distance d, which reaches several percent in the range of their ratio values $\Delta d/d = 0.1$.

In order to eliminate the effect of the leakage impedance of the supply cable, reaching tens of pF and GΩ units per 1 m of its length, *active shielding* is used to connect the capacitive sensors to the evaluation circuits, in which there is a *triaxial cable* with an internal and external shielding (Figure 2.27).

Assuming ideal characteristics of the op-amp connected in the voltage follower function with transfer 1, the cable's internal shield is at the potential of the sense electrode E_1, and therefore the leakage impedances between the cable's internal shield and its inner conductor do not apply. The external shielding of the triaxial cable, including the shielding cover of the sensor, is grounded together with the E_3 electrode. The condition for the correct operation of this method of shielding is the use of a floating, i.e., ungrounded power supply of the operational amplifier.

To increase the sensitivity of capacitive sensors with a variable distance of plane-parallel electrodes, a differential arrangement of these sensors is

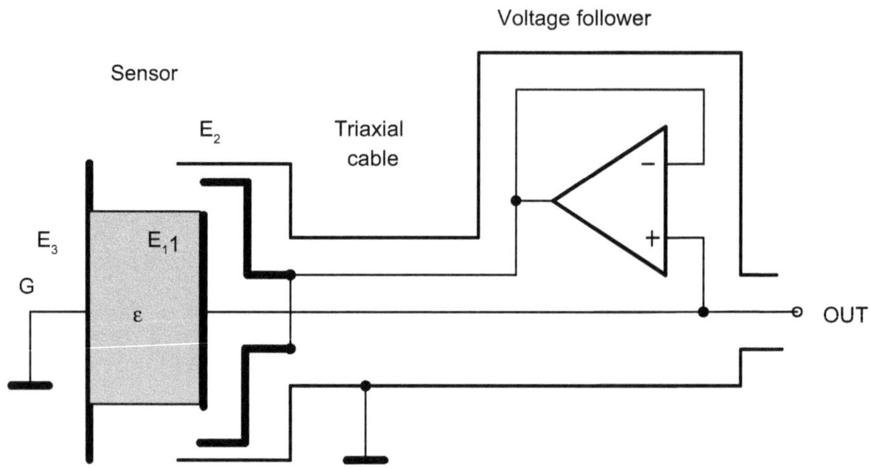

Figure 2.27 Suppression of the influence of the supply cable capacity by active shielding.

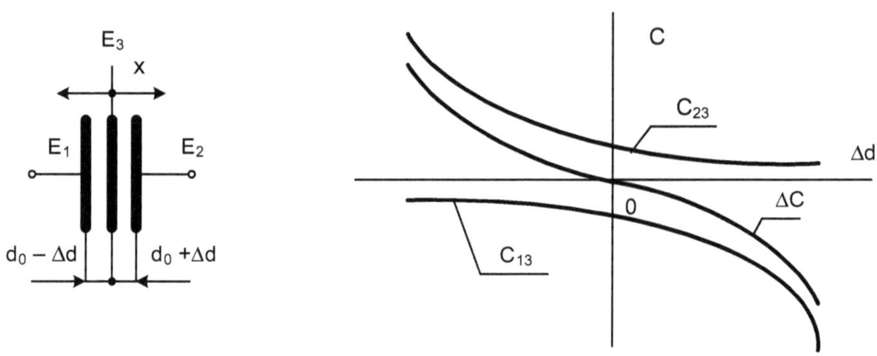

a) arrangement of the sensor b) conversion characteristic of the sensor

Figure 2.28 Differential capacitive position sensors with variable electrode distance.

used, in which the electrode E_3 moves between two fixed electrodes E_1 and E_2 (Figure 2.28).

Because it applies to the inter-electrode capacitances of these sensors

$$C_{13} = \varepsilon \cdot \frac{S}{d_0 - \Delta d}, \qquad C_{23} = \varepsilon \cdot \frac{S}{d_0 + \Delta d}, \tag{2.29 a, b}$$

their sum and difference are determined by the equations

$$C_{13} - C_{23} = \varepsilon \cdot S \frac{2 \cdot \Delta d}{(d_0 + \Delta d) \cdot (d_0 - \Delta d)}, \tag{2.30 a, b}$$

$$C_{13} + C_{23} = \varepsilon \cdot S \frac{2 \cdot \Delta d_0}{(d_0 + \Delta d) \cdot (d_0 - \Delta d)}.$$

If only the difference of these partial sensor capacities is evaluated

$$C_{13} - C_{23} \approx \varepsilon \cdot S\left[1 + 2\left(\frac{\Delta d}{d}\right)^2\right], \tag{2.31}$$

then this sensor has twice the sensitivity compared to a simple sensor, but its dependence of the change in capacitance on the proportional change in distance Δd is approximately quadratic.

If we relate the differential capacitance to the sum of the partial capacitances of the differential sensor

$$\frac{C_{13} - C_{23}}{C_{13} + C_{23}} = \frac{\Delta d}{d_0}, \tag{2.32}$$

then the proportional difference of its partial capacities is directly proportional to the proportional change in the distances of their electrodes. Compared to the previous arrangement, differential capacitance sensors have less non-linearity and suppress the effect of temperature on the change in their dimensions and permittivity of the dielectric. Therefore, these sensors are used to detect small position changes in the range of tenths to units of mm. The micromechanical structures of *micro-electronic mechanical systems* (MEMS) of capacitive sensors make it possible to sense changes in lengths up to tens of μm.

Capacitive position sensors with a variable overlapping area of plane-parallel electrodes in the simple and differential arrangements in Figure 2.29 have inter-electrode capacities equal to

$$C_{12} \approx C_0\left(1 + \frac{\Delta l}{l_m}\right), \qquad \Delta C = C_{13} - C_{23} \approx C_0\left[1 + 2\left(\frac{\Delta d}{d}\right)^2\right], \tag{2.33 a, b}$$

where C_0 are the maximum values of the inter-electrode capacitances of the sensors at their maximum overlap, i.e., at $\Delta l = 0$.

Similar to the previous case, a linear dependence of its proportional differential capacitance on the proportional change in the position of its internal movable electrode can also be achieved with the differential design of the capacitive sensor with its electrodes overlapping

$$\frac{C_{13} - C_{23}}{C_{13} + C_{23}} = \frac{\Delta l}{d_0}. \tag{2.34}$$

Capacitive position sensors with a variable *position of inserted insulator*, the permittivity of which is different from the permittivity of air in simple and differential arrangements, are shown in Figure 2.30.

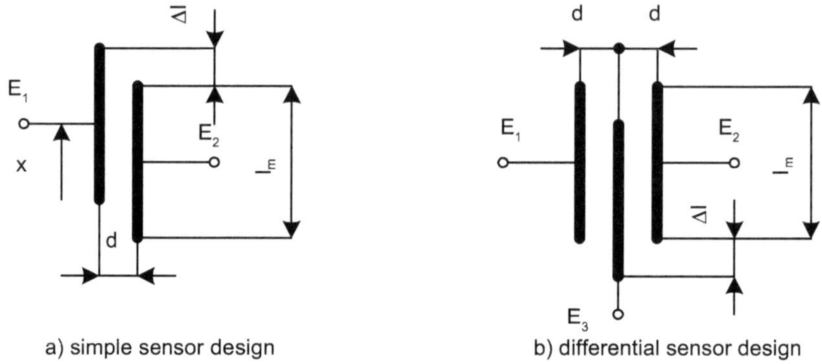

Figure 2.29 Capacitive position sensors with variable overlap of plate electrodes.

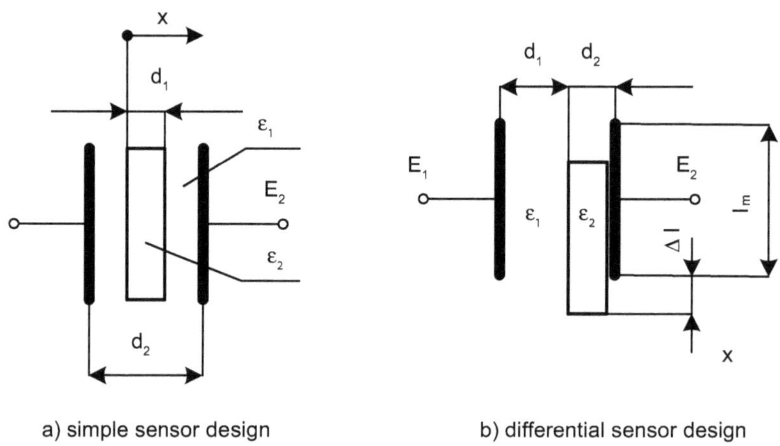

Figure 2.30 Capacitive position sensors with variable position of inserted insulator.

It applies to the inter-electrode capacitances of these sensors

$$C_{12} \approx \frac{\varepsilon_1 \cdot S}{d_2 - d_1(x)}\left(1 + \frac{\varepsilon_1}{c}\right), \qquad C_{12} = \frac{\varepsilon_1 \cdot S}{d_1 + d_2}\left[1 + \frac{\Delta l}{l_m} \cdot \frac{1 - \dfrac{\varepsilon_1}{\varepsilon_2}}{\dfrac{d_1}{d_2} + \dfrac{\varepsilon_1}{\varepsilon_2}}\right], \tag{2.35 a, b}$$

The most frequently used moving plates are Teflon plates with a relative permittivity of approximately 2.1, glass plates with a relative permittivity in the range of 4–6, and mica plates with a relative permittivity in the range of 7–11.

The circuit for evaluating changes in the capacities of differential capacitive position sensors in Figure 2.31 consists of an alternating voltage

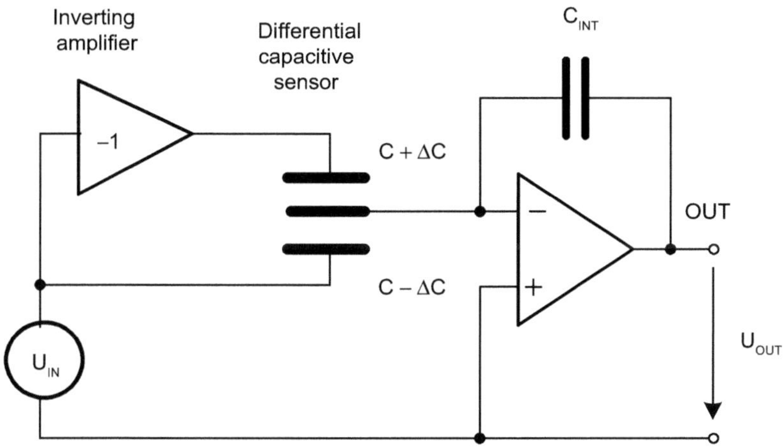

Figure 2.31 Differential capacitive position sensor circuit.

source U_{IN}, an inverting amplifier, and an inverting integrating amplifier with an integrating capacitor C_i.

When the capacitances of the differential capacitance sensor change, ΔC applies to the output voltage of the circuit

$$U_{OUT} = 2 \cdot \frac{\Delta C}{C_i} \cdot U_{IN}. \tag{2.36}$$

A typical circuit for processing the signals of differential capacitance sensors is, for example, the AS7150 circuit from Analog Devices, which enables the digitization of capacitance changes in two channels in the range of 0–13 pF with a resolution of up to 1fF (10^{-16} F).

Capacitive sensors with plane-parallel electrodes are also used to measure inclination (*capacitive inclinometer*). These sensors are made up of a fixed circular electrode E_1 of diameter D, above which is placed a movable circular electrode E_3 of the same diameter, tilted relative to the fixed electrode (Figure 2.32).

Assuming that the electrode distances d_1, d_2 << D, the capacitance between them is

$$C_{13} = \frac{\varepsilon \cdot \pi \cdot D^2}{2(d_1 + d_2)}. \tag{2.37}$$

The slope of the electrodes is then approximately determined by their inter-electrode capacitance

$$\text{INL} = \frac{d_1 - d_2}{D} \approx \frac{\varepsilon \cdot \pi \cdot D}{4 \cdot C_{13}}. \tag{2.38}$$

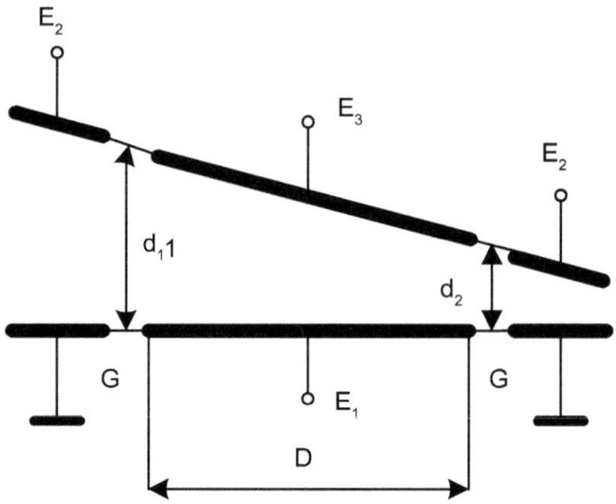

Figure 2.32 Capacitive sensors of the inclination.

To suppress the effect of interfering signals, a coaxial shielding electrode E_2 in the shape of a conductive annulus is placed around the perimeter of the sensing electrode E_1, thus creating an equipotential shielding of the sensor, since the voltage difference between this electrode and the ground is ideally zero.

For *two-dimensional non-contact position measurement*, and/or tilt, capacitive sensors with four fixed electrodes E_1–E_4, over which one moving electrode E_0 moves, are used (Figure 2.33).

If the moving electrode has a square shape with edge a, then when it is deviated from the x- and y-axes of electrodes E_1–E_4, the differences in inter-electrode capacities are

$$C_{20} - C_{10} = 2\frac{\varepsilon}{a}\Delta x\left(\frac{a}{2} + \Delta y\right), \qquad C_{40} - C_{10} = 2\frac{\varepsilon}{a}\Delta y\left(\frac{a}{2} + \Delta x\right). \tag{2.39 a, b}$$

These sensors are used, for example, to center moving objects, which is conditioned by the equality of inter-electrode capacitances C_{10}–C_{40}.

Two-dimensional capacitive sensors can also be used to measure tilt in two directions. In this case, the moving electrode E_5 floats in a liquid whose density is greater than the density of the material from which the electrode carrier is made. This electrode then remains in a horizontal position and changes in the tilt of the fixed electrodes E_1–E_4 cause changes in the inter-electrode capacitances of the sensor, which are evaluated by a pair of capacitance sensors.

Capacitive sensors can also be used to detect the *rotation angle*. A typical representative of these sensors is, for example, a sensor consisting of four

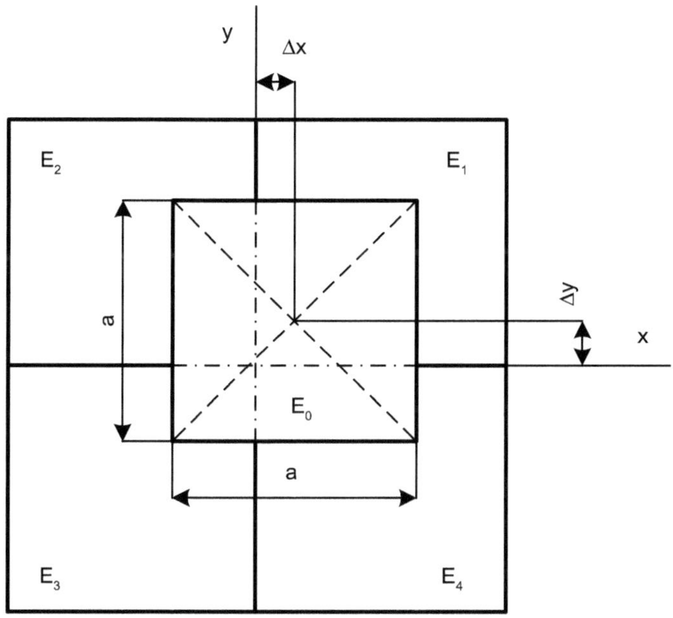

Figure 2.33 Two dimensional capacitive position sensor.

symmetrical fixed electrodes E_1–E_4 and a moving electrode E_0, formed by a half section with a diameter r (Figure 2.34).

Since the area of a segment of a circular segment is

$$S(\alpha) = \frac{\pi \cdot r^2}{360} \cdot \alpha = \frac{1}{2} r^2 \cdot \mathrm{arc}(\alpha). \tag{2.40}$$

where α (°) is the angle of rotation of the electrode E_0 with respect to the x-axis of the fixed electrodes, valid in the range of the angle of rotation of the movable electrode 0°–90° for inter-electrode capacities

$$C_{10} = \varepsilon \frac{S_{10}}{d} = \varepsilon \frac{r^2}{2 \cdot d} \mathrm{arc}(\alpha), \tag{2.41 a}$$

$$C_{30} = \varepsilon \frac{S_{30}}{d} = \varepsilon \frac{r^2}{d} \left[\pi - \frac{\mathrm{arc}(\alpha)}{2} \right], \tag{2.41 b}$$

$$C_{20} = C_{40} = \varepsilon \frac{\pi \cdot r^2}{d}, \tag{2.41 c}$$

where ε is the permittivity of the insulator between the electrodes and d is their distance.

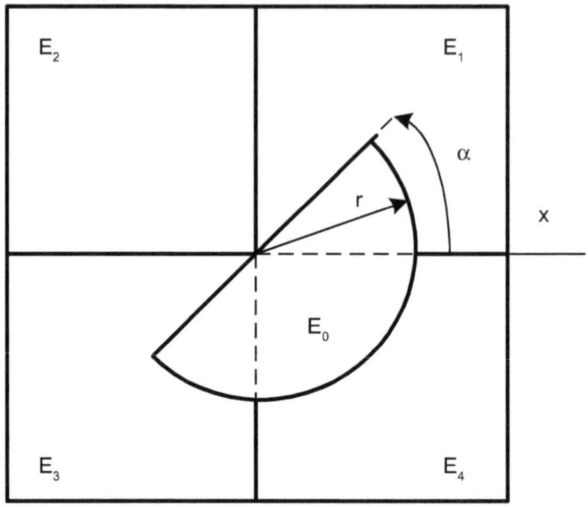

Figure 2.34 Capacitive rotation angle sensor.

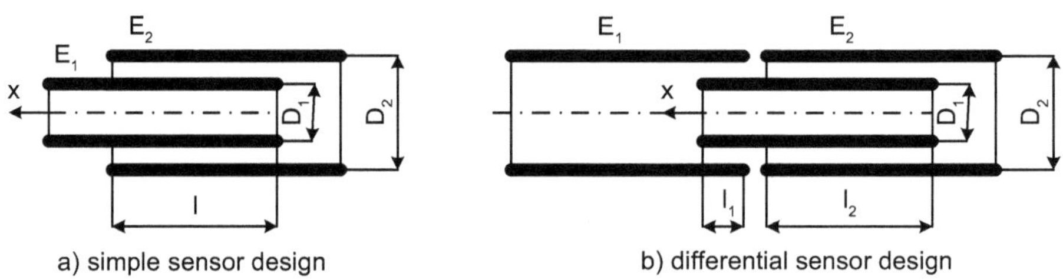

Figure 2.35 Capacitive position sensors with cylindrical electrodes.

Capacitive sensors with concentric cylindrical electrodes in simple and differential arrangement have inter-electrode capacities directly proportional to the mutual insertion lengths l, respectively, of the difference in the insertion lengths $l_1 - l_2$ of the inner electrodes with a diameter of D_1 into the outer electrodes with an inner diameter of D_2 (Figure 2.35),

$$C_{12} = \varepsilon \frac{2 \cdot \pi \cdot l}{\ln \frac{D_1}{D_2}}, \qquad \Delta C = C_{13} - C_{23} \approx 2 \cdot \varepsilon \frac{2 \cdot \pi}{\ln \frac{D_1}{D_2}} (l_2 - l_1), \quad (2.42\text{ a, b})$$

where ε is the permittivity of the insulator located between the electrodes.

Capacitive sensors are used not only for direct measurement of longitudinal and circular motion but also for measuring mechanical deflection of deformation members of mechanical stress sensors, deflection of flexible membranes of pressure sensors and height sensors of liquid levels and bulk materials.

If the permittivities of the insulators of the sensing capacitors have negligible losses, it is sufficient to determine the changes in their capacities only by determining their reactances. In the case of non-negligible dielectric losses of the dielectric capacitors, it is necessary to use *synchronous detection* to process their signals, and to determine the position of the sensors, only *imaginary components* of the measured impedances should be used.

2.1.7 Optoelectronic incremental and code sensors

Optoelectronic incremental sensors are made up of metal or glass rulers or disks, on which geometrically equidistant lines with a spacing of up to 0.1 mm are marked. These lines are illuminated by LED diodes and the reflexively reflected light rays are filtered by two shutters CL_1 and CL_2, and the spacing of the holes is shifted by ¼ of the line spacing. The optical signals passed through these apertures are converted in the photodiode sensors into the corresponding pulse signals d_1 and d_2, which are used to control the flip-flop whose output signal Q controls the direction of pulse counting in the reversible counter. The sampling frequency of the counter is derived from the pulse signal d_3 of the third photodiode sensor through the triggered clock pulse generator, whose frequency fs is multiplied by the PLL (*phase locked loop*) by the division ratio N of the frequency divider connected to its feedback.

To achieve higher resolution of incremental sensors, interpolation methods are used based on the interpolation of the triggering edges of pulses d_1 and d_2. The most frequently used interpolations in 4–256 time equidistant values make it possible to increase the resolution of incremental longitudinal motion sensors up to several μm and circular motion sensors up to several thousandths of a degree.

Figure 2.36 shows the location of the apertures of the incremental longitudinal movement sensor and the circuit for processing pulsed photodiode signals.

Optical code sensors are used to determine the longitudinal or circular movement of a *Grey's code* with a change of only one of the bits in successive codewords. This method of coding prevents the occurrence of concurrency hazards that arise when using binary (double) coding. Table 2.4 shows the conversion of a four-bit *Greye's code* to a four-bit binary code.

To convert the *Greye's code* into a *binary code*, logic circuits of the exclusive logical sum EXOR (*Exclusive Or*) are used, connected according to Figure 2.37.

Photodiode reflective sensors are also most often used to detect code marks, and the number of code paths, determined by the number of Greye's

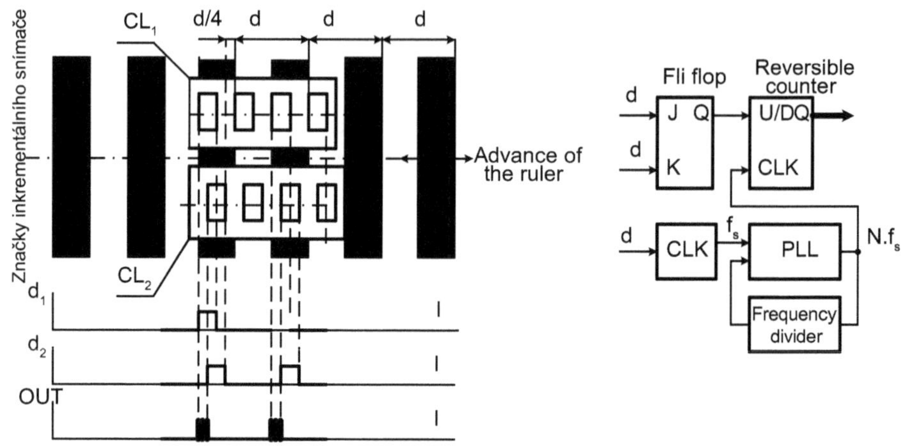

a) sensor aperture arrangement b) sensor evaluation circuit

Figure 2.36 Incremental sensor of longitudinal movement.

Table 2.4 Conversion of 4-bit Greye's to 4-bit binary code.

Decadal equivalent	0	1	2	3	4	5	6	7
Grey's code	0000	0001	0011	0010	0110	0111	0101	0100
Binary code	0000	0001	0010	0011	0100	0101	0110	0111
Decadal equivalent	8	9	10	11	12	13	14	15
Grey's code	1100	1101	1111	1110	1010	1011	1001	1000
Decadal equivalent	1000	1001	1010	1011	1100	1101	1110	1111

code bits, ranges from 8 to 16, which corresponds to a proportional length resolution of 2^{-8} to 2^{-16} and an angular resolution of 1.4° to $5 \cdot 510^{-3\circ}$.

Figure 2.38 shows the wheels of an incremental circular motion sensor, a 4-bit binary-coded sensor, and a 4-bit *Greye's-coded sensor.*

2.1.8 Continuous optoelectronic position sensors

Photodiode sensors using one or several discrete *quadrature photodiodes*, *one-* or *two-dimensional* photodiodes, charge-coupled line, and image photodiode sensors are used for optoelectronic position sensors.

A *one-dimensional quadrature photodiode position sensor* with two identical photodiodes that are irradiated with a light beam of a circular shape, the center of the axis of which is deviated from the axis of the photodiodes by a change in displacement Δx, is shown in Figure 2.39(a).

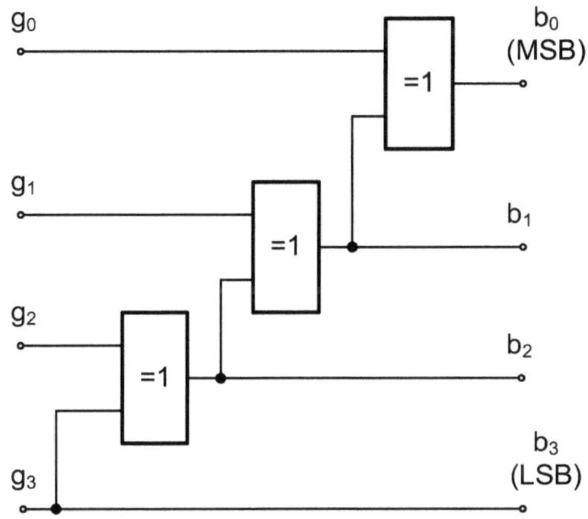

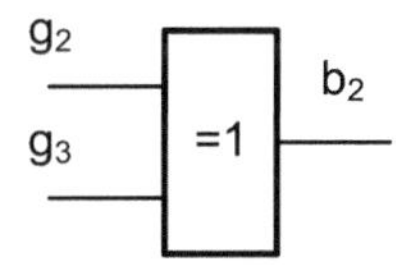

g_2	g_3	b_2
0	0	0
0	1	1
1	0	1
1	1	0

a) Gray to bin codfer b) EX-OR logic table

Figure 2.37 Four bit Greye's code to four bit binary code converter.

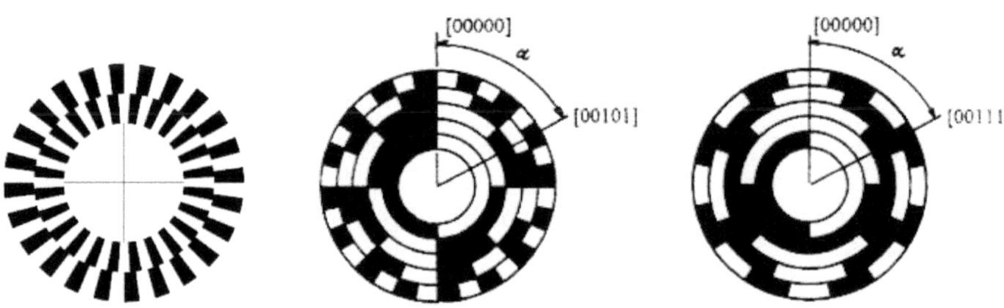

a) incremental disc b) binary disc c) Greye's code disc

Figure 2.38 Discs of circular motion sensors.

Assuming that the light sensitivity of both photodiodes is the same, the center of the circular trace of the light beam incident on them is deflected in the Δx direction

$$\Delta x = \frac{i_{FD1} - i_{FD2}}{i_{FD1} + i_{FD2}}, \qquad (2.43)$$

where i_{FDi}, i = 1,2, are the photodiode currents.

A *two-dimensional quadrature photodiode position sensor* with four centrally located identical photodiodes that are irradiated by a light beam is shown in Figure 2.39(b).

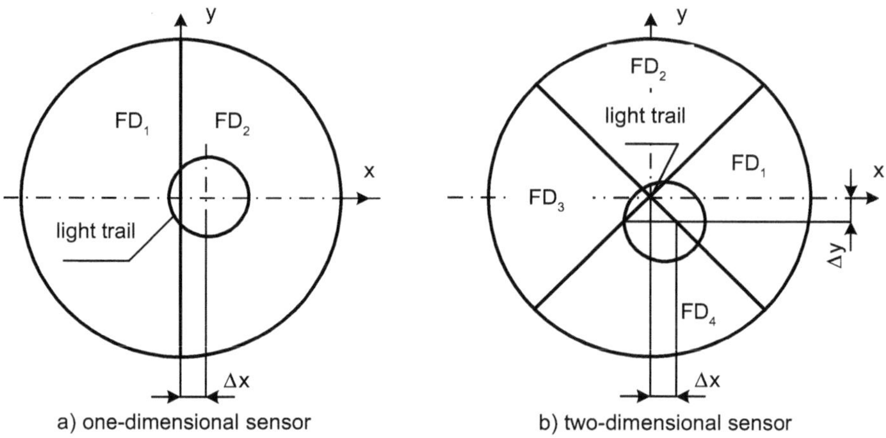

Figure 2.39 Quadrature photodiode position sensors.

Assuming that the light sensitivity of all four photodiodes is the same, the center of the circular trace of the light beam incident on them is deflected in the Δx and Δy directions

$$\Delta x = \frac{i_{FD1} - i_{FD3}}{i_{FD1} + i_{FD3}}, \qquad \Delta y = \frac{i_{FD2} - i_{FD4}}{i_{FD2} + i_{FD4}}. \qquad (2.44\text{ a, b})$$

Photodiode currents are converted to the corresponding voltage in current amplifiers for both types of sensors.

Photodiode PSD sensors are made up of large-area one-dimensional or two-dimensional photodiodes, on the surface of which a light beam falls (Figure 2.40).

By its impact on the P layer, free charge carriers are generated in the sensing Si layer, which are transferred to the opposite P and N layers, which causes the generation of current I_0 between them. This current is divided between sub-currents i_1 and i_2 depending on the position x of the light beam according to

$$\frac{i_1}{I_0} = \frac{L - x}{L}, \qquad \frac{i_2}{I_0} = \frac{x}{L}. \qquad (2.45\text{ a, b})$$

The position of the light beam x can be determined from the ratio of the difference and the sum of these partial currents.

$$\frac{i_1 - i_2}{i_1 + i_2} = \frac{i_1 - i_2}{I_0} = 1 - 2\frac{x}{L}. \qquad (2.46)$$

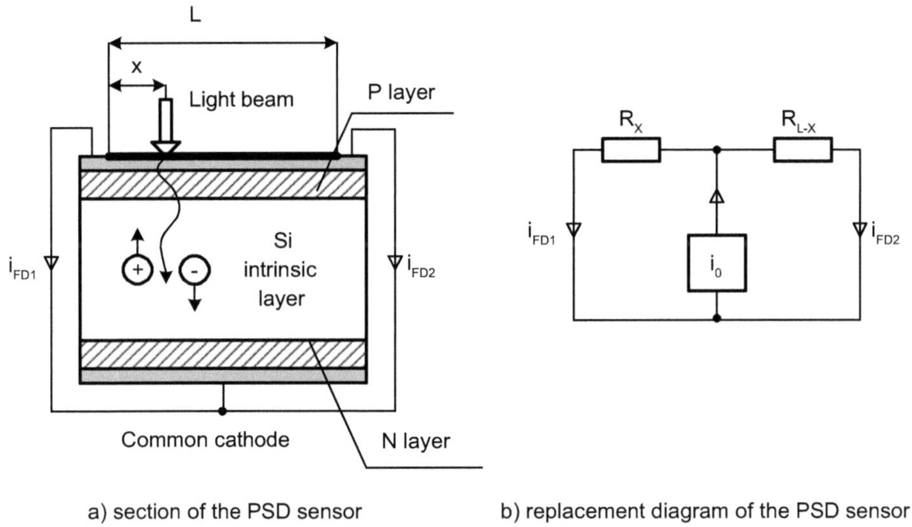

Figure 2.40 One dimensional PSD sensor.

Two-dimensional position sensors PSD also work on a similar principle, in which the currents i_1, i_2 and i_3, i_4 are sensed from the opposite electrodes E_1, E_2, E_3, and E_4, from which, based on the equations

$$\frac{i_1 - i_2}{i_1 + i_2} = 1 - 2\frac{\Delta x}{L}, \qquad \frac{i_3 - i_4}{i_3 + i_4} = 1 - 2\frac{\Delta y}{L} \tag{2.47 a, b}$$

determined the positions of the light beam relative to the axis PSD of the sensor (Figure 2.41).

In both types of photodiode sensors, amplitude-modulated *monochromatic laser* beams are used to irradiate the photodiodes, which makes it possible to significantly suppress the influence of external light interference when using synchronous demodulation of signals from these sensors. In addition, by using proportional methods of signal processing, the effects of temperature dependence of photodiode sensitivity are significantly suppressed in these sensors. The basic proportional length resolution of these sensors without additional optics, given by the signal-to-noise ratio of the photodiode sensors, is up to 10 μm.

To convert photodiode currents into voltage, photodiode amplifiers are used, in which the photodiode is polarized in the closing direction. In this case, its current is directly proportional to the intensity of its illumination (Figure 2.42).

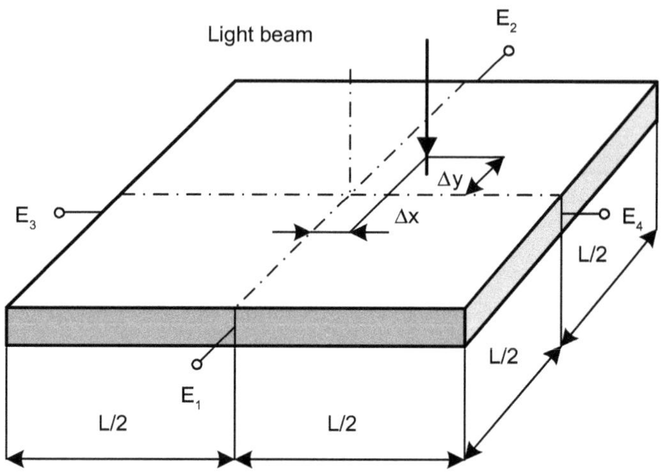

Figure 2.41 Two dimensional position PSD sensor.

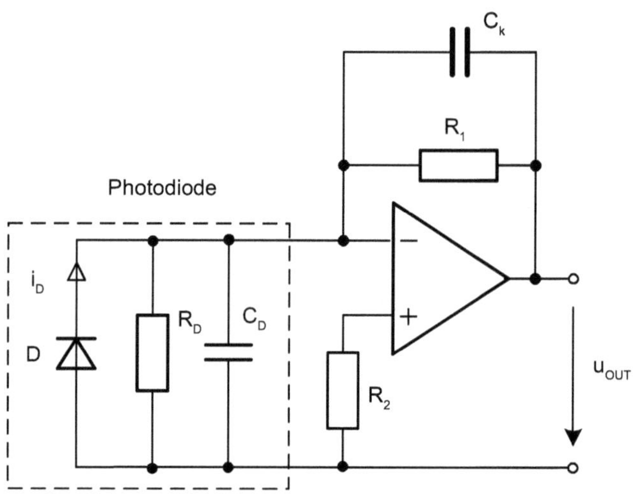

Figure 2.42 Photodiode amplifier.

The influence of input currents of operational amplifiers on their output voltage is suppressed by choosing the same resistances of resistors $R_1 = R_2$. With this option, only the differences in the input currents of the operational amplifiers are applied. The influence of the parasitic capacitances of the photodiodes in the closing C_D is compensated by the correction capacitor C_k, for the capacity of which the

$$C_k = \frac{R_D}{R_1} \cdot C_D, \tag{2.48}$$

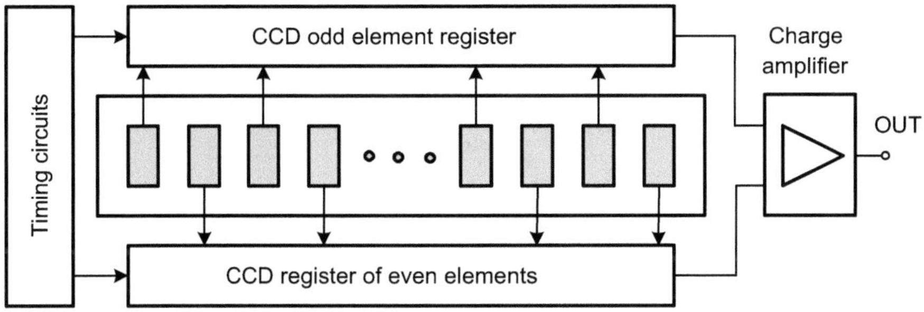

Figure 2.43 Line CCD sensor.

where R_D is the internal resistance of the photodiode in the closing direction of its polarization.

Charge-coupled line and image photodiode sensors use MOS (*metal oxide semiconductor*) technology, which allows the realization of charge-coupled structures with switched CCD capacitors (*charge capacitor devices*) with a high integration density.

The basic element of these sensors is a *photosensitive capacitor*, one transparent electrode of which is irradiated with a light beam. This capacitor generates a charge between the electrodes proportional to the energy of the incident part of the light beam. CCD shift registers are used to transfer the resulting charges to the charge amplifiers that convert the charge to voltage. When charges are moved, they overflow between individual capacitors (blooming), which causes a decrease in the charge transfer efficiency CTE charge transfer efficiency, reaching values of up to 0.99.

Line CCD sensors are made up of a series of photosensitive capacitors (*pixels*) with dimensions of 7 × 7 μm to 100 × 100 μm, the number of which ranges from 128 to 6000, which corresponds to the active length of these sensors in the range of 4–60 mm. The sampling frequency of line sensors reaches units to tens of MS/s, or even hundreds of MS/s.

In line CCD sensors, a two-phase system is usually used to transfer charges from photosensitive capacitors, consisting of temporal interleaving of odd and even samples of the charges of these capacitors (Figure 2.43).

Area CCD sensors are formed by a matrix of photosensitive capacitors with a total number of several tens to hundreds of MPX (*mega pixels*) whose areas are in the range of several to hundreds of mm^2. To transfer charges from the photosensitive capacitors of these sensors, an interline or frame method is used (Figure 2.44).

Interline transfer (ILT) transfers charges from the parallel capacitors of each row to the corresponding elements of the *vertical shift registers*. From

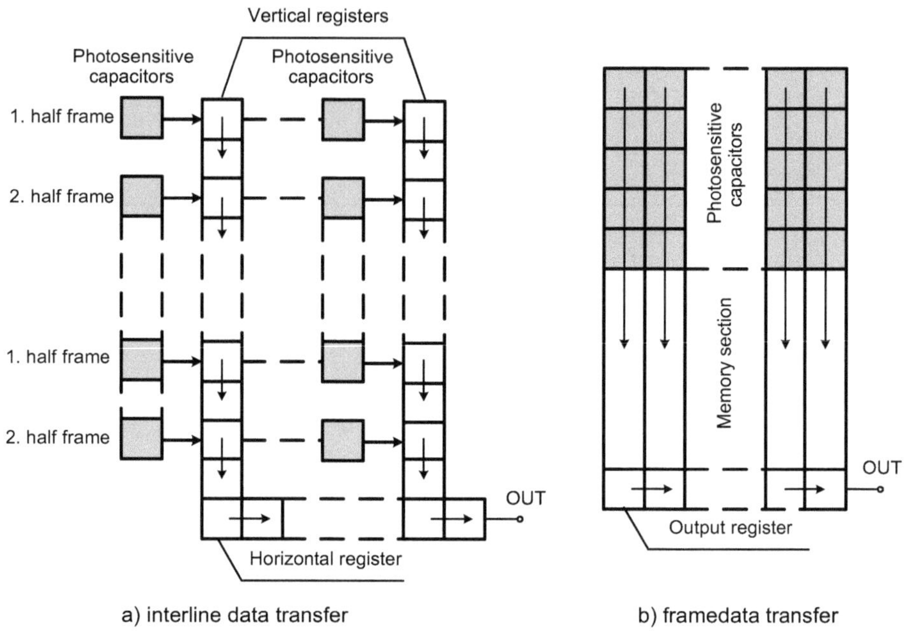

Figure 2.44 Data transfer in area CCD sensors.

the horizontal shift register, the charge obtained in this way is fed to the *charge amplifier*, where it is converted into a sequence of voltage pulses. With this method of transmission, the lines are interlaced, i.e., odd lines of photosensitive capacitors form the first half-frame and even line the second half-frame.

Frame transfer (FT) has the sensor's photosensitive capacitors separated from its memory part. In this method of transmission, individual half-frames are transferred by four-phase control pulses to the parallel vertical registers of the memory part of the sensor, from which the half-frame data is transferred to the horizontal output register of the sensor. To increase the transmission speed, this method of data transmission uses the principle of *time sharing*, during which the captured image is exposed and the data of the previous image is digitized in one time period.

To measure more distant moving objects with optical position sensors, the *triangulation principle* of measurement is most often used, in which the measured object is irradiated with a modulated laser beam, the surface of which has a non-zero reflection coefficient, and the reflection of this beam is concentrated by a coupler-type lens onto the photosensitive part of the optical sensor (Figure 2.45).

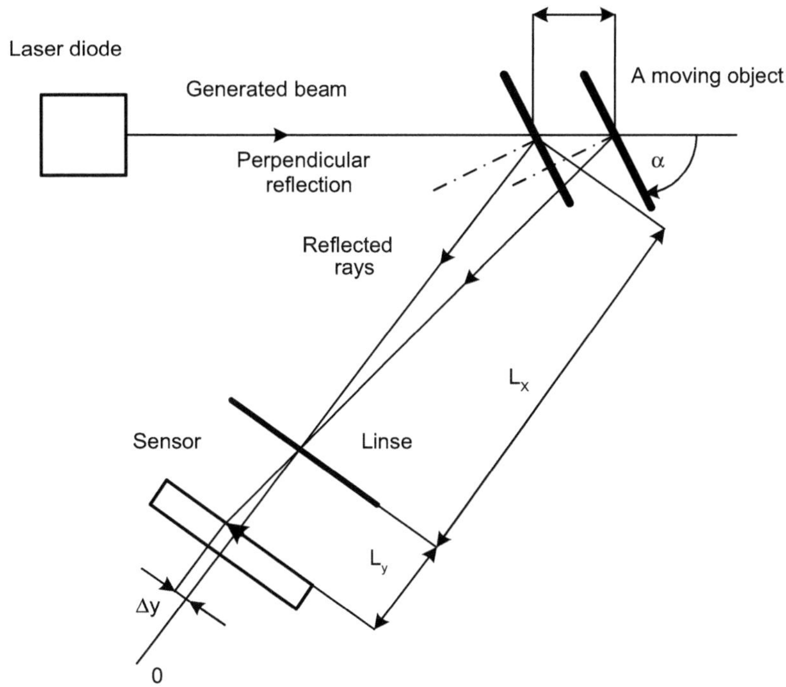

Figure 2.45 Triangulation principle of measuring more distant moving objects.

If this assumption cannot be met or the surface of the measured object is rough or wavy, a reflective surface is glued to the measured object. Assuming that the reflected beam forms an angle α with the axis of the reflective surface, and the axes of the lens and the sensor are perpendicular to this axis, the change in displacement of the measured object Δx corresponds to the change in the position Δy of the reflected beam on the photosensitive layer of a one-dimensional PSD or CCD line optical sensor

$$\Delta y = \sin\alpha \cdot \frac{L_x}{L_y} \cdot \Delta_x, \tag{2.49}$$

where L_x and L_y are the axial distances between the axes of the moving object, the lens, and the sensor.

In simpler cases with which to measure movement or speeds, use *triangulation reflective systems* with one or two photodiodes that work as reflective optical barriers.

Another field of use of semiconductor position sensors is *optoelectronic interferometers*, which use the interference of light waves of a reflected light beam from a moving object. Based on the theory of oscillations, it can be

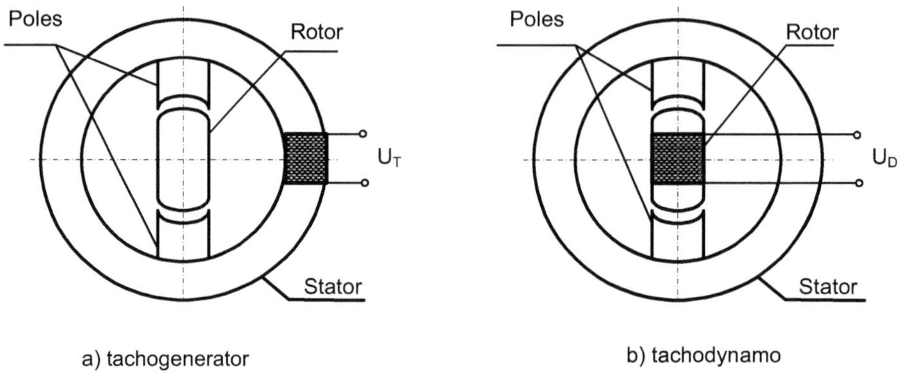

Figure 2.46 Tachometric speed sensors.

deduced that the minimum and maximum values of the reflected light intensities correspond to changes in the position of the measured object

$$\Delta x_1 = N \cdot \frac{\lambda}{2}, \qquad \Delta x_2 = \left(N + \frac{1}{2}\right) \cdot \frac{\lambda}{2}, \qquad (2.50\text{ a, b})$$

where $N = 1, 2, \ldots$ are natural numbers and λ is the wavelength of radiation.

Due to their high spectral sensitivity, especially in the area of infrared radiation with a wavelength of 850 nm, photodiode sensors are also used for non-contact measurement of temperatures and temperature fields (see Section 2.8.5).

2.2 Speed Sensors

Tachometric, inductive, and optical sensors are used to detect the revolutions of the rotating shafts.

2.2.1 Tachometric speed sensors

Tachometric speed sensors are tachogenerators and tachodynamometers (Figure 2.46).

The *tachogenerator* consists of a rotor with prominent poles, made of permanent magnets, whose circular motion induces alternating voltage in the stator coils of the tachogenerator, the effective value and frequency of which are directly proportional to the number of revolutions of the rotor n and the number of its poles N

$$U_T = k_G \cdot n \cdot M, \qquad f = 2 \cdot M \cdot n, \qquad (2.51\text{ a, b})$$

where k_G is the voltage conversion constant of the tachogenerator.

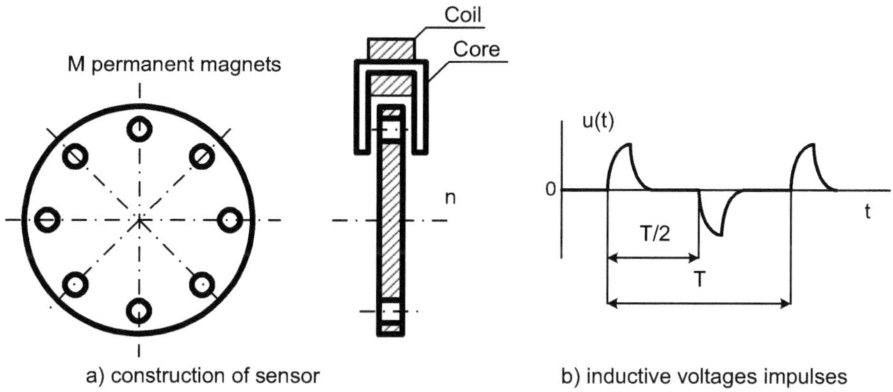

Figure 2.47 Inductive speed sensors.

The *tachodynamo* consists of a stator made of a permanent magnet, in whose direct magnetic field an alternating voltage is induced in the rotor coils, which is rectified by means of a commutator to a direct voltage of medium value

$$U_D = k_D \cdot n \cdot M, \tag{2.52}$$

where k_D is the tachodynamo voltage conversion constant.

2.2.2 Inductive speed sensors

Inductive speed sensors with a sensing coil are formed by a disc with M permanent magnets fixed at equidistant around its circumference. By rotating the disk, their induction fluxes change in the vicinity of these permanent magnets, which causes voltage pulses in their sensing coils with a frequency determined by the number of permanent magnets M and the revolutions n of the rotating disk (Figure 2.47)

$$f = \frac{1}{T} = 2 \cdot M \cdot n. \tag{2.53}$$

Instead of sensing coils, *Hall probes* are also used to indicate changes in the magnetic field of permanent magnets, whose output voltage is proportional to changes in the magnetic field near these magnets.

2.2.3 Optoelectronic speed sensors

According to the method of optical speed sensing, we distinguish between optical sensors with a passing light beam and reflective optical sensors (Figure 2.48).

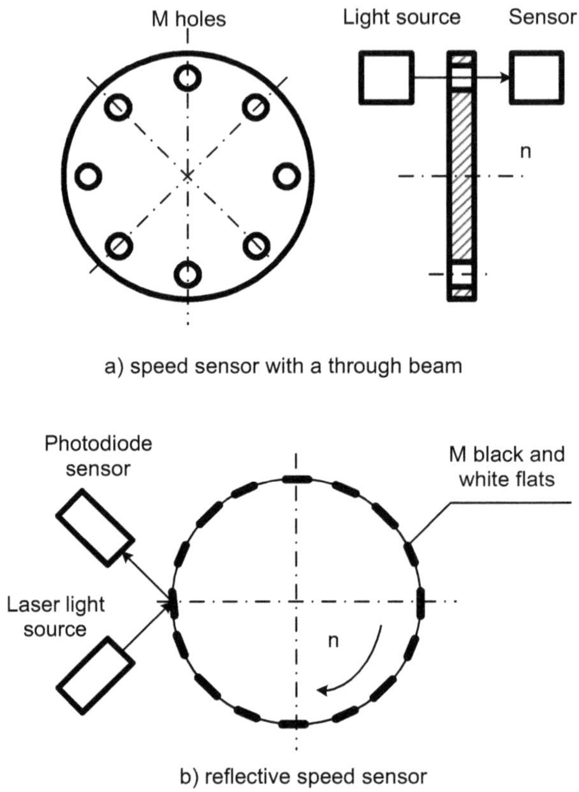

Figure 2.48 Optical speed sensors.

Optical speed sensors with a passing light beam are made up of an optical system containing a light beam source, which is most often a *light emitting diode* (LED) or a *laser diode*, whose light rays pass through the holes of the disk whose revolutions are being measured. The transmitted light rays are sensed by a fast photodiode whose current in the closing direction is frequency modulated

$$f = M \cdot n, \tag{2.54}$$

determined by the revolutions of the disc n and the number of its holes or translucent parts M.

Reflective optical speed sensors use the reflection of a light beam, generated by a laser diode, from evenly distributed reflective surfaces located on the circumference of the shaft, e.g., in the form of glued *zebra tape*. Reflex

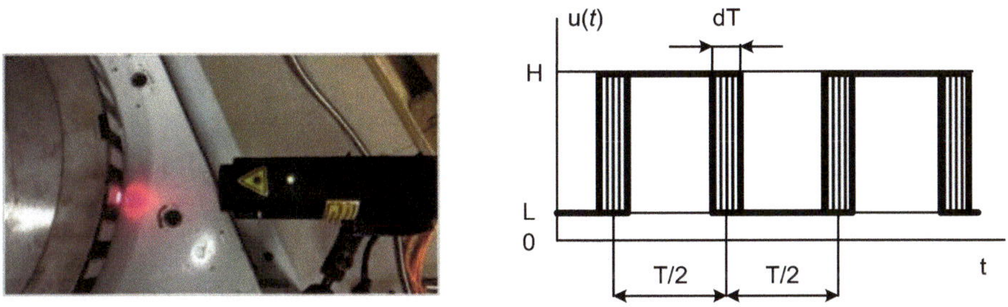

a) reflective laser speed sensor b) time course of the signal from the photodiode sensor

Figure 2.49 Measurement of shaft torsion and its torsional oscillation by the reflection optical method.

speed sensors are also used to measure the torsion of shafts and their torsional oscillation, determined by the proportional changes in times $\Delta T/T$, where

$$T = \frac{1}{n \cdot N} \tag{2.55}$$

is the time of the period of the impulse current signal of the photodiode, n is the number of its revolutions, N is the number of black and white marks on the shaft, and ΔT is the time change of time T, given by the torsional oscillation of the shaft (Figure 2.49).

2.3 Sensors of mechanical stress and force

Tensometric, piezoelectric, magnetoelastic, resonance, and photoelastic measuring principles are most often used to measure mechanical stress in technical practice.

2.3.1 Tensometric sensors of mechanical stress

Wire, foil, and semiconductor *strain gauges* are used to measure mechanical stress and torque. Special epoxy is used for gluing the strain gauge sensors to the investigated structures, which guarantee high adhesion and minimal slippage of the sensors against the structure.

Metal strain gauges are produced in the form of resistance wires or grids from constantan Cu(57) Ni(43), karma Ni(73) Cr(20), nichrome Ni(80) Cr(20), and platinum and tungsten alloys Pt(92) W (8), or these metal alloys are steamed onto elastic mats.

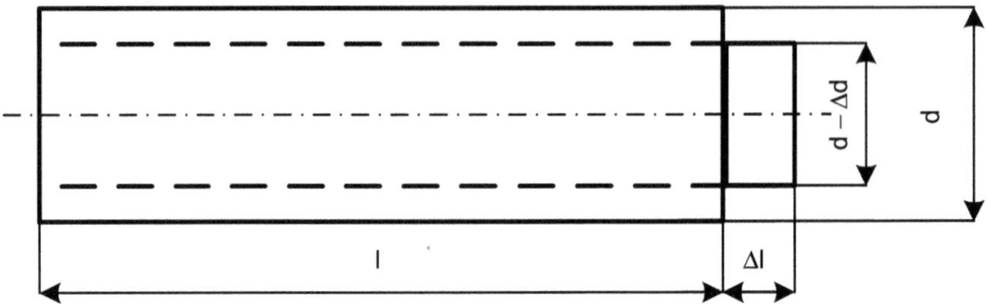

Figure 2.50 Conductor stress and its deformation.

If the conductor is of length l, cross section S, resistivity ρ, and resistance

$$R = \rho \frac{l}{S} \tag{2.56}$$

mechanically stressed in tension, then the proportional change in its resistance is determined by the proportional changes in its length $\Delta l/l$, cross section $\Delta S/S$, and resistivity $\Delta\rho/\rho$ (Figure 2.50)

$$\frac{\Delta R}{R} = \frac{\Delta l}{l} - \frac{\Delta S}{S} + \frac{\Delta \rho}{\rho}$$

Assuming that there are no structural changes due to the stress of the conductor material, which can be assumed for changes in the length of the conductor $\varepsilon = \Delta l/l < 10^{-3}$, the proportional change in the resistance of the conductor related to the proportional change in its length can be expressed in the form

$$\frac{\Delta R}{R} / \frac{\Delta l}{l} = k_\varepsilon = 1 + 2 \cdot \mu + \frac{\Delta \rho}{\rho} / \frac{\Delta l}{l} = 1 + 2 \cdot \mu + \pi_e \cdot E, \tag{2.57}$$

where k_ε is the coefficient of deformation sensitivity, μ is *Poisson's constant*, π_e is the piezoresistive factor, and E is *Yong's modulus of elasticity.*

Important properties of strain gauges are changes in their stress sensitivity $dK_\varepsilon/d\varepsilon$, the temperature dependence of their resistance $\alpha_R = dR/d\theta$, and the temperature dependence of the deformation sensitivity coefficient $\alpha_{k\varepsilon} = dk_\varepsilon/d\theta$.

Metal strain gauges, most often made of nichrome or platinum and tungsten alloys, have sensitivity $dk_\varepsilon/d\varepsilon = 10^{-3}$, deformation sensitivity $k_\varepsilon = 2.1$, resistance temperature coefficient $\alpha_R = 10^{-6}$/K, and temperature coefficient of deformation sensitivity $\alpha_K = 5 \cdot 10^{-6}$/K. *Semiconductor silicon strain gauges* have sensitivity $dk_\varepsilon/d\varepsilon = 10^{-2}$, deformation sensitivity $k_\varepsilon = 125$, resistance temperature coefficient $\alpha_R = 10^{-5}$/K, and temperature coefficient of deformation sensitivity $\alpha_K = 2 \cdot 10^{-6}$/K.

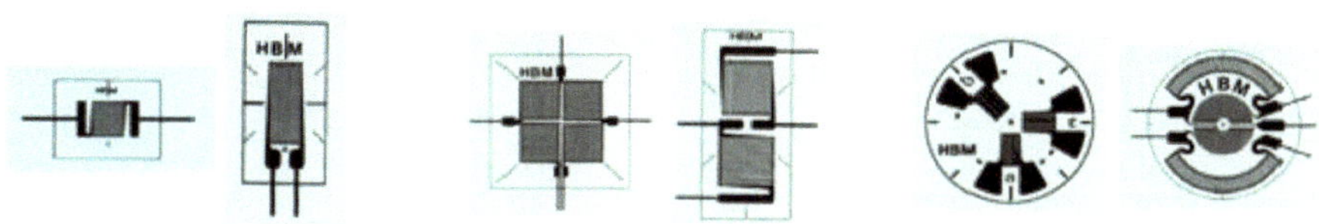

Figure 2.51 Types of strain gauges.

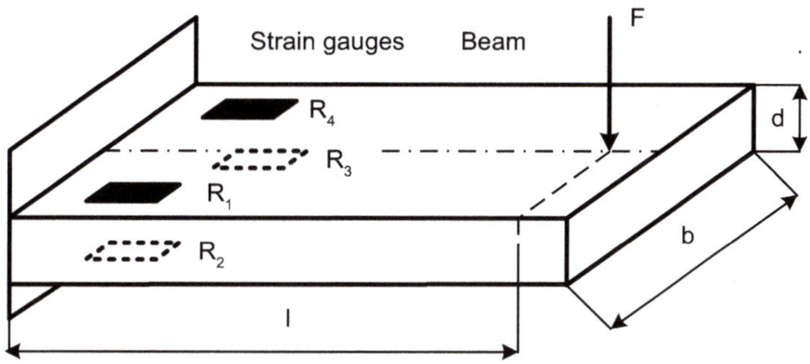

Figure 2.52 Deformation element with embedded beam.

Single-axis tensometers are used to measure longitudinal strain, and strain gauge crosses or rosette gauges are used to measure transverse strain (Figure 2.51).

Various shapes of deformation members are used to measure force, whose deformations are detected by strain gauges. Up to forces of tens of kN, deformation members with a *threaded beam* are most often used, which is stressed at its free end by a force F acting at a distance l from its threaded point (Figure 2.52).

The resulting deformation of the beam material in the vicinity of its insertion is determined by the *bending moment* M_o and the *cross-sectional modulus* in bending W_o of its profile

$$\varepsilon = \frac{M_0}{E \cdot W_0} = \frac{F \cdot l \cdot 6}{E \cdot b \cdot d}. \tag{2.58}$$

When the beam is loaded with a force F, a pair of strain gauges with resistances R_1 and R_4, located on the upper part of the beam, are subjected to pressure, and their resistances therefore increase. A pair of strain gauges of resistances R_2 and R_3 are stressed with pressure, and their resistances therefore decrease by the same differences ΔR as those of R_1 and R_4 but with the opposite sign.

In practice, these prototype strain gauges are connected to the bridge so that the change in resistance of the strain gauges $\Delta R/R$ corresponds

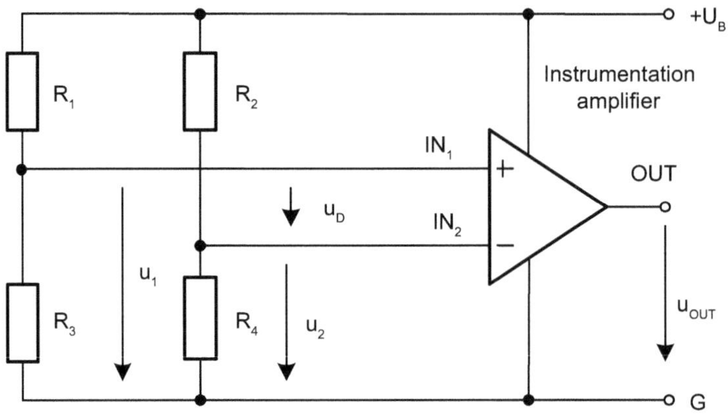

Figure 2.53 Connecting the strain gauges to the resistance bridge.

to the maximum change in the differential voltage of the bridge $\Delta u_D/u$ (Figure 2.53).

In addition, the bridge arrangement of strain gauges significantly reduces the influence of their temperature dependence on the accuracy of determining mechanical stress.

If to amplify the differential voltage of the bridge, a differential instrumentation amplifier with differential amplification G_D of the differential voltage $u_c = (u_1 + u_2)/2$ and common amplification G_C of the common voltage $u_c = (u_1 + u_2)/2$ is used, then for the output voltage of the instrumentation amplifier

$$u_{\mathrm{OUT}} = G_D \cdot u_D + G_C \cdot u_C = \frac{\Delta R}{R} U_B + G_C \cdot \frac{U_B}{2}. \tag{2.59}$$

For example, with a proportional change in the resistance of strain gauges of 5% and a supply voltage of 5 V in the bridge, the difference voltage of the bridge is 50 mV and the concordant voltage is 2.5 V. If an instrument amplifier with differential gain $G_D = 100$ and concordant gain $G_C = 10^{-3}$ is used to amplify the differential voltage of the bridge, then the error component of the voltage 2.5 mV is caused by the concordant voltage of the strain gauge bridge of 2.5 V.

In more demanding applications, instead of DC voltage excitation of resistance bridges, their AC current supply is used, which eliminates the occurrence of errors caused by thermoelectric voltages due to the existence of different metal materials of their supply cables and different temperatures at the place of their connection. For example, in the connection of a NiCr strain gauge with a copper conductor, a thermoelectric voltage of tens of μV/°C is generated.

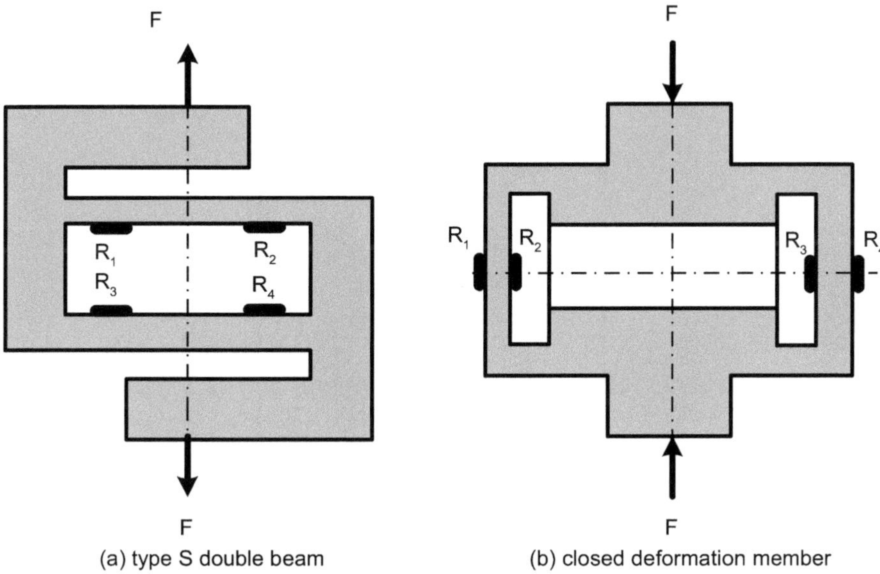

Figure 2.54 Deformation elements for measuring higher forces.

To measure greater forces in the range of tens of kN to MN units, either deformation members consisting of double beams of type *S*, stressed to bend, or closed deformation members, stressed to tension and compression (Figure 2.54), are used.

In both types of deformation members, four strain gauges are again used to measure the mechanical stress of their flexible parts, of which strain gauges with resistance R_1 and R_4 are stressed to tension and strain gauges with resistance R_2 and R_3 are stressed to pressure.

A frequently used deformation member in the form of a cuboid or cylinder, axially loaded with the measured force *F*, can measure forces in the range of up to several tens of MN (Figure 2.55).

For this deformation member, a pair of strain gauges with resistances R_2 and R_4 sense the longitudinal deformation $\varepsilon_{\text{long}}$ and a pair of strain gauges with resistances R_1 and R_3 sense its transverse deformation ε_{tr}, which is smaller in the ratio of *Poissson's constant* μ of the material of the deformation component with respect to the longitudinal deformation. For the differential voltage of a strain gauge bridge supplied by U_B, the following applies

$$u_D = k_\varepsilon \cdot U_B \cdot \varepsilon_{\text{long}} \cdot \left(1 + \frac{1}{\mu}\right), \tag{2.60}$$

where k_ε is the strain sensitivity coefficient of the strain gauges.

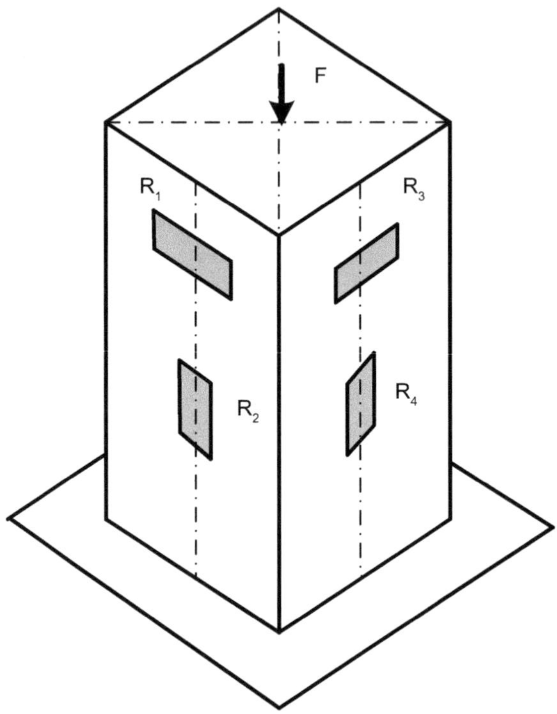

Figure 2.55 Cuboid shaped deformation element for measuring higher forces.

2.3.2 Piezoelectric sensors of mechanical stress

These sensors use the *piezoelectric effect*, consisting in the polarization of some crystalline and polycrystalline piezoelectrics, and thus for the subsequent generation of electric charge on electrodes located perpendicular or parallel to the direction of their mechanical stress (Figure 2.56).

If the cut of a piezoelectric crystal is stressed by the force F_x perpendicular to its electrodes, then the charge on its electrodes does not depend on the dimensions of the crystal and is determined only by the piezoelectric constant k_{pz}

$$q_x = k_{pz} \cdot F_x. \tag{2.61}$$

The most common piezoelectrics are, for example, silica SiO_2, barium titanate $BaTIO_3$, lead titanate $PbTiO_3$, and zirconium titanate $PbZrO_3$, whose proportional permittivity and piezoelectric constants are given in Table 2.5.

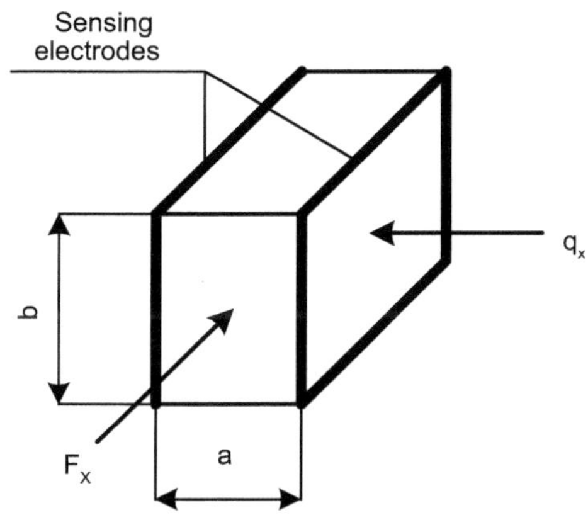

Figure 2.56 Piezoelectric crystal cut.

Table 2.5 Piezoelectric constants of the most common piezoelectrics.

Piezoelectric	**Relative permittivity** ε_r	k_{pz} **(pC/N)**
SiO_2	4.5	2.3
$BaTiO_3$	10^3	374
$PbTiO_3$	200	51
$PbZrO_3$	$1.6 \cdot 10^3$	374

To convert charge to voltage, piezoelectric transducers use charge amplifiers consisting of an inverting connection of an operational amplifier in whose feedback branch the C_i integration capacitor is connected (Figure 2.57).

When replacing the piezoelectric transducer with a current source i_s with an internal resistance R_s and a capacitance C_s, assuming the ideal properties of an operational amplifier with infinite differential gain, zero input residual voltage, and current, the output voltage of the charge amplifier is

$$u_{\mathrm{OUT}}(t) = -\frac{1}{C_i}\int i_s(t)\cdot dt = -\frac{Q}{C_i} = -\frac{k_{\mathrm{pz}}\cdot F_x}{C_i}, \tag{2.62}$$

where Q is the charge on the electrodes of the piezoelectric sensor.

Ideally, the output voltage of the integration amplifier is not affected by the internal capacitance and resistance of the sensor, including the leakage capacity and the leakage resistance of the supply cable.

Grounding of the piezo sensor housing and the charge amplifier input at different locations can therefore cause undesirable interference by the flow of earth currents through a common section of conductor between different

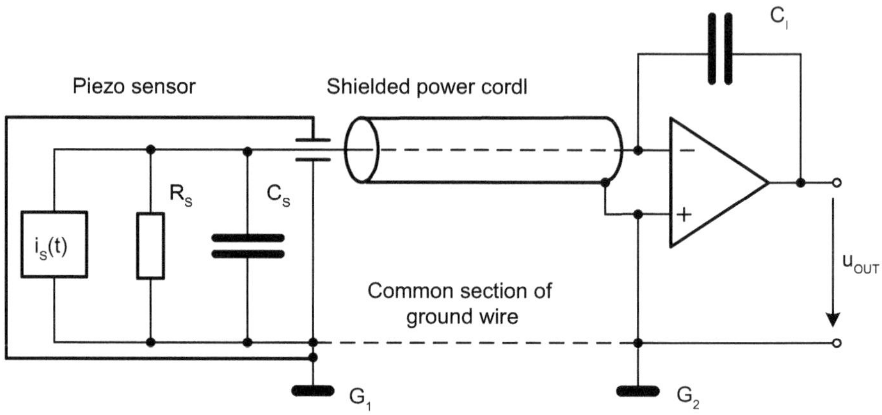

Figure 2.57 Connection of the piezo sensor to the amplifier.

countries G_1 and G_2. This deficiency is eliminated by the connection of the piezo sensor to the *differential amplifier* according to Figure 2.58.

In addition to the A_1 integration amplifier with the C_{i1} integration capacitor, an inverting amplifier A_2 with –1 transmission is used for this connection, which introduces negative integration feedback into the non-inverting input of the Z_1 amplifier with the C_{i2} integration capacitor. This treatment ensures that none of the sensor electrodes are grounded.

A shielded pair with twisted conductors is used to connect the piezo sensor to the differential charge amplifier. The sensor shielding, including the twist pair shielding, is grounded at the location of the G_1 differential integration amplifier, which may be different from its G_2 signal country, provided the power supply of the differential integration amplifier is isolated.

To eliminate the influence of the temperature dependence of the piezoelectric constant a pair of sensors is used, which are exposed to the opposite direction of stress. The charges of these sensors are converted by the A_1 and A_2 integration amplifiers to voltages, the difference between which is amplified by the A_3 difference amplifier (Figure 2.59).

Assuming identical properties of the piezo transducer, i.e., their piezoelectric constants, conformity of capacities of integration capacitors $C_{i1} = C_{i2}$, and paired conformity of resistors $R_1 = R_2$ and $R_3 = R_4$, the output voltage of the differential integration amplifier is

$$u_{\mathrm{OUT}} = 2 \cdot k_{\mathrm{pz}} \cdot \frac{R_3}{R_1} \cdot \frac{F}{C_i}. \tag{2.63}$$

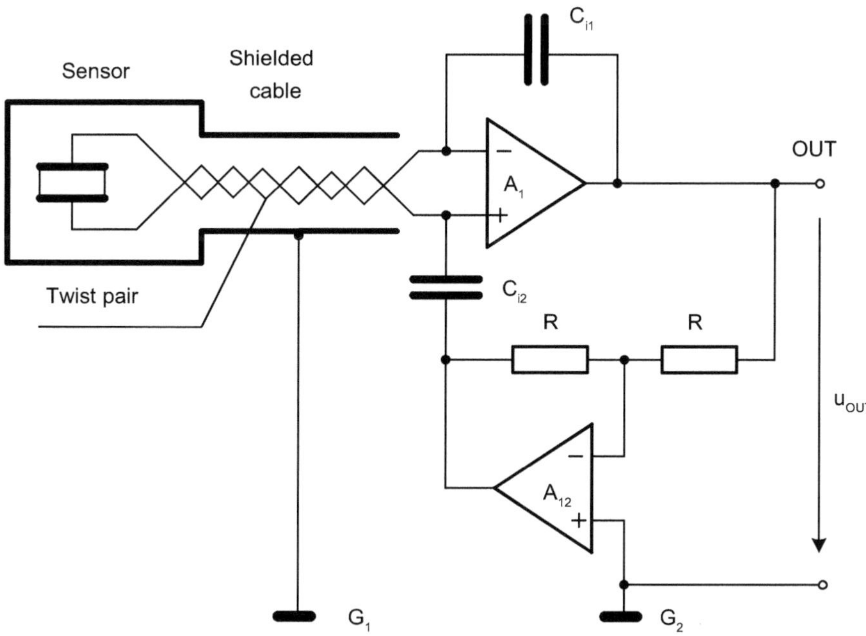

Figure 2.58 Connection of the piezo sensor to the differential charge amplifier.

2.3.3 Magnetoelastic load cells

They use changes in the permeability of some ferromagnetic materials, e.g. FeNi, caused by their mechanical deformation

$$\frac{\Delta\mu}{\mu(0)} = \lambda \cdot \mu(0) \cdot \sigma. \tag{2.64}$$

where $\Delta\mu = \mu(0) - \mu(\sigma)$ is the difference between the permeabilities of unstressed and stressed ferromagnetics by deformation σ and λ is the coefficients of magnetostriction.

Since the *coefficient of magnetostriction* is proportional to the ratio of squares of magnetic induction of stressed and unstressed ferromagnetics, it applies to the proportional change in the permeability of the stressed ferromagnetics

$$\frac{\Delta\mu}{\mu(0)} = k \cdot \left[\frac{B(\sigma)}{B(0)}\right]^2 \cdot \mu(0) \cdot \sigma, \tag{2.65}$$

This change in permeability causes a change in the inductance of the coil in the closed magnetic circuit of the sensor, which is converted in the impedance

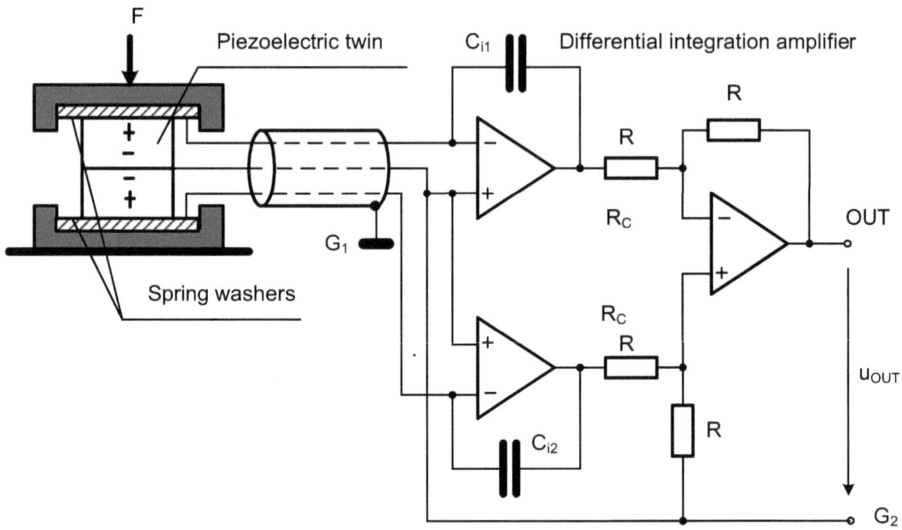

Figure 2.59 Connection of a pair of piezo sensors to a differential charge amplifier.

converter into a change in the voltage amplitude of the alternately excited current coil.

Suitable magnetostrictive materials are materials with a crystalline structure whose magnetostriction factor ranges from 10^{-u} to 10^{-0}. Promising magnetoelastic materials are, for example, *amorphous magnetic materials* such as compounds of iron, beryllium, and silicon in the percentage composition Fe(80) B(14) Si(6), whose magnetorestriction factor is $3 \cdot 10^{-5}$. In addition, these materials allow mechanical stress up to $\sigma_{max} = 1.5 \cdot 10^3$ Pa.

Magnetoelastic force sensors consist of a ferromagnetic cylindrical core in which the winding of the cylindrical coil is stored. Deformation of the walls of the core surrounding the coil increases the permeability of ferromagnetics and thus increases the inductance of the coil (Figure 2.60).

The *transformer magnetoelastic sensors* consist of a primary winding, which is excited by alternating current with frequencies of several tens of kHz, and a secondary winding in which alternating currents are induced, the amplitude of which is directly proportional to the change in the permeability of the ferromagnetic core. The transformer principle is also used by magnetoelastic sensors, consisting of a ring of amorphous magnetic materials, on which the primary and secondary coils of the resulting transformers are wound.

String resonant force sensors are used to determine the strength the change in the resonant frequency of a string (wire), which depends on its length l, the magnitude of its tension force F, and density ρ

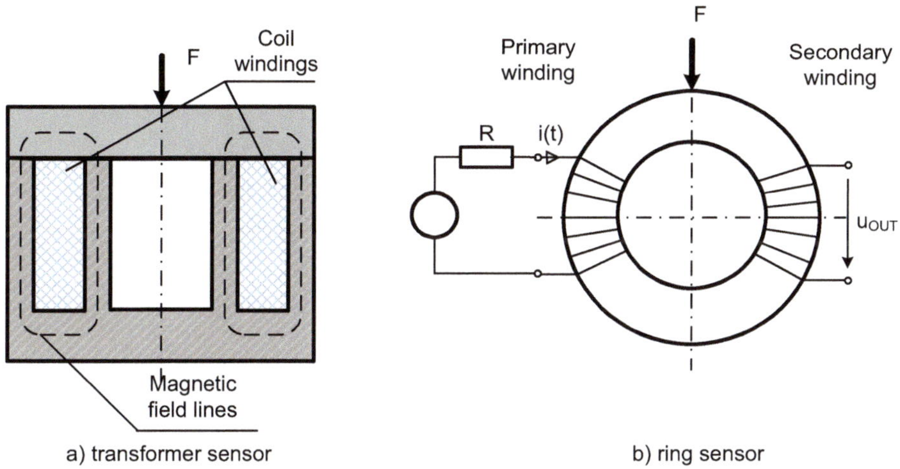

Figure 2.60 Magnetoelastic force sensors.

$$f_r = \frac{N}{2 \cdot l}\sqrt{\frac{F}{\rho}}, \tag{2.66}$$

where N is a natural number.

These sensors, usually consisting of a pair of strings, are loaded with a measured force and vibrated by an alternating electromagnetic field. From the maximum deflection of such excited strings, sensed by the sensing coil, the resonant frequency of the strings is determined. Thanks to the high quality factor of this electromechanical resonance system, string force transducers have a high linearity in the order of 10^{-F} and very good time stability.

2.3.4 Micromechanical force sensors

They work on the principle of propagation of the *surface acoustic wave* (SAW), which propagates from the ridge structure of the electromechanical transmitter to the place of the electromechanical receiver. These sensors use micromechanical technology MEMS (*micro electro mechanical systems*; Figure 2.61).

The comb shapes of the electrodes, formed on the base plate of quartz or similar semiconductor, cause deformation due to the indirect piezoelectric effect on the transmitter side of the sensor. The resulting *Rayleigh-type* surface waves propagate to the receiving structure of the transducer.

Based on the direct *piezoelectric effect*, the resulting waves are transformed into an electrical signal. The minimum wave attenuation between the transmitting and receiving piezoelectric converter occurs when the

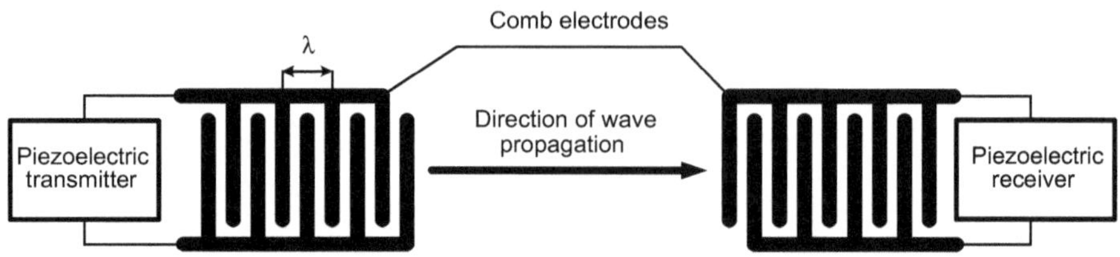

Figure 2.61 Resonant sensor with surface acoustic wave.

pitch between the teeth d of the converter corresponds to half the wavelength $\lambda/2$.

By connecting this sensor to the amplifier feedback and by phase shift φ, an electromechanical harmonic generator with a frequency is created when the oscillation condition is met

$$f_{\text{osc}} = \frac{1}{\Delta t}\frac{2 \cdot \pi \cdot N - \varphi}{2 \cdot \pi} = \frac{v_{\text{SAW}}}{l}\left(n - \frac{\varphi}{2 \cdot \pi}\right), \tag{2.67}$$

where v_{PAV} is the velocity of the wave and $\lambda = v_{\text{SAW}}/f$ is its wavelength.

As with string sensors, the frequency of oscillations in SAW sensors is determined by integral multiples n of the fundamental frequency of the electromechanical resonator.

Micromechanical resonance sensors are produced in various forms, e.g., in the shape of a silicon prism, double beam, or diaphragm. Piezoelectric actuators are used to excite of oscillations. For oscillation sensing, piezoelectric, strain gauge, and capacitive pressure sensors.

The wave propagation speed of silicon structures is approximately 10^3 m/s. When micromechanical sensors with silicon membrane are used, the velocity of wave propagation is inversely proportional to its density ρ

$$v = \frac{\alpha}{\sqrt{\rho}}, \tag{2.68}$$

where α is the shape constant of the membrane, determined by its diameter and thickness.

2.3.5 Torque and transmitted power sensors

The classic way of measuring torque and transmitted power is to measure the deformation of the inserted deformation member between the engine and the load (brake).

The analysis of mechanical stresses on a shaft or deformation component shows that the tangent components of the mechanical stress are oriented at angles of 45° to the shaft or deformation component axes (Figure 2.62).

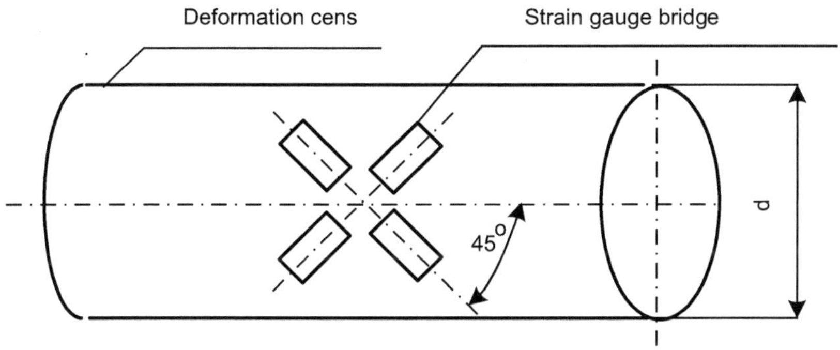

Figure 2.62 Location of strain guage for sensing shaft torque.

If the circular shaft (elastic member) is stressed by the *torque* M_k, then the tangent stress e_t on the shaft surface is defined by the equation

$$\varepsilon_t = \frac{M_k}{W_k}, \tag{2.69}$$

where W_k is the cross-sectional modulus in torsion.

For a *solid shaft* with diameter d and for a *hollow shaft* with an inner diameter d and an outside diameter D, the moduli of elasticity are defined by the equations

$$W_k(d) = \frac{\pi \cdot d^3}{16}, \qquad W_k(d,D) = \frac{\pi}{16 \cdot D}(D^4 - d^4). \tag{2.70 a, b}$$

The transmitted power between the motor and the load is then determined by torque and speed

$$P = M_k \cdot \omega. \tag{2.71}$$

The torque can also be determined from the torsional stress on the shaft or deformation component, which is defined by the equation based on *Hooke's law*

$$\varphi = \frac{M_k}{G\langle I_p} \cdot l[\text{rad}], \tag{2.72}$$

where l is the length of the shaft, G is the *shear modulus* of the material, and I_p is the polar torque of the shaft profile (Figure 2.63).

For polar moments of circular and inter-circle shaft profiles,

$$I_p(d) = \frac{\pi}{32}d^4, \qquad I_p(d,D) = \frac{\pi}{32}(D^4 - d^4). \tag{2.73 a, b}$$

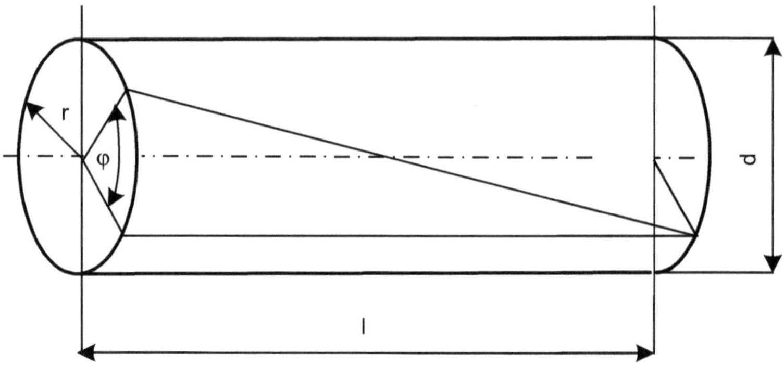

Figure 2.63 Torsion angle of solid shaft.

In the context of the definition of torsion, the *relative torsion* of a shaft is also defined, given by its torsional angle φ related to the length *l* of the shaft, which is expressed either in curve (radians) or in degrees

$$\alpha = \frac{\varphi}{l} = \frac{M_k}{G \cdot I_p}[\text{rad}] = \frac{M_k}{G \cdot I_p} \cdot \frac{180}{\pi}\ [° / \text{m}]. \tag{2.74}$$

Magnetic sensors with a *Hall probe* or retroreflective optical sensors of the light beam reflection speed from evenly distributed reflective facets located on the shaft, e.g., in the form of glued *zebra tape*, are most often used to measure the speed or torsion of the shaft.

Torque sensors using magnetic reluctance changes tubes made of ferromagnetic material consist of two rows of slots. If the axes of these slits are oriented in the direction of shear stresses, i.e., in the direction of 45° in relation to the axis of the tube, then the measured torque transmitted by this tube causes the slits to increase in one row and to shrink in the other row of slits (Figure 2.64).

If the tube is magnetized by an alternating magnetic field through a primary winding with the number of turns N_1, then the differential voltage of symmetric secondary windings of equal numbers of turns $N_2 = N_3$, located symmetrically around the tube, corresponds to the difference in the magnitude of the magnetic resistances of the slots of the pipe deformed by the transmitted torque.

Magnetic anisotropy of *amorphous magnetic materials* is also used to measure torque. These sensors consist of two mechanically pre-stressed films made of cobalt–silicon–beryllium alloy in the percentage composition Co(75) Si(15) Be(10), which are placed on the shaft so that the direction of

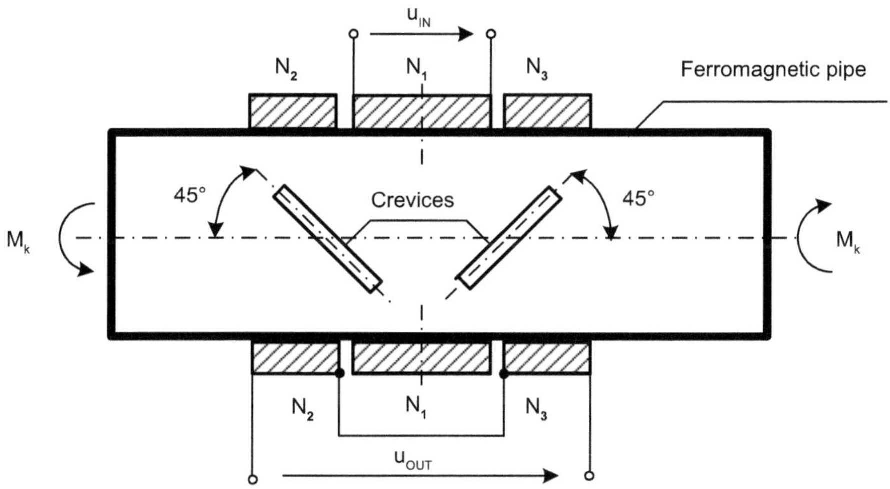

Figure 2.64 Torque sensor with ferromagnetic tube.

their anisotropy forms an angle of 45° with the shaft axis, which ensures sensing of the tangential stress of the shaft σ in two perpendicular axes (Figure 2.65).

On the shaft, there is a primary winding of a transformer with the number of turns N_1, which is excited by an AC source, and two symmetrical secondary windings with the same number of turns $N_2 = N_3$, which are connected antiserially. If the transmitted torque is zero, then the magnetic foils are not mechanically stressed and the differential voltage of the secondary windings is zero.

If the shaft is stressed by non-zero torque, then the *magnetic foils* are mechanically stressed, which causes a change in their permeability and the differential voltage of the secondary windings is directly proportional to the sensed torque. Synchronous detection is used to rectify the AC differential secondary voltage, synchronized by the AC excitation source.

Torque sensors using magnetic anisotropy of the shaft material work on a similar principle. These sensors use two coils with ferromagnetic cores, located above the shaft in two mutually perpendicular directions, shifted by 45° relative to the shaft axis.

Since the magnetic anisotropy of steel, from which shafts are usually made, is approximately two orders of magnitude smaller compared to the magnetic anisotropy of CoSiBe alloys, these sensors are less accurate and sensitive compared to foil sensors.

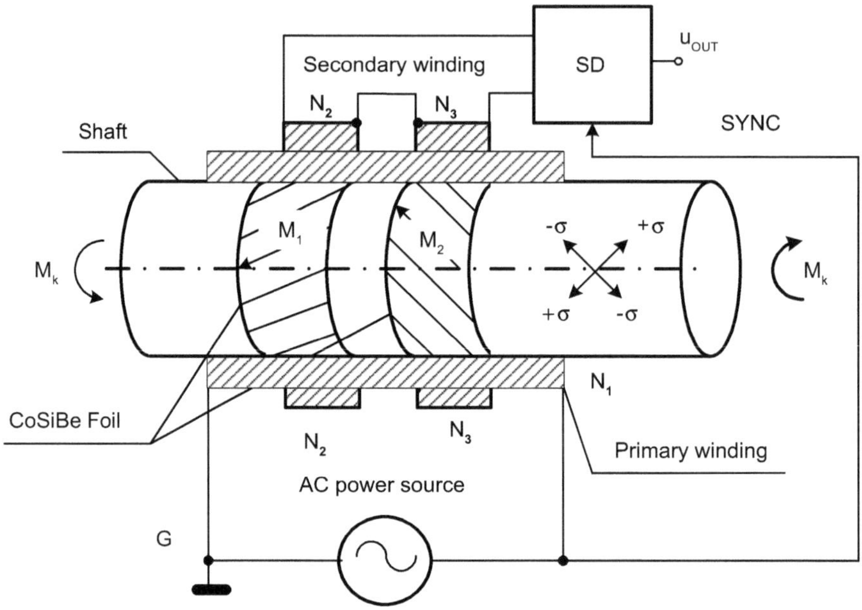

Figure 2.65 Torque sensor with magnetic foil.

Torque sensors using the *Wiedamann effect* consist of a ferromagnetic cylinder (tube) through which alternating current flows. This creates a circular magnetic field with an H_C intensity vector whose direction coincides with the coil axis strung on the cylinder (Figure 2.66).

If the ferromagnetic head pulley is not subjected to torque stress, then the induced voltage in the telecoil is zero. When the cylinder is subjected to torque $M_k = F{\cdot}L$, the magnetic field strength vector $\boldsymbol{H}_C$ is rotated by the torsion angle α and the resulting component of magnetic field strength $\boldsymbol{H}_l$, perpendicular to the telecoil surface, induces an alternating voltage proportional to the sensed torque.

2.4 Pressure Sensors

Diaphragm sensors are most commonly used to sense the pressure of gaseous or liquid substances, while strain gauge, capacitive, and piezoelectric sensors are used to measure the deformation and deflection of their membranes.

In terms of function, we distinguish *absolute pressure sensors*, measuring pressure against zero vacuum pressure, *relative pressure sensors*, measuring pressure relative to ambient pressure, and *differential pressure sensors*, measuring pressure difference.

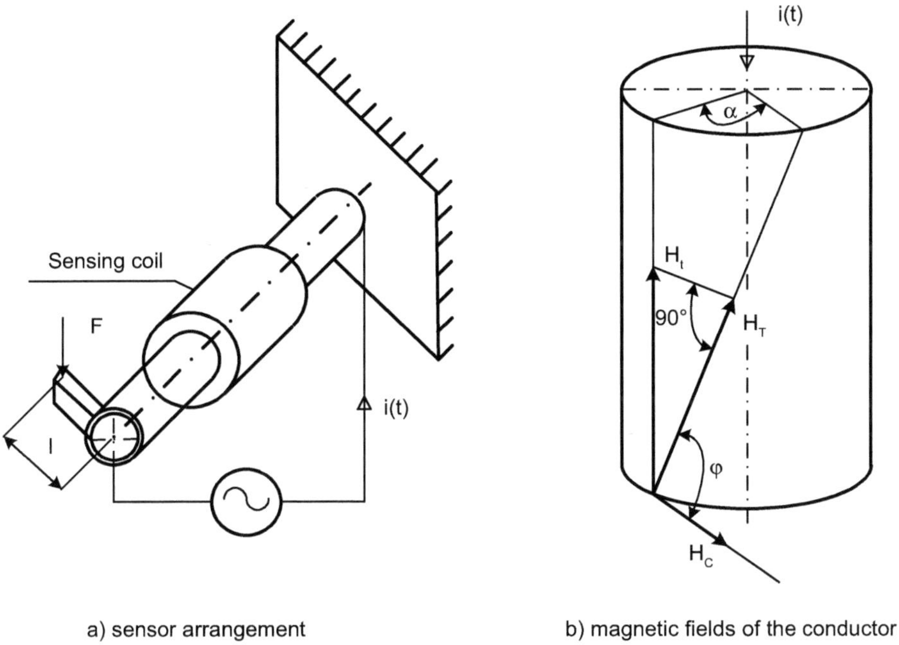

Figure 2.66 Torque sensor using the Wiedemann effect.

2.4.1 Strain gauge pressure sensors

A *differential pressure sensor* with a diaphragm whose deflection is determined by pressure difference $\Delta p = p_1 - p_2$ is shown in Figure 2.67.

Assuming that the deflection of the membrane s is negligible with respect to its thickness h, the radial and tangential components of its mechanical stress can be determined at a distance r_2 from its axis

$$\sigma_r = \frac{3}{4}\Delta p \frac{z \cdot r_1^2}{h^3}\left[1 + \mu - (3 + \mu)\frac{r_2^2}{r_1^2}\right],$$

$$\sigma_t = \frac{3}{4}\Delta p \frac{z \cdot r_1^2}{h^3}\left[1 + \mu - (3 \cdot \mu + 1)\frac{r_2^2}{r_1^2}\right], \qquad (2.75\text{ a, b})$$

where z = 0 specifies the neutral plane lying in the middle of the membrane thickness and μ is the *Poisson constant* of the membrane material.

It follows from the above equations that on the neutral plane $z = 0$, both stress components are zero, and on both surfaces of the membranes, i.e. for $z = \pm h/2$, on radius r_2, both stresses are proportional to the ratio $(r_1/r_2)^2$.

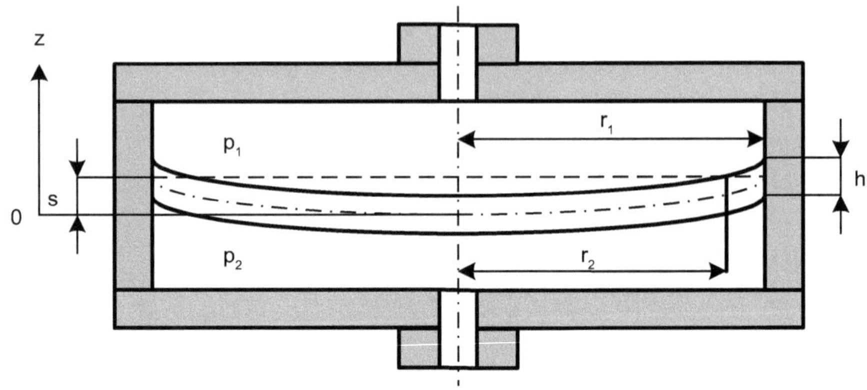

Figure 2.67 Diaphragm differential pressure sensor.

A suitable strain gauge probe for determining these stresses is, for example, a *foil strain gauge sensor* in the shape of a rosette, consisting of two sensors located on its edge, measuring the radial stress of the diaphragm, and two spiral sensors for measuring the tangential component of the mechanical stress on the membrane.

These four strain gauge sensors are connected to the bridge, and when determining the deformation sensitivity of the bridge, given by the ratio of the change in its differential stress to the change in pressure difference $\Delta u_d/\Delta p$, it must be taken into account that the radial and tangential components of the mechanical stress on the membrane are determined by the relations

$$\varepsilon_r = \frac{\sigma_r - \mu \cdot \sigma_t}{E}, \qquad \varepsilon_t = \frac{\sigma_t + \mu \cdot \sigma_r}{E}. \qquad (2.76\ \text{a, b})$$

where E is the *Yong modulus of elasticity* in the tensile of the membrane material.

For higher pressures, tens to hundreds of MPa, *tube sensors* are also used instead of membrane sensors, on the surface of which four strain gauges are placed according to Figure 2.68(a).

The maximum deformation sensitivity of the bridge can be achieved when the strain gauges are positioned so that they detect only the tangent components of the deformation of the tube. If strain gauges can also detect the axial components of the tube deformation, for thin tubes with a wall thickness $d < 0.1 \cdot r$, the ratio between axial and tangential deformation is $\varepsilon_a/\varepsilon_t \approx 0.5$.

Modified tubular pressure sensors consisting of a deformation tube or a part of a deformable pipe that is wrapped with strain gauge wire or strain gauge foil are used for continuous measurement of higher pressures of gases and liquids in the pipeline (Figure 2.68(b)).

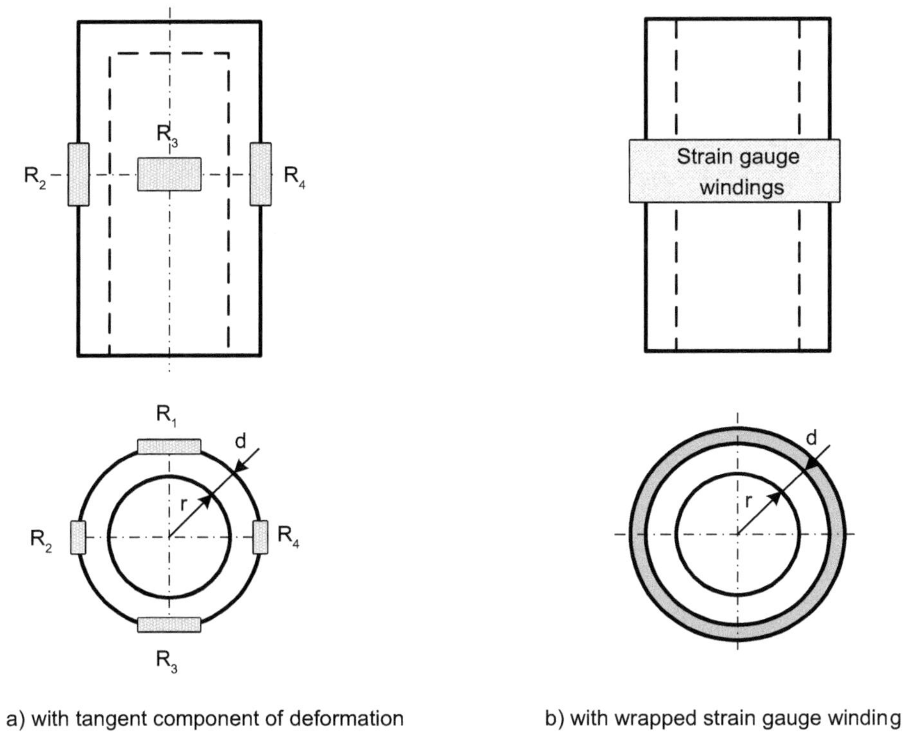

Figure 2.68 Tubular pressure sensor.

A change in the pressure in the tube causes a change in its diameter and thus in the circuit, which causes a change in the resistance of the strain gauge winding or strain gauge film.

The *monolithic integrated silicon diaphragm pressure sensors* in Figure 2.69 have piezoresistors directly on the diaphragm with a thickness of several μm implemented by diffusion or ion technology, whose resistance varies with their mechanical stress. Piezoresistors PR_1–PR_4 are again connected to the resistance bridge, which practically eliminates their considerable temperature resistance dependence. These compact sensors allow you to measure pressures in the range of several kPa to tens of MPa with an accuracy of up to 0.5%.

2.4.2 Capacitive pressure sensors

They use capacitance changes between the conductive membrane E_3 and two fixed electrodes E_2 and E_3 (Figure 2.70) to measure pressure.

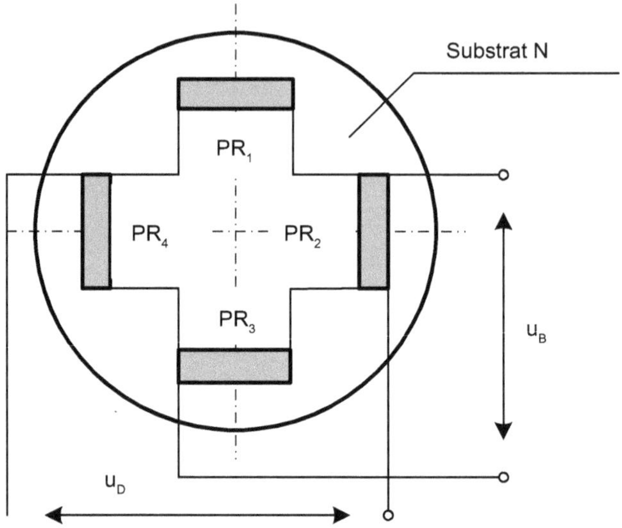

Figure 2.69 Location of piezoresistors on silicon membrane.

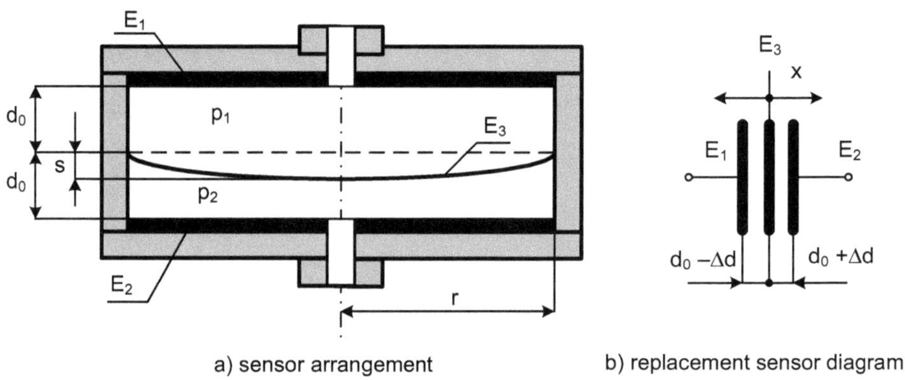

Figure 2.70 Capacitive pressure sensor.

Assuming that the deflection with respect to its diameter r is negligibly small, i.e., that $s << r$, its deflection is directly proportional to the difference in pressures in the space above and below it. The mean value of their distances is then equal to half the deflection of the membrane $d = s/2$. Under these assumptions, the inter-electrode capacitances of this differential capacitor

$$C_{13} = \varepsilon_1 \cdot \frac{\pi \cdot r^2}{d_0 + \Delta d}, \qquad C_{23} = \varepsilon_2 \frac{\pi \cdot r^2}{d_0 - \Delta d}, \qquad (2.77\ \text{a, b})$$

where ε_1, ε_2 are the dielectric permittivities between the electrodes E_1, E_3 and E_2, E_3 and d_0 are the distances between the diaphragm and the fixed electrodes of the sensor at zero differential pressure.

The difference in inter-electrode capacities then determines the difference in measured pressures

$$\Delta p = k_s(C_{13} - C_{23}) = k_s \frac{\Delta d}{d_0}, \tag{2.78}$$

where k_s is the transducer conversion constant, given by the geometric dimensions of the membrane and its stiffness.

2.4.3 Resonant pressure sensors

These sensors use SAW (*surface acoustic wave*) technology, described in Section 2.2.4, with the difference that two IDM *delay lines* are located on the silicon membrane with a piezoelectric layer, one of which is sensing and the other compensating (Figure 2.71).

These delay lines form harmonic signal generators with A_1, A_2 amplifiers, whose frequencies are determined by the time delay of the delay lines, which depends on their mechanical stress. After multiplying both signals and filtration by a low-pass filter, the frequency difference of the two generators is directly proportional to the measured pressure.

2.5 Vibration Sensors

They are designed to measure the path, speed, and acceleration of vibrating objects. According to the reference point to which these quantities relate, we distinguish between absolute and relative sensors. *Absolute sensors* determine the values of vibrational quantities to the Earth's gravitational field. *Relative sensors* determine these values to a reference point that is not identical to Earth.

The principle of all these sensors is to sense the motion of a seismic body with mass m with respect to a reference body of mass M (Figure 2.72).

The motion of a body of mass m is defined by the differential equation of motion

$$m\frac{d^2y(t)}{dt^2} + b\frac{dy}{dt} + k \cdot y = F_b = M \cdot \frac{d^2x(t)}{dt^2}, \tag{2.79}$$

where y is the deflection of a seismic body of mass m, dy/dt is its velocity, d^2y/dt^2 is its acceleration with respect to a reference object of mass M, b is

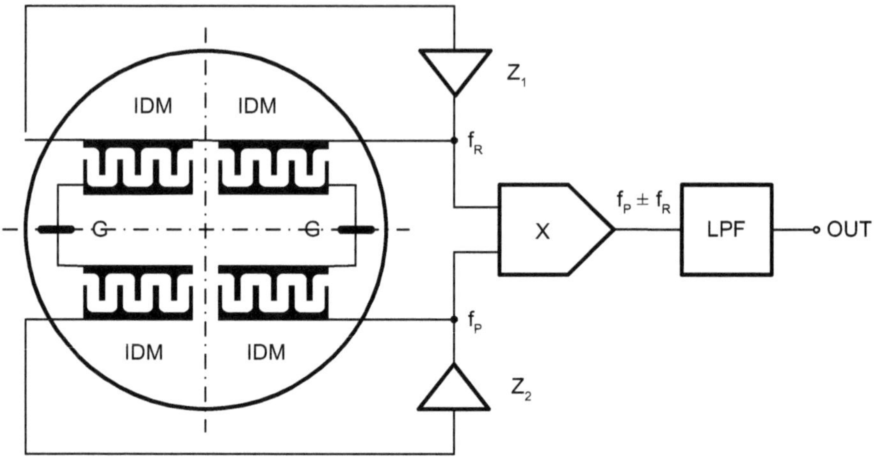

Figure 2.71 Resonant pressure transmitter.

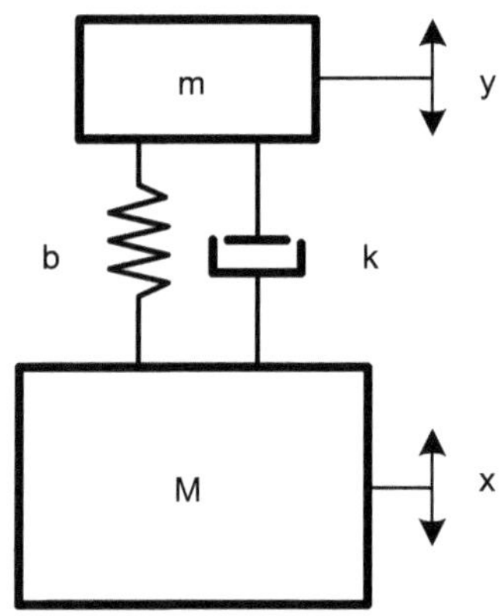

Figure 2.72 Principle of absolute vibration sensor.

the damping coefficient, k is the stiffness of the spring element, and F_b is the force acting on a seismic body of mass m.

Suitable selection of parameters m, b, and k can then create sensors of vibrometry, speed, and acceleration.

Vibrometers are characterized by a large mass of the seismic body m, comparable to the mass M of the reference body, low damping b, and low stiffness to the spring element. For these sensors, the equation of motion approximately applies

$$y(t) \approx x(t). \tag{2.80}$$

The deflection of a seismic body with mass m is then approximately equal to the deflection of the reference body in these sensors

$$m \cdot \frac{d^2y(t)}{dt^2} \approx M \cdot \frac{d^2x(t)}{dt^2}. \tag{2.81}$$

Velocimeters are characterized by a significantly small seismic mass m, a large damping b, and a low stiffness to the spring element. For these sensors, the equation of motion approximately applies

$$b \cdot \frac{dy}{dt} \approx M \cdot \frac{d^2x(t)}{dt^2}. \tag{2.82}$$

Therefore, the deflection of a seismic body of mass m is proportional to the speed of movement of the reference body in these sensors

$$y \approx \frac{M}{b} \cdot \frac{dx(t)}{dt}. \tag{2.83}$$

Accelerometers have considerable spring stiffness and a small mass of the seismic body m. For these sensors, the equation of motion approximately applies

$$ky \approx M \cdot \frac{d^2x(t)}{dt^2}. \tag{2.84}$$

The deflection of a seismic body with mass m is then proportional to the acceleration of the reference body for these sensors

$$y \approx \frac{M}{k} \cdot \frac{d^2x(t)}{dt^2}. \tag{2.85}$$

In connection with the application of acceleration sensors, the influence of their weight on the determination of the acceleration of oscillating parts of machines must also be taken into account. If the weight of the accelerometer m_a is not negligible to the weight of the machine, then its oscillation speed is valid

$$a_s = \frac{m_s + m_a}{m_s} \cdot a_m, \tag{2.86}$$

where a_m is the measured acceleration of the accelerometer.

2.5.1 Electrodynamic oscillation rate sensors

These sensors contain a *permanent magnet* that moves inside the telecoil on springs or flexible diaphragms (Figure 2.73).

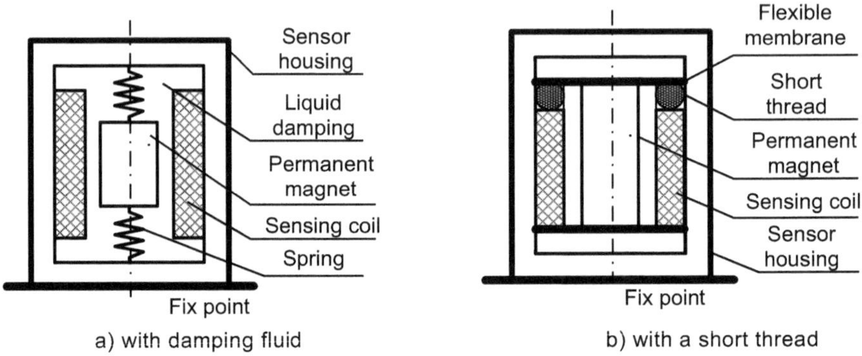

Figure 2.73 Electrodynamic oscillation rate sensors.

To dampen the movement of a permanent magnet, either a damping fluid in which the magnet moves or a short-term thread is used, in which the movement of the permanent magnet induces a current acting with a force opposite to the force caused by the movement of the permanent magnet. In both cases, the movement of the magnet induces tension in the telecoil

$$u(t) = B \cdot l \cdot v, \tag{2.87}$$

where B is the induction of the magnetic field of the permanent magnet, l is the active length of the conductor moving in this field, and v is the speed of movement of the permanent magnet.

Electrodynamic sensors are essentially *absolute motion sensors*, but since the induced voltage in their coil is proportional to the *speed of movement* of a permanent magnet (seismic body), they are speed sensors. Electrodynamic sensors are able to measure oscillation speed in the range of several mm/s to several m/s in the frequency range of several Hz to several kHz. By integrating the voltage of these sensors, information about their movement can be obtained.

2.5.2 Piezoelectric acceleration sensors

Piezoelectric accelerometers consist of a seismic body that is attached through one end to the sensor housing via the spring and damping layer and the other end is attached to the piezoelectric sensor (Figure 2.74).

If acceleration a acts on a seismic body with mass m, then the resulting force F, acting on the piezoelectric sensor, creates a charge between its electrodes

$$q(t) = k_{pz} \cdot F(t) = k_{pz} \cdot p \cdot m \cdot a(t), \tag{2.88}$$

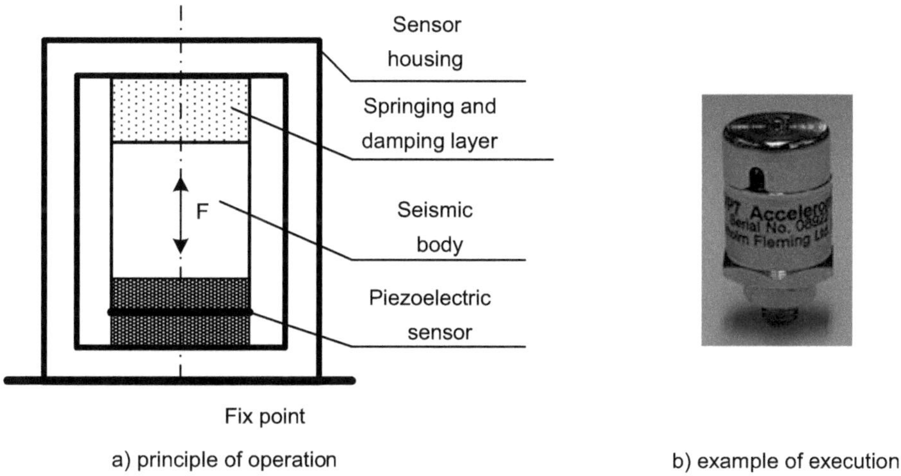

Figure 2.74 Piezoelectric accelerometer.

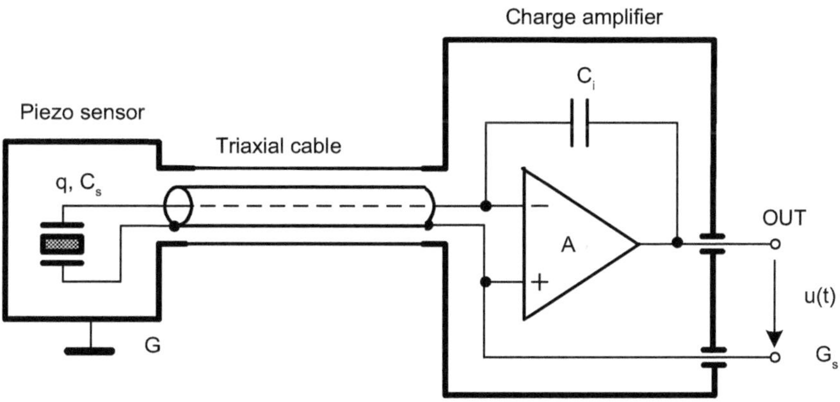

Figure 2.75 Connection of the piezoelectric sensor to the charge amplifier.

where k_{pz} is the piezoelectric constant of the piezo transducer. The resulting charge is converted to voltage by a charge amplifier consisting of an operational amplifier with a feedback capacitor C_i (Figure2.75),

$$u(t) = \frac{q(t)}{C_s} = k_{pz} \cdot \frac{m}{C_s} \cdot a(t). \tag{2.89}$$

Assuming that the operational amplifier A used has a sufficiently large differential gain and the input residual voltage is approximately zero, the effect of leakage resistance and capacitance between the inner conductor and its casing, forming the signal ground of the G_s sensor, is not applied.

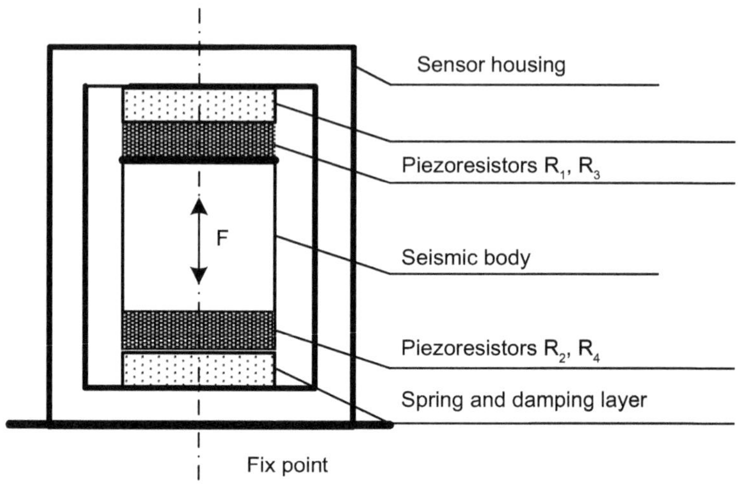

Figure 2.76 Accelerometer with piezoresistors.

To connect the piezoelectric sensor to the hub amplifier, special triaxial cables with minimal *piezoelectric* and *triboelectric effects* are used, which would cause the formation of disturbing charges under mechanical stress. Modern piezoelectric accelerometers, referred to as ICP (*integrated circuit piezoelectric*), have a hub amplifier located directly in the accelerometer housing and therefore do not require their connection with special and expensive cables.

Piezoelectric accelerometers allow you to measure acceleration in the range of 1–50 $m \cdot s^{-2}$, their sensitivity reaches several tens to hundreds of mV/ms^{-2}, and their frequency span ranges from several Hz to tens of kHz.

Piezoresistive accelerometers also consist of a seismic body that is attached to the sensor housing via springing and damping layers (Figure 2.76).

Between the spring and damping layers, two pairs of *piezoresistors* with identical properties and nominal values are deposited, while these pairs are alternately mechanically stressed for tension and pressure by the time change of the measured acceleration. These piezoresistors are connected to a bridge whose differential voltage is valid

$$u_d(t) = k_{\text{pr}} \cdot \frac{dR}{R} \cdot U_B, \tag{2.90}$$

where k_{pr} is the sensitivity of the piezoresistor, dR/R is the relative change in its resistance, and U_B is the supply voltage of the bridge (Figure 2.77).

Since the change in resistance of piezoresistors is directly proportional to their mechanical stress, the differential stress of the bridge is

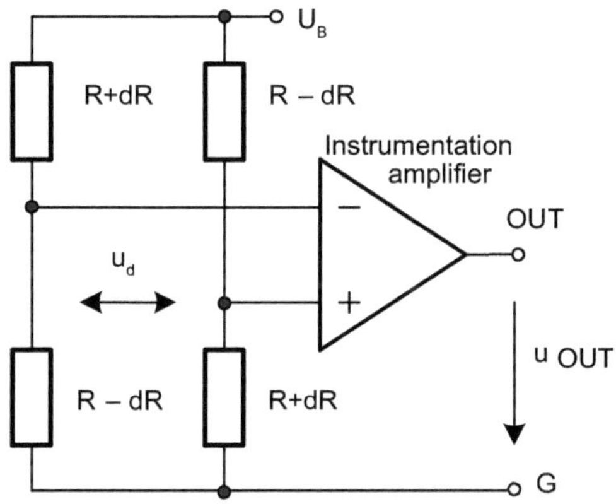

Figure 2.77 Connecting a bridge with piezoresistors to an instrument amplifier.

$$u_d(t) = U_n \cdot k_{pr} \cdot F(t) = U_B \cdot k_{pr} \cdot m \cdot a(t), \tag{2.91}$$

where U_B is the supply voltage and k_{pr} is the sensitivity of piezoresistive sensors.

The differential voltage, reaching a maximum of several tens of mV, is amplified by the instrument amplifier to the desired value. In addition, by connecting piezoresistors to the bridge, their temperature dependence is eliminated.

2.5.3 Micromechanical acceleration sensors

Capacitive micromechanical accelerometers use micromechanical technology MEMS (micro electro mechanical systems) to measure acceleration. This technology uses electrodes to which seismic bodies and *differential ridge capacitors* are attached to measure motion (Figure 2.78).

A change in measured acceleration causes changes in the distances of the $d \pm \Delta d$ differential capacitors from their central electrodes, causing changes in their capacities $C \pm \Delta C$. These capacitance changes are converted into voltage changes by the *charge amplifier*,

$$\Delta u = 2 \cdot \frac{\Delta C}{C_i} \cdot U, \tag{2.92}$$

where C_i is the integration capacitance of the hub amplifier.

Capacitive accelerometers are manufactured in uniaxial, two-axis, or three-axis design, which allows their use even in vector measurement of vibrometric quantities. Capacitive accelerometers allow you to measure

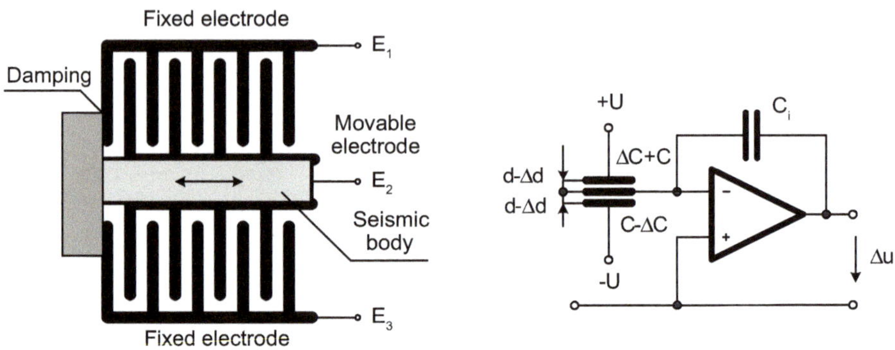

a) internal layout b) accelerometer wiring

Figure 2.78 MEMS accelerometer.

a) accelerometer for sticking b) accelerometer for mounting

Figure 2.79 Three-axis accelerometers MEMS.

acceleration in the range of several to hundreds of $m \cdot s^{-2}$, their sensitivity reaches several tens to hundreds of mV/ms^{-2}, and their frequency range is several tens of kHz.

Figure 2.79 shows typical three-axis accelerometers used to measure vibrations in industrial practice.

To attach the sensors, either stud screws are most often used or the sensors are glued with a special *cyanoacrylate adhesive*. Some sensors are also equipped with permanent magnets, which allow their easy connection to ferromagnetic materials.

2.6 Liquid and Gas Flow Sensors

According to the method of measurement, we distinguish *volume* and *mass sensors of liquid* flow. Assuming laminar fluid flow, positive displacement sensors determine the volume of fluid flowed ΔV per unit of time Δt, which

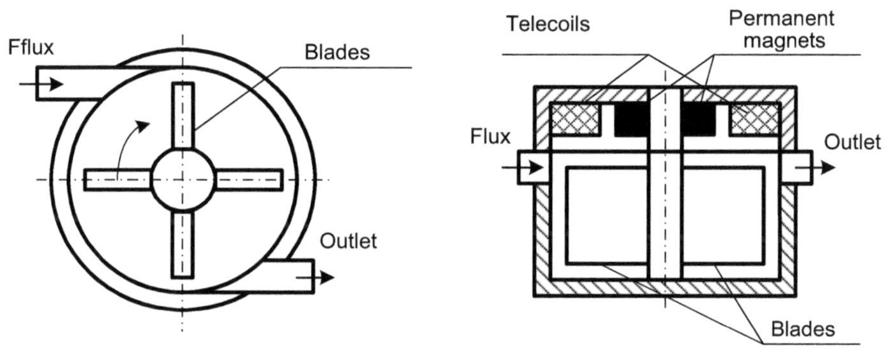

a) principle of the flow meter b) section of the flowmeter

Figure 2.80 Turbine flow meter.

can be expressed by the product of the mean velocity in its flow and the cross-sectional area S through which the fluid flows

$$Q_V = \frac{\Delta V}{\Delta t} = v \cdot S. \tag{2.93}$$

Mass flow sensors determine the volume of liquid flowed from the ratio of changes in its mass Δm flowing through a cross section of area S per unit of time Δt

$$Q_m = \frac{\Delta m}{\Delta t} = \rho \cdot v \cdot S, \tag{2.94}$$

where ρ is the density of the liquid flowed.

Turbine, induction, orifice plate, and ultrasonic sensors are used to measure volumetric flow. To measure the mass flow, sensors working on the principle of *Coriolis force* and thermal sensors are used.

2.6.1 Turbine flow sensors

They determine the volumetric flow of the liquid from the number of revolutions of the turbine, which is inserted into its stream. Permanent magnets are placed on the turbine axis, whose movement in the vicinity of the sensing coils induces current pulses, which are read by a counter (Figure 2.80).

The volumetric flow rate of the liquid is determined by the number of revolutions of the turbine n and the volume of liquid ΔV passed through the sensor corresponding to one turbine revolution

$$Q_v = k_V \cdot n \cdot \Delta V\ . \tag{2.95}$$

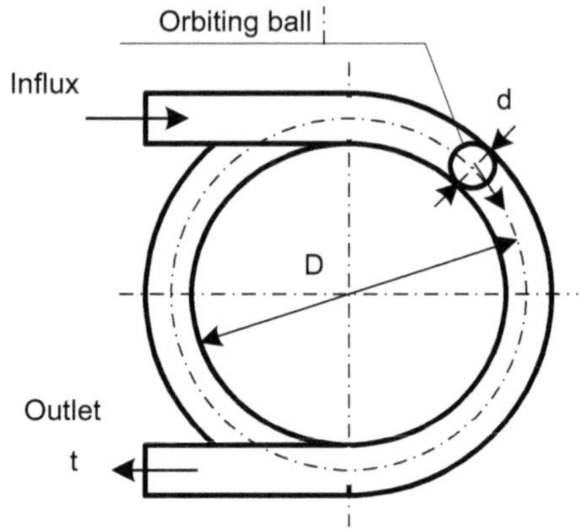

Figure 2.81 Flowmeter with rotating ball.

where k_V is the conversion constant of the flow meter, ranging from 0.9 to 0.95.

Turnless flow meters measure the volume of liquids flowed from the number of circulators of a rotating ball of diameter *d*, which moves in a circular pipe of mean diameter *D*, through which the measured liquid flows (Figure 2.81).

The volume flow of the liquid is determined by the number of ball circulations and the volume of the circular pipe ΔV in which the ball circulates

$$Q_v = k_V \cdot n \cdot \Delta V = k_v \cdot n \frac{\pi \cdot d^2}{4} \pi \cdot D, \qquad (2.96)$$

where k_V is the conversion constant of the flow meter, ranging from 0.95 to 0.99.

The number of ball orbits is again sensed by a magnetic or optical speed sensor.

2.6.2 Inductive flow sensors

Inductive volume flow sensors consist of a permanent magnet between the poles of which a non-magnetic pipeline of inner diameter *d* is inserted, through which a conductive liquid whose volume is measured flows at an average speed (Figure 2.82).

When passing ions with a charge *q* in a conductive liquid flowing at a speed in the magnetic field of a permanent magnet with induction *B*, these ions are deflected by the force F_1 to the walls of the non-conductive pipe, on the surface of which the electrodes E_1 and E_2 are located. The charges on the

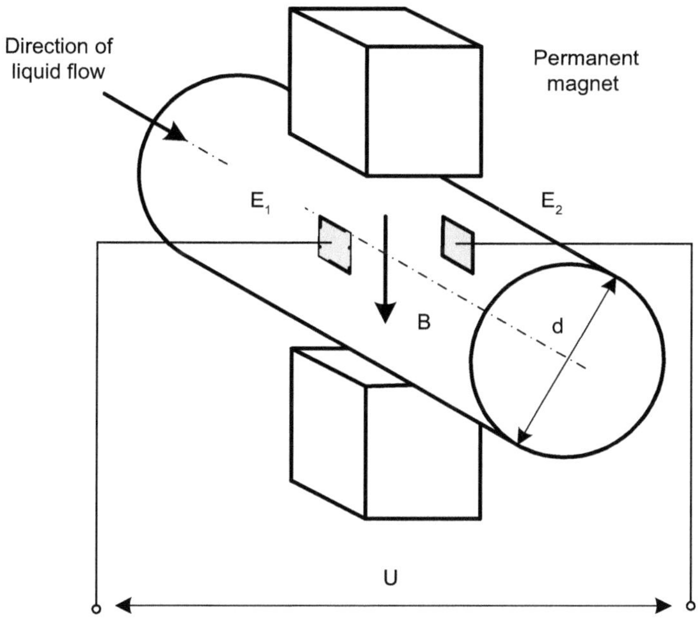

Figure 2.82 Induction flow meter.

electrodes between them create an electric field of intensity E acting on the ions with the force F_2. From the equality of these forces

$$F_1 = q \cdot v \cdot B = F_2 = q \cdot E = q \cdot E\frac{U}{d} \tag{2.97}$$

open-circuit voltage between sensor electrodes can be determined

$$U = d \cdot B \cdot v, \tag{2.98}$$

from which the volumetric flow rate of liquids can be determined

$$Q_v = v \cdot S = \frac{U}{d \cdot B}\frac{\pi \cdot d^2}{4}. \tag{2.99}$$

The internal resistance of this voltage source is determined by the relative conductivity of the liquid σ and the area of the S_E sensing electrodes

$$R_i = \frac{d}{\sigma \cdot S_E}. \tag{2.100}$$

When measuring liquids with low ionization, the voltage at the sensing electrodes reaches only a few tens of mV. In order to eliminate the influence of interfering DC voltages caused, for example, by thermoelectric voltages on the contacts of different metals at their different temperatures, an alternating magnetic field is used instead of a permanent magnet, which causes alternating voltage on the sensing electrodes.

These alternating voltages are amplified by AC amplifiers and then rectified by synchronous detectors whose control signals are derived from the alternating excitation currents of electromagnets. By using synchronous detection, any alternating components of interfering voltages whose frequency is not identical to the frequency of the excitation current of electromagnets are also suppressed. Thus, flow measurement accuracy of up to several percentages can be achieved in the range of units up to hundreds of *l/s*.

2.6.3 Throttle flow sensors

They use the equality of static and kinetic energies of flowing fluids in front of and after the throttles to measure volumetric flow. When measuring in a horizontal pipeline, *Bernoulli's equation* applies

$$p_1 + \frac{1}{2}\rho \cdot v_1^2 = p_2 + \frac{1}{2}\rho \cdot v_2^2, \tag{2.101}$$

where v_1, v_2 are the fluid flow velocities, p_1, p_2 are the static pressures in the cross sections S_1, S_2, before and at the constriction, and S_1, S_2 is the fluid density (Figure 2.83).

If the throttle is a *ring orifice plate* with diameter d, then the volumetric flow of the flowing fluid can be determined from the pressure difference p_1 before the orifice plate and the pressure p_2 after the orifice plate based on *Bernoulli's equation*

$$Q_v = \frac{\pi \cdot d^2}{4}\sqrt{2 \cdot \frac{p_1 - p_2}{\rho}}. \tag{2.102}$$

Venturi tubes with variable profile allow, compared to orifice plate throttles, to measure the flow rates of these substances in a wider range of their flow rates and thus smaller pressure differences with an accuracy of up to 1%.

Pitot tubes detect both static and dynamic pressure p_c through a curved tube oriented against the direction of flow of liquids or gases, and a tube located in the wall of the pipe, sensing only the static pressure p_s. From the difference of these pressures, it is possible to determine the flow velocity of the media and their volumetric flow

$$v = \sqrt{\frac{p_c - p_s}{2 \cdot \rho}}, \qquad Q_v = S \cdot v = \frac{\pi \cdot d^2}{4}\sqrt{\frac{p_c - p_s}{2 \cdot \rho}}, \tag{2.103}$$

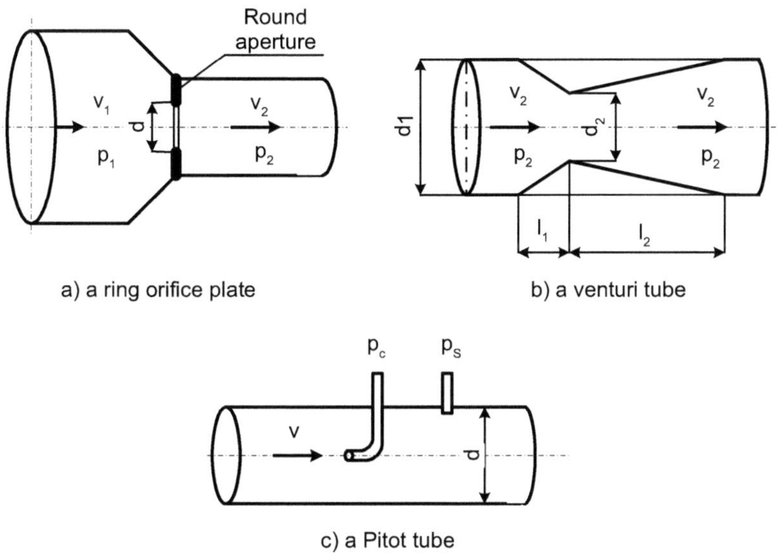

Figure 2.83 Throttle flow measurement.

Burst flow and volumetric sensors of liquids and gases use the force F applied to a circular plate with an area S whose axis coincides with the direction of the flowing liquid or gas (Figure 2.84).

Because it applies to this power

$$F = k_d \cdot S \cdot \frac{\rho}{2} v^2, \tag{2.104}$$

where k_d is the shape constant of the circular plate.

The volume flow rate of liquids, determined by the equation

$$Q_v = S \cdot v = S \sqrt{\frac{2 \cdot F}{k_d \cdot S \cdot \rho}}. \tag{2.105}$$

These sensors are suitable for measuring not only the flow of liquids but also gases with high viscosity with an achievable accuracy of several percentages.

2.6.4 Ultrasonic flow sensors

They work on the principle of measuring the difference in propagation times of an ultrasound signal through the medium whose volume flow is being measured. The ultrasonic signal, generated by the transmitter, whose axis

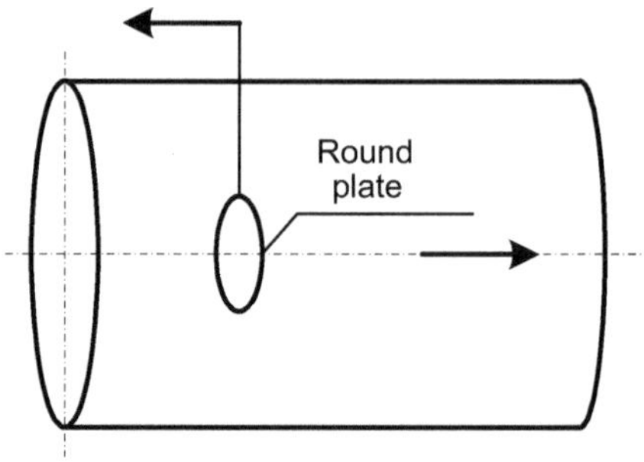

Figure 2.84 Impact volume flow sensor with circular plate.

makes an angle α with the axis of the liquid flow, propagates to the reflector at a speed

$$v_1 = c_0 + v \cdot \cos\alpha. \tag{2.106}$$

The reflected signal travels from this reflector to the ultrasound receiver at speed

$$v_2 = c_0 - v \cdot \cos\alpha, \tag{2.107}$$

where v is the mean speed of the liquid flow and c_0 is the propagation speed of the ultrasonic signal in it (Figure 2.85).

Because it applies to the transit times of these signals between the ultrasound transmitter and the receiver

$$\Delta t_1 = \frac{L}{c_0 + v \cdot \cos\alpha}, \qquad \Delta t_2 = \frac{L}{c_0 - v \cdot \cos\alpha}. \tag{2.108 a, b}$$

is their difference

$$\Delta t_2 - \Delta t_1 = 2 \cdot L \cdot \frac{v \cdot \cos\alpha}{c_0^2 - v^2 \cdot \cos^2\alpha}. \tag{2.109}$$

Assuming that the mean velocity of the liquid flow is significantly smaller than the propagation speed of the ultrasonic echo in it, i.e., that in $v << c_0$, the liquid flow velocity can be expressed by the approximate relation

$$v \approx \frac{c_0^2}{2 \cdot L \cdot \cos\alpha}\left(\Delta t_2 - \Delta t_1\right). \tag{2.110}$$

Since the propagation speed of the ultrasonic signal depends not only on the physico-chemical composition of the liquid but also on its temperature, the product of times is used to exclude this influence

$$\Delta t_1 \cdot \Delta t_2 = \frac{L^2}{c_0^2 - v^2 \cdot \cos^2\alpha}, \tag{2.111}$$

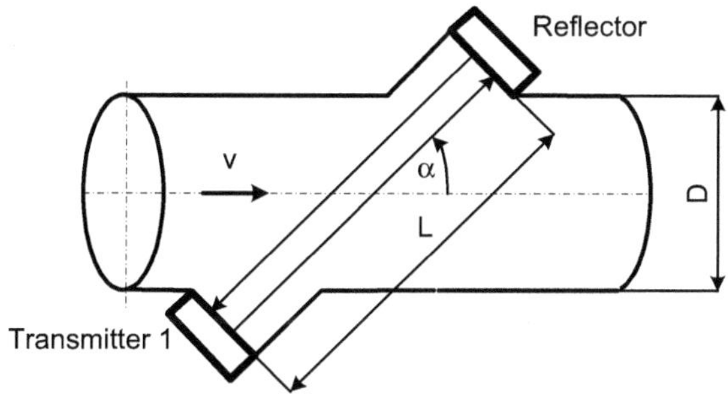

a) with single reflection

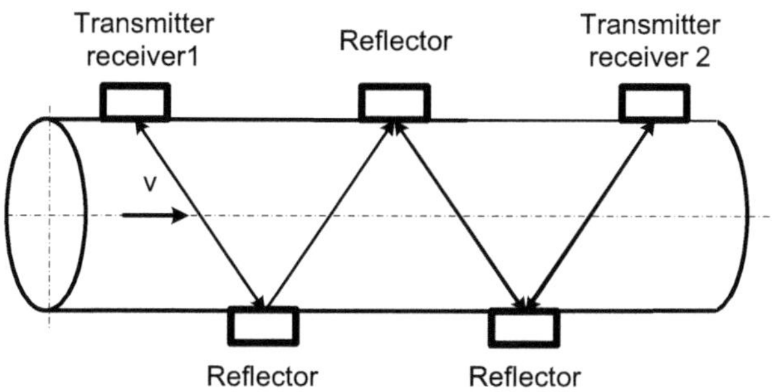

b) with multiple reflections

Figure 2.85 Ultrasonic volume flow sensors.

which makes it possible to determine the mean velocity of the liquid flow

$$v = \frac{c_0^2}{2 \cdot L \cdot \cos\alpha} \cdot \frac{\Delta t_2 - \Delta t_1}{\Delta t_1 \cdot \Delta t_2}. \quad (2.112)$$

Sensors with multiple reflections of ultrasonic signals are used to increase the resolution and the related accuracy of determining the flow rate of liquids. When using an n-fold reflection, assuming statically independent measurements, the mean square deviation of the measurement can be theoretically reduced in the ratio $1/\sqrt{n}$ compared to the measurement with one reflection.

2.6.5 Flow sensors using the Coriolis force

Sensors using the *Coriolis force* are used to measure the *mass volume of the flow*, which acts on a rectilinearly moving body of mass m at speed v in a system rotating at angular speed ω

$$F_C = m \cdot v \cdot \omega = m \cdot a_c, \quad (2.113)$$

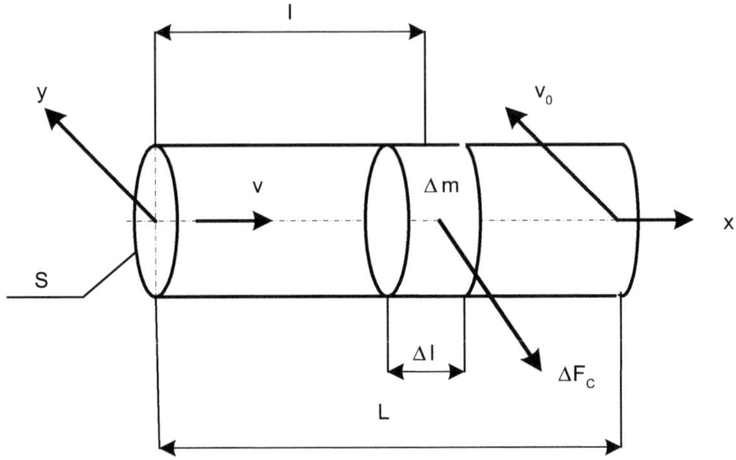

Figure 2.86 The principle of operation of the volume flow mass sensor.

where a_c is the *Coriolis acceleration.*

If a liquid flows in a pipe with a velocity v that rotates at an angular velocity ω, then the *Coriolis force* acts on each element of the liquid of mass Δm

$$\Delta F_C = 2 \cdot \Delta m \cdot v \cdot \omega \tag{2.114}$$

in the direction perpendicular to the plane of the vector of this velocity (Figure 2.86).

Since the speed of liquid flow in a pipe is determined by the change in its path per unit time $v = \Delta l / \Delta t$, the change in the *Coriolis force* acting on the pipe is directly proportional to the mass flow rate Q_m of the flowing liquid

$$\Delta F_C = \frac{\Delta m}{\Delta t} 2 \cdot \Delta l = 2 \cdot Q_m \cdot \omega \cdot \Delta l. \tag{2.115}$$

Tensometric or piezoelectric sensors are most often used to measure the *Coriolis force* in mass flow meters. Sensors using the *Coriolis force* make it possible to measure mass flow in the range of several hundred g/s to hundreds of kg/s with an accuracy of several tenths of a percent in the temperature range of –40 to 150 °C.

2.6.6 Thermal flow sensors

These sensors use heat exchange between its source and the flowing liquid, whose mass flow rate we measure. These so-called *thermoanemometers*

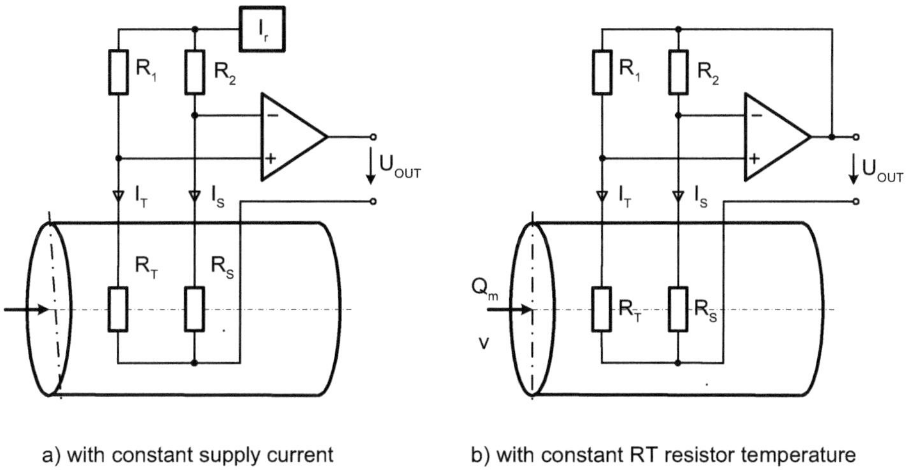

Figure 2.87 Thermal flow sensors.

consist of a heated Pt resistor R_T and a sensing Pt resistor R_S, while the ratio of their nominal resistances is, for example, 1/100. Therefore, a significantly higher current I_T flows through the resistor R_T than the current I_S flowing through the resistor R_S. Both resistors, immersed in the flowing liquid, form a resistance bridge with identical resistors $R_1 = R_2$. These sensors can work either with a constant supply current or with a constant temperature of the heated resistor (Figure 2.87).

For thermal flow sensors with a *constant supply current*, as the flow rate increases, the resistor R_T is cooled significantly more than the resistor R_s, which causes the resistance bridge to become unbalanced. In thermal flow sensors with a constant temperature of the heated resistor, the bridge is powered by the output current of the differential amplifier. In both cases, the output voltage of the differential amplifier determines the mass flow rate of the liquid

$$u_{\mathrm{OUT}}^2 = a + b \cdot \sqrt{Q_m}, \tag{2.116}$$

where a is the sensor constant, taking into account the heat transfer from the resistors to the liquid and vice versa from the liquid to the resistors by conduction and radiation at zero liquid flow rate. The constant b depends on the geometry of the R_T and R_s resistors, the density, viscosity, thermal conductivity, and specific heat capacity of the liquid.

To measure small flows, differential temperature flow sensors are used, which are formed by a pair of sensing resistors R_{s1} and R_{s2} and one heating resistor R_T, while all these resistors are placed perpendicular to the direction of flow in the pipe through which the liquid flows (Figure 2.88).

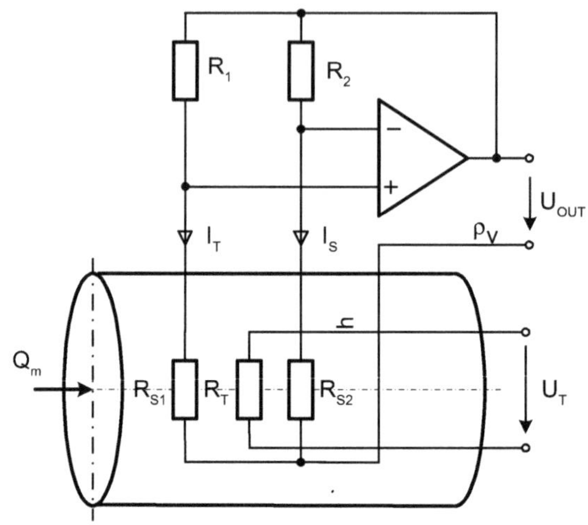

Figure 2.88 Differential temperature flow sensor.

Similar to the previous case, the sensing resistors are connected to the resistor bridge. Assuming that the resistances of sensors R_s and R_{s2} are directly proportional to their temperature, it applies to the measured liquid mass flow

$$Q_m = \frac{A \cdot c_P}{\Phi_q}\left(R_{s2} - R_{s1}\right), \tag{2.117}$$

where A is the geometric and thermal constant of the sensor, c_p is the specific heat capacity of the liquid, and Φ_q is the heat flux of the heating resistor.

A variant of the temperature differential temperature sensor is a sensor of *small flows* with a branch in the form of a thin-walled tube with good thermal conductivity, on which two resistance temperature sensors R_{s1} and R_{s2} are wound and between which the heating winding RT is placed (Figure 2.89).

With these flow sensors, mass flow rates of liquids in the range of a few g/h can be measured with an accuracy of up to a few tenths of a percent.

2.7 Level Sensors

In addition to float and hydrostatic sensors, capacitive, ultrasonic, radar, and optical sensors are used to measure the level of liquids and bulk materials.

2.7.1 Float level sensors

They are formed by a float whose specific volume density is less than the specific volume density of the liquid whose level we are measuring. The vertical

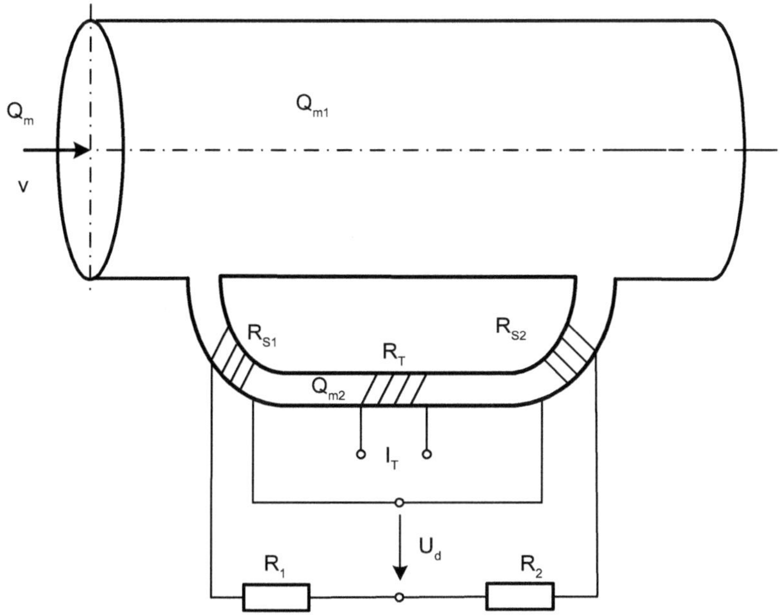

Figure 2.89 Differential temperature sensor of small flows.

movement of the float is sensed either by a potentiometric, inductive, capacitive, or incremental optical movement sensor (Figure 2.90).

2.7.2 Hydrostatic level sensors

To measure the level of liquids, these sensors use either buoyancy force sensors that act on a submerged body in a liquid whose specific volume density ρk is greater than the specific volume density of the environment above the liquid ρ_v, or the difference in hydrostatic pressures in the liquid p_k and above its surface (Figure 2.91).

In the first case, it acts on a submerged body in the shape of a cylinder with cross section S, height l, and mass m, and when immersed in a liquid for a length h, the buoyant force is

$$F_{vz} = m \cdot g - \rho_k \cdot S \cdot h - \rho_v \cdot S(h - l), \tag{2.118}$$

where $g = 9.81\ \mathrm{m \cdot s^{-2}}$ is the acceleration of the Earth.

In the second case, the immersion depth of the body h_x is determined by the pressure difference

$$p_k - p_v = \rho_k \cdot g \cdot h_x, \tag{2.119}$$

which is measured by a differential pressure sensor.

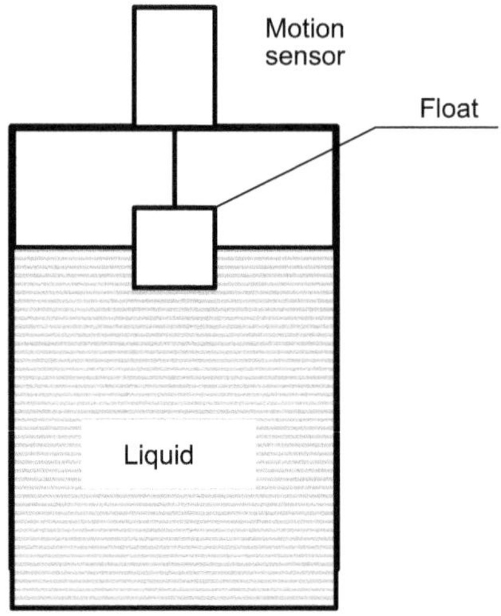

Figure 2.90 Float level sensor.

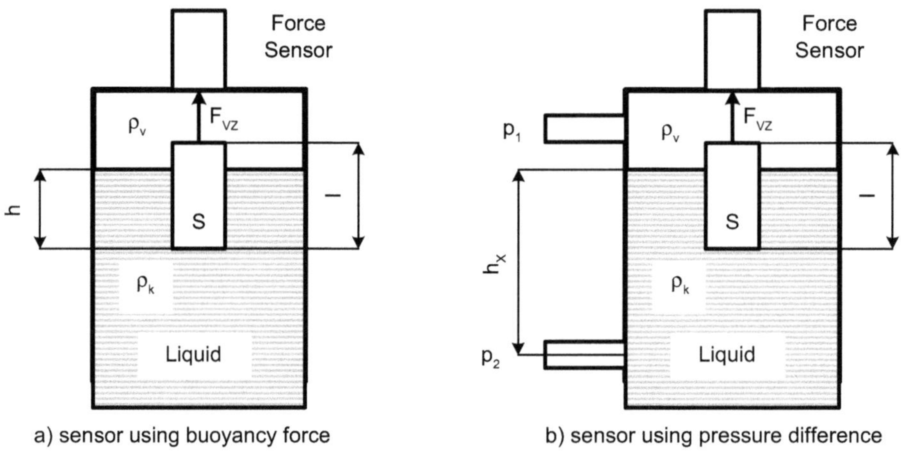

Figure 2.91 Hydrostatic level sensors.

2.7.3 Capacitive level sensors

The essence of these sensors is a capacitor with a variable capacity, which is formed by the sum of the capacities between the non-immersed and immersed parts of the electrodes. A necessary condition for the use of these capacitive sensors is the different permittivity of the environment above the liquid level and the permittivity of the liquid. Capacitive sensors with plate, circular, and coaxial capacitors are used to sense the water level.

Capacitive level sensor with a plate capacitor consists of plane-parallel electrodes E_1 and E_2 with width b and distance d (Figure 2.92).

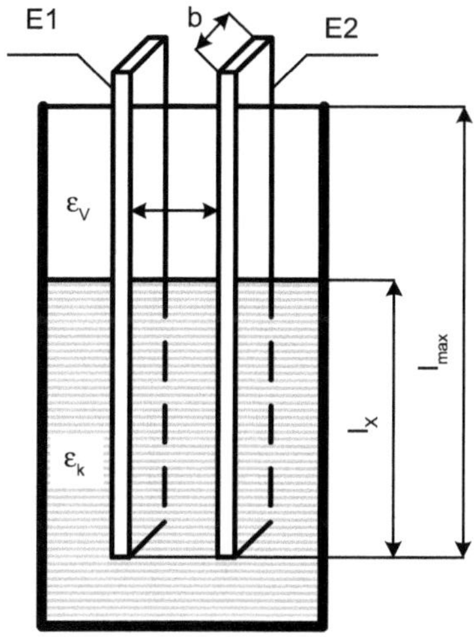

Figure 2.92 Capacitive level sensor with plate capacitor.

The capacitance between the electrodes of this sensor is determined by the sum of the capacitances of the immersed and non-immersed parts of the plate capacitors

$$C_{12} = \frac{\varepsilon_k \cdot a \cdot l_x}{d} + \frac{\varepsilon_v \cdot a \cdot \left(l_{\max} - l_x\right)}{d}, \tag{2.120}$$

where ε_k is the permittivity of the liquid, ε_v is the permittivity of the environment above the liquid level, l_x is the height of the level from the lower edge of the electrodes, and $l_{\max}$ is the range of the capacitive sensor.

A *capacitive level sensor* with *one rod electrode* consists of a circular electrode with a diameter of *D*, while the second electrode is formed by the conductive walls of the container (Figure 2.93(a)).

The capacitive level sensor in Figure 2.93(b) with a pair of rod electrodes of diameter *D* and pitch *d* has an inter-electrode capacitance

$$C_{12} = \frac{\pi \cdot \left[\varepsilon_v \cdot \left(l_{\max} - l_x\right) + \varepsilon_r \cdot l_x\right]}{\ln\left[\frac{d}{D} \cdot \left(1 + \sqrt{1 - \frac{D^2}{d^2}}\right)\right]}. \tag{2.121}$$

Capacitive level sensor with *coaxial circular of electrodes* in Figure 2.94 and consists of an inner circular electrode with a diameter of D_1 and an outer electrode with a diameter of D_2. It applies to the inter-electrode capacitance of this sensor

$$C_{12} = \frac{(\varepsilon_v - \varepsilon_k) \cdot l_{\max} - \varepsilon_k \cdot l_x}{\ln \frac{D_1}{D_2}}. \tag{2.122}$$

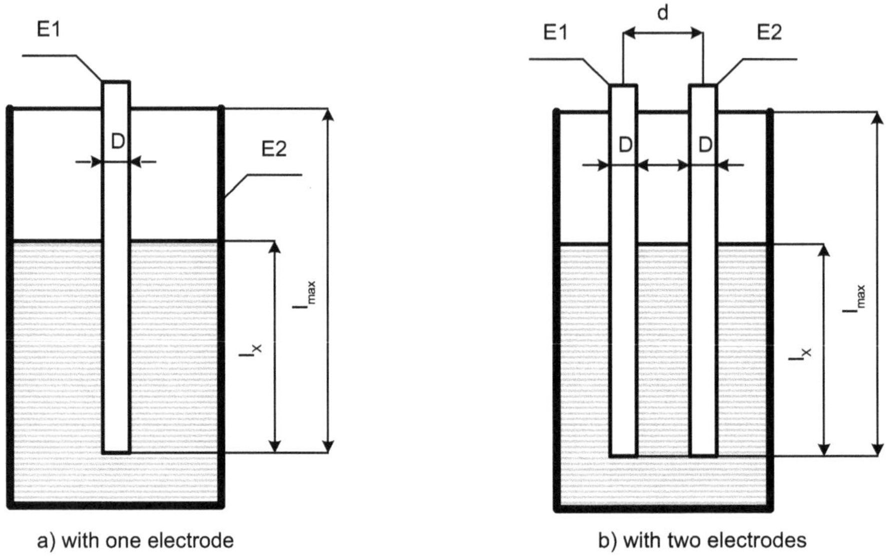

Figure 2.93 Capacitive level sensors with circular electrodes.

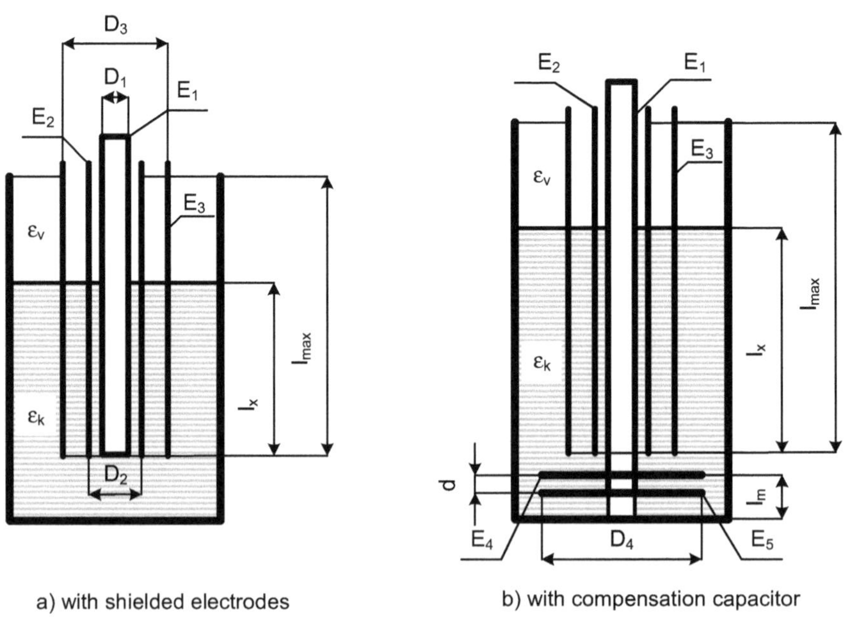

Figure 2.94 Capacitive level sensors with coaxial electrodes.

In order to suppress the influence of external interference, capacitive sensors are usually supplemented with an additional *shielding electrode*, which is galvanically connected to the metal shell of the container. Therefore, capacitive sensors are connected to evaluation circuits using *triaxial cables* with *double shielding*, while active shielding is used to suppress the effect of capacitance between the sensing electrodes and the shielding electrode (see Section 2.1.6, Figure 2.27).

The sensitivity of the sensor $\Delta C_{12}/\Delta l_x$ depends on the ratio of the permittivity of the liquid to the permittivity of the external environment $\varepsilon_k/\varepsilon_v$. For example, for diesel and other liquid hydrocarbons, this ratio ranges from 2.15 to 2.50 at a temperature of 20 °C, and its temperature change is approximately $1.8 \cdot 10^{-3}$/°C. This temperature dependence can be partially compensated by another compensating plate capacitor, which is located at the bottom of the container (Figure 2.94(b)). In addition, its use allows determining the minimum height of the level in the container, l_{min}.

If the liquid is conductive, Teflon or glass coatings are used to insulate the sensing electrodes, which, due to their thickness of a few tenths of an mm, practically do not affect the metal–electrode capacities of the sensors.

2.7.4 Ultrasonic level sensors

They consist of an *ultrasonic transmitter* generating ultrasonic pulses in the frequency range of several tens to hundreds of kHz, which are detected by an ultrasonic receiver after reflection from the surface of the liquid (Figure 2.95).

Assuming that ultrasonic waves propagate in the environment above the surface of a liquid or bulk material at a speed c_u, then the time delay ΔT_x between the transmitted and reflected ultrasonic pulse of the ultrasonic transmitter and receiver UVZ determines the height of the level of the measured medium

$$l_x = l_{\text{max}} - c_u \cdot \frac{\Delta T_x}{2}. \tag{2.123}$$

The speed of propagation of ultrasound in a gaseous environment

$$c = \sqrt{k_a \cdot R \cdot \frac{\Theta}{M}} \tag{2.124}$$

depends on the adiabatic coefficient k_a, the gas constant R, the absolute temperature Θ, and the molecular weight of the gas M.

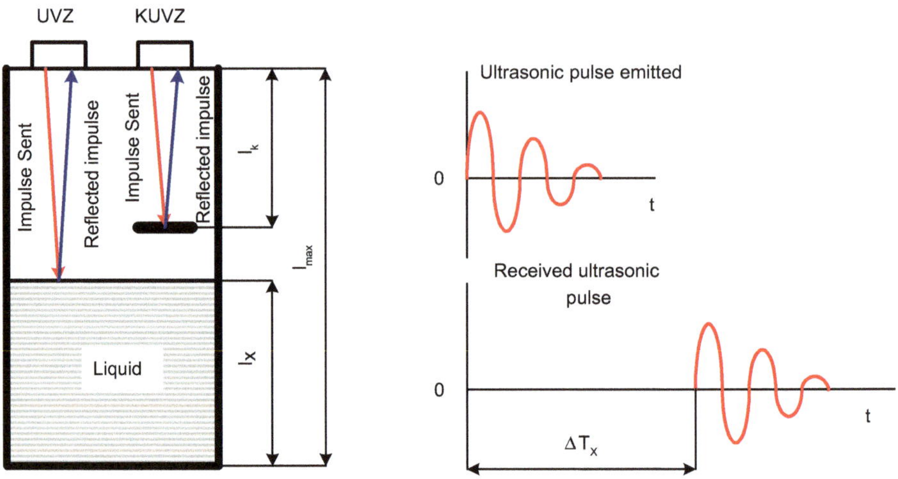

a) arrangement of the sensor b) time course of the sensor pulses

Figure 2.95 Ultrasonic level sensor.

The speed of sound propagation in dry air at a temperature of 0 °C is 332 m/s. With increasing temperature, this rate increases non-linearly and at a temperature of 100 °C, it is already 17% greater than at a temperature of 0 °C. The influence of temperature fluctuations on the output signal of ultrasonic level meters is therefore compensated for with these sensors by adding a reference reflection surface, distant by the reference distance l_k. Because it applies to the corrected level height

$$l_k = c_u \cdot \frac{\Delta T_k}{2}, \tag{2.125}$$

where ΔT_k is the time delay between the sent and reflected ultrasonic pulse of the corrective ultrasonic receiver-transmitter KUVZ, applies to the corrected level height of the ultrasonic sensor

$$l_{xk} = l_{\max} - \frac{\Delta T_x}{\Delta T_k} l_x. \tag{2.126}$$

The quality of the ultrasound signal reflection is significantly influenced by the properties of the surface of the measured medium. It is important for the function of the sensors that the reflected wave has sufficient energy. If the measured medium is a foaming liquid or loose material consisting of granules, less energy will be reflected from the surface of their surface. Therefore, to achieve the same measuring range, it is necessary to use a larger initial energy of the transmitted signal.

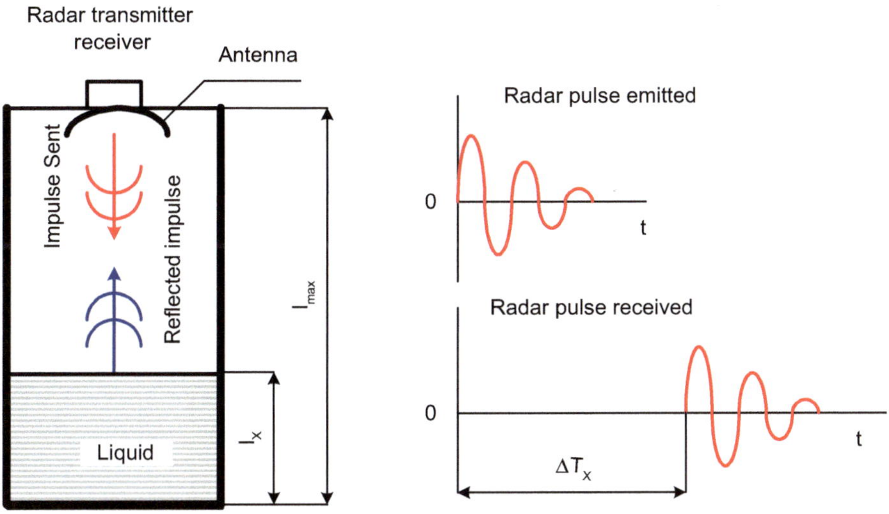

a) arrangement of the sensor b) time course of the sensor pulses

Figure 2.96 Level sensor with impulse radar signal.

2.7.5 Radar level sensors

They work on a similar principle to ultrasonic sensors, with the difference that instead of an ultrasonic signal, electromagnetic waves with a frequency of several GHz are used to measure the water level. We distinguish between sensors with pulsed and frequency-swept radar signals.

In the case of sensors with a *pulsed radar signal*, short radar pulses are sent by the radar transmitter, which are partly reflected back on the surface and partly passed into the measured medium, whose surface height is being measured (Figure 2.96).

The level height of the measured medium is determined from the time delay ΔT_x between the sent and received radar wave

$$l_x = l_{max} - c_r \cdot \frac{\Delta T_x}{2}, \tag{2.127}$$

where c_r is the speed of propagation of the radar wave through the environment above the measured surface. This is determined by its wavelength l and the frequency of electromagnetic waves f

$$c_r = \lambda \cdot f. \tag{2.128}$$

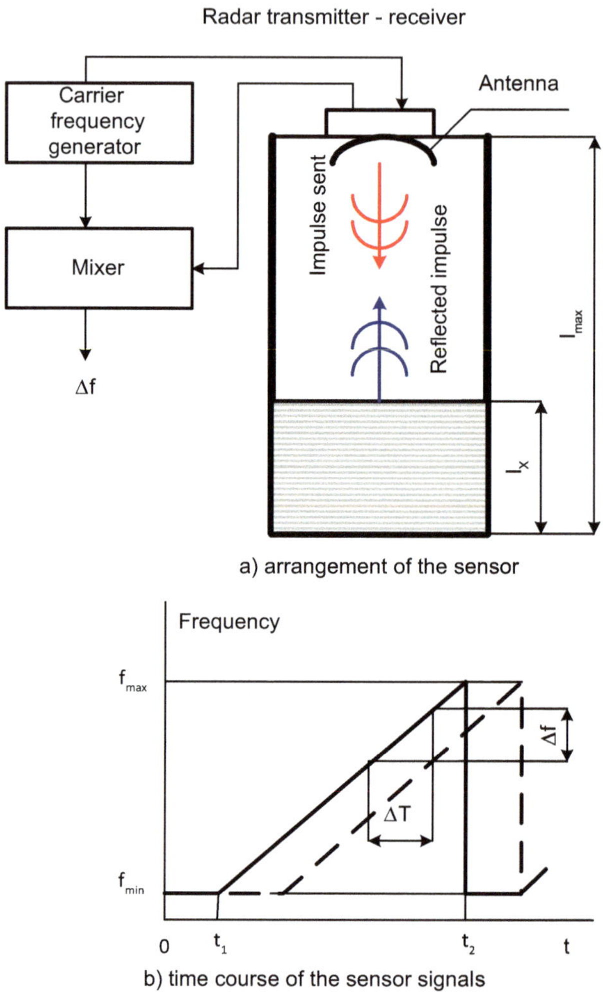

Figure 2.97 Level sensor by frequency-swept radar signal.

In sensors with a *frequency-swept continuous radar signal*, referred to as FMCW (frequency-modulated continuous wave), electromagnetic waves are transmitted continuously, while the frequency of the transmitted signal f_v is modulated by a modulation signal of frequency f_m, which changes linearly with time (Figure 2.97).

The received signal is mixed with the transmitted signal in the mixer, and the resulting intermediate frequency signal with frequency Δf determines the distance of the radar transmitter-receiver from the measured level l_x

$$l_x = l_{\max} - \frac{c_r}{2 \cdot \Delta f}. \tag{2.129}$$

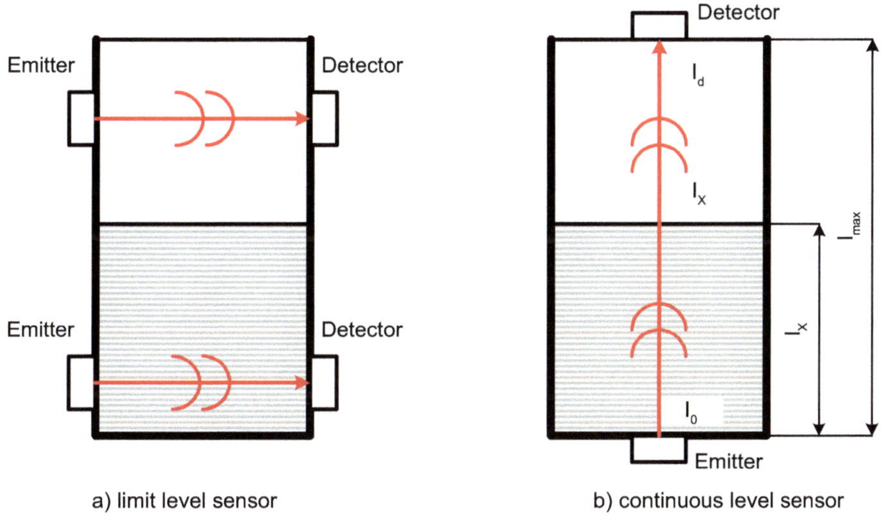

Figure 2.98 Level sensors with radioactive emitters.

A special group of these microwave sensors is TDR sensors (*time domain reflection*), in which microwave pulses are conducted through a coaxial cable into the measured liquid. The intensity of the signal reflected from its surface depends on its permittivity. Again, the time difference between the sent and received impulses is evaluated. The advantage of these sensors is that the pulses propagate along the line and therefore are not dampened by the external environment. They also allow measuring the height of not only liquids but also powder and granular materials.

2.7.6 Level sensors with radioactive emitters

These sensors use the weakening of the intensity of ionizing radiation when it passes through the measured medium. So that these sensors do not have to be calibrated often, the source of ionizing gamma radiation is the isotopes Co 60, Cs 137 with a long half-life. *Geiger–Müller detectors* or scintillation detectors with a *photomultiplier* are used to detect radiation.

Limit level sensor with a pair of radioactive emitters and detectors in Figure 2.98 allows discontinuous measurement of the level height of liquids, bulk substances, and viscous media.

The continuous level sensor in Figure 2.98(b) determines its height l_x from the ratio of the intensity of ionizing radiation above the level I_x, which for a gaseous environment is practically equal to the intensity of radiation at the location of the detector I_d, to the intensity of radiation at the location of the emitter I_0

$$\frac{I_x}{I_0} = e^{-\mu \cdot l_x}, \tag{2.130}$$

where μ is the attenuation coefficient, which depends on the atomic number and density of the medium and on the wavelength of the radiation. It then applies to the height of the water level

$$l_x = -\frac{1}{\mu} \ln\left(\frac{I_x}{I_0}\right). \tag{2.131}$$

Since the attenuation coefficient cannot be precisely determined, it is used similarly to the ultrasonic height sensors of their calibration at a known height of the measured medium.

Optical level sensors work on a similar principle and can only be used for translucent or slightly cloudy liquids.

2.8 Sensors of Thermal Quantities

Touch and *non-contact sensors* are used to measure thermal quantities (temperature, thermal radiation, and thermal energy). Touch temperature sensors include resistive metal and polycrystalline resistance sensors, semiconductor, and thermocouple sensors. Total and spectral pyrometers and thermal imaging cameras are intended for non-contact temperature measurement.

2.8.1 Resistance metal temperature sensors

Resistance temperature sensors include metal platinum, nickel, and polycrystalline sensors.

Platinum temperature sensors consist of a resistance wire with a diameter of several hundred mm or a platinum layer applied to a ceramic or glass plate. Platinum sensors are manufactured in the range of nominal values from 100 Ω to 100 kΩ in accuracy classes of 0.125 and 0.25 °C. Their resistances are defined in the temperature range of –200 °C to +850 °C by the equation

$$R_{Pt}(\vartheta) = R_{Pt}(0) \cdot \left(1 + \alpha_{Pt} \cdot \vartheta + \beta_{Pt} \cdot \vartheta^2 + \gamma_{Pt} \cdot \vartheta^3\right), \tag{2.132}$$

where $R_{Pt}(0)$ is the sensor resistance at temperature $\vartheta = 0$ °C, $\alpha_{Pt} = 3.91 \cdot 10^{-3}\,K^{-1}$, and $\beta_{Pt} = -5.8 \cdot 10^{-7}\,K^{-2}$. For temperatures $\vartheta > 0$ °C, the coefficient < 0 °C is $\gamma_{Pt} = -4.273 \cdot 10^{-12}\,K^{-3}$.

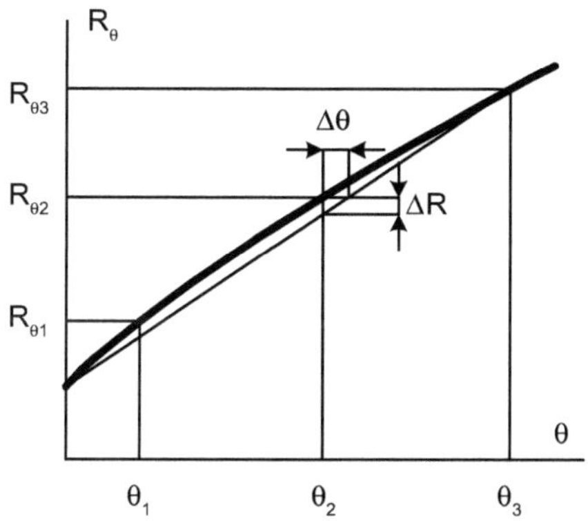

Figure 2.99 Dependence of Pt sensor resistance on temperature.

For example, in the temperature range from 0 to 100 °C, the relative non-linearity of the temperature resistance dependence of platinum sensors reaches $1.425 \cdot 10^{-3}$ (Figure 2.99).

In less demanding applications, nickel temperature sensors are used, which are manufactured with nominal resistances of 100 Ω to 20 kΩ with an accuracy of ±2 °C. Their temperature resistance dependence in the temperature range of –60 to 180 °C is defined by a polynomial

$$R_{Ni}(\vartheta) = R_{Ni}(25) \cdot \left(1 + \alpha_{Ni} \cdot \vartheta + \beta_{Ni} \cdot \vartheta^2\right), \tag{2.133}$$

where $R_{Ni}(25)$ is the resistance of the sensor at a temperature of 25 °C, $\alpha_{Ni} = 5.49 \cdot 10^{-3}\ \mathrm{K}^{-1}$, and $\beta_{Ni} = 6.80 \cdot 10^{-6}\ \mathrm{K}^{-2}$. For temperatures $\vartheta < 0$ °C, the coefficient $\gamma_{Ni} = 0$ and for $\vartheta > 0$ °C, $\gamma_{Ni} = 9.24 \cdot 10^{-9}\ \mathrm{K}^{-3}$.

The non-linearities of these sensors can be reduced by using tabular or computational correction of their temperature dependences.

In less demanding applications, their *two-wire connection* is used to evaluate changes in the resistances of these sensors, in which the influence of the resistances of their inputs on the measurement accuracy is applied. In more demanding applications, a *four-wire connection* is therefore used, in which its current terminals are connected to the supply current source, and their resistance is determined from the voltage drop on the sensor's voltage terminals.

Figure 2.100 shows an example of a circuit connection with a four-wire connection of a Pt100 resistance temperature sensor with a nominal resistance of 100 Ω at a temperature of 0 °C.

The sensor is powered by a current $I_r = U_r/R_0$ and the voltage drop at its voltage terminals is amplified by an instrument amplifier with matched resistors R_2, R_3, and R_4. It applies to the output voltage of the amplifier

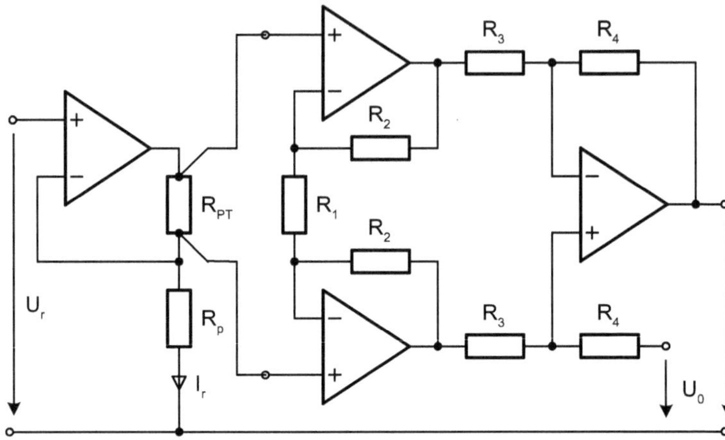

Figure 2.100 Connection of a platinum sensor with an instrument amplifier.

$$U_2 = \frac{R_{\text{Pt}}}{R_0} \cdot \frac{R_4}{R_3}\left(1 + \frac{R_2}{R_1}\right)U_r + U_0. \qquad (2.134)$$

If, for example, the reference voltage source has a value of $U_r = 2.5$ V and its working resistor has a resistance of $R_p = 2.5$ kΩ, then a current of 1 mA flows through the sensor. When requiring the output voltage of the amplifier to vary between 0 and 1 V as the sensor temperature changes between 0 and 100 °C, then the differential gain of the instrumentation amplifier is 26.3. With identical resistances $R_2 = R_3 = R_4 = 100$ kΩ, the resistor has a resistance of $R_1 = 7.9$ kΩ. Since at the sensor temperature 0 °C the voltage drop across it is 100 mV, the voltage source has a value of $U_0 = 2.63$ V.

The *bridge connection* of the amplifier in Figure 2.101 allows the measurement of the temperature difference. The connection is formed by a resistance bridge with a pair of identical Pt sensors and a pair of temperature-independent identical resistors $R_1 = R_2$, whose resistances are identical to the resistances of the $R_{\text{Pt}}(0)$ sensors at 0 °C.

Bridge differential voltage

$$u_2 - u_1 \approx \frac{U_r}{4R_0}\left(R_{\text{Pt2}} - R_{\text{Pt1}}\right) \qquad (2.135)$$

is amplified by an asymmetric *instrumentation amplifier* whose output voltage is

$$u_{OUT} = \frac{U_r}{2 \cdot R_0}\left(1 + \frac{R_4}{R_1}\right) \cdot R_{Pt}(0) \cdot \alpha_{Pt}\left(\vartheta_2 - \vartheta_1\right). \qquad (2.136)$$

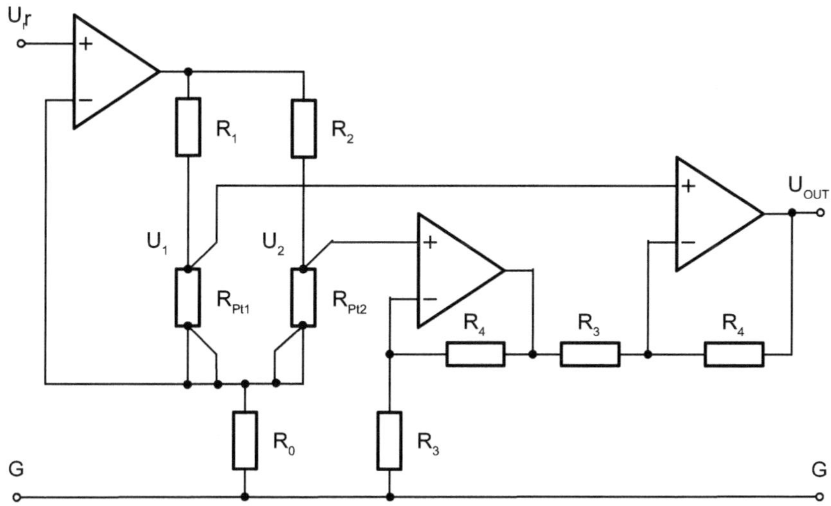

Figure 2.101 Bridge connection of two Pt sensors.

If, for example, the reference voltage is U_r = 2.5 V, R_o = 1 kΩ and $R_{\text{Pt1}}(0) = R_{\text{Pt2}}(0) = R_1 = R_2 = 100\ \Omega$, then the difference in sensor temperatures of 100 °C corresponds to the differential voltage bridge $u_1 - u_2$ = 24.4 mV. To amplify this voltage to 1 V, the differential gain of the instrument amplifier corresponds to 40.9 and the ratio of resistors $R_4/R_3 = 39.9$.

2.8.2 Polycrystalline resistance temperature sensors

Polycrystalline temperature sensors NTC (*negative temperature coefficient*), also referred to as negastors, are usually produced by powder technology from a mixture of metal oxides, e.g., $Fe_2O_3 + TiO_2$, $MnO + C0_2$. Their resistance dependence on temperature is defined by the equation

$$R_{\text{NTC}}(\Theta_1) = R_{\text{NTC}}(\Theta_2) \cdot e^{B\left(\frac{1}{\Theta_1} - \frac{1}{\Theta_2}\right)}, \tag{2.137}$$

where $R_{\text{NTC}}(\Theta_1)$ and $R_{\text{NTC}}(\Theta_2)$ are the supports of the sensor at absolute temperatures Θ_1, Θ_2 and B is the material constant of the sensor, reaching values of 1500–7000 K. This constant can be determined by knowing the resistances $R_{\text{NTC}}(\Theta_1)$ and $R_{\text{NTC}}(\Theta_2)$ by

$$B = \frac{\Theta_1 \cdot \Theta_2}{\Theta_2 - \Theta_1} \ln\left(\frac{R_{\text{NTC}}(\Theta_1)}{R_{\text{NTC}}(\Theta_2)}\right). \tag{2.138}$$

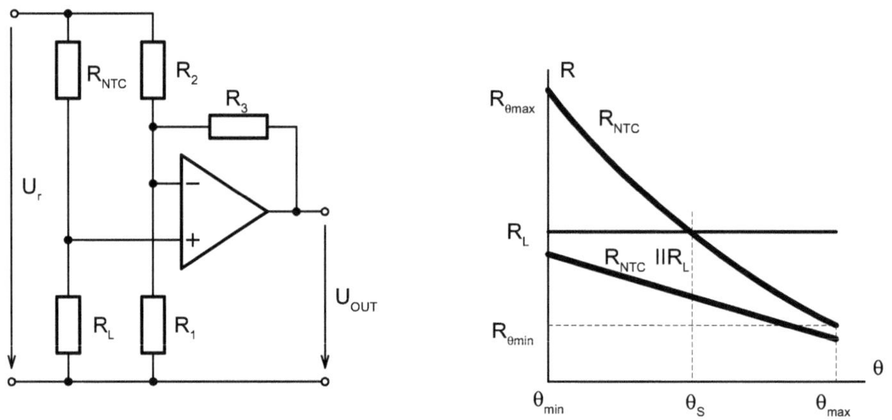

a) connection of the sensor b) conversion characteristic of the sensor

Figure 2.102 Linearization of NTC temperature sensor.

The average temperature coefficient of the sensor in the temperature range $\Theta_1 = 291.15$ K and $\Theta_2 = 358.15$ K is

$$\alpha = -\frac{B}{\Theta^2}. \tag{2.139}$$

These sensors are usually connected to a resistance bridge according to Figure 2.102.

To linearize the non-linear dependence of resistance on temperature, a linearization resistor R_L is used for these sensors, the resistance of which is chosen based on the *tangent linearization* at the inflection point of the temperature resistance dependence of the sensor according to the relation

$$R_L = \frac{B - \Theta_s}{B + 2 \cdot \Theta_s} \cdot R_{\mathrm{NTCs}}\left(\Theta_s\right) = R_{\mathrm{NTCs}}\left(\Theta_s\right), \tag{2.140}$$

where

$$R_{\mathrm{NTCs}}\left(\Theta_s\right) = \frac{R_{\mathrm{NTC}}\left(\Theta_{\mathrm{min}}\right) + R_{\mathrm{NTC}}\left(\Theta_{\mathrm{max}}\right)}{2} \tag{2.141}$$

is the mean value of the sensor resistances in the temperature range Θ_{min} and Θ_{max}.

When using tangent linearization, temperature measurement accuracy of several percentages can be achieved in the temperature range of –100 to 450 °C.

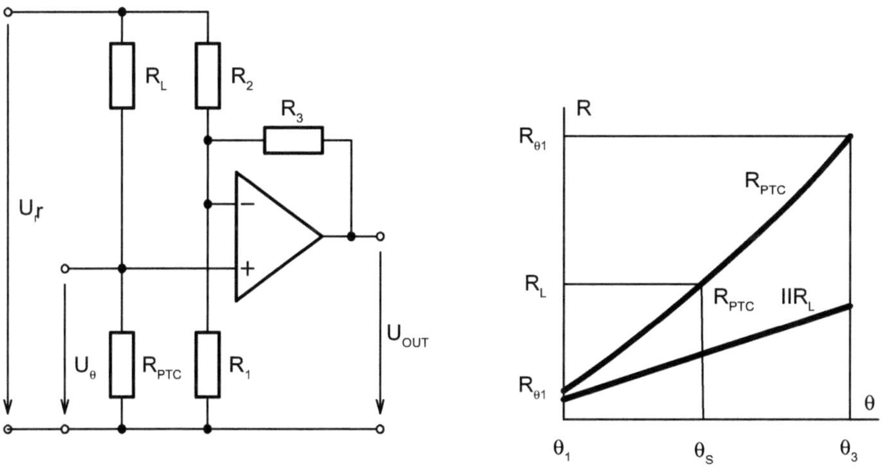

a) connection of the sensor b) conversion characteristic of the sensor

Figure 2.103 Linearization of PTC temperature sensor.

Polycrystalline PTC (*positive temperature coefficient*) temperature sensors, also referred to as positors, are manufactured from polycrystalline ferroelectric ceramics, e.g., $BaTiO_3$. Their resistance dependence on temperature is defined by the equation

$$R_{\mathrm{PTC}}(\vartheta) = R_{\mathrm{PTC}}(25^{\circ})\left[1 + 7.95 \cdot 10^{-3}(\vartheta - 25) + 1.95 \cdot 10^{-5}(\vartheta - 25)^{2}\right], \tag{2.142}$$

where R_{PTC}(25 °C) is the resistance of the sensor at a temperature of 25 °C.

These sensors are also connected to resistance bridges, while tangent linearization is again used to linearize them (Figure 2.103).

2.8.3 Semiconductor temperature sensors

These sensors, referred to as BG (*band gap*), use the temperature dependence of the UBE voltage of the BE transition of a bipolar transistor, connected in the function of a *transistor diode*. The temperature variation of the BE junction voltage is determined by

$$\frac{dU_{\mathrm{BE}}}{d\vartheta} = \frac{k}{q_e} \cdot \frac{U_{\mathrm{BE}}}{U_T} - 2.2\left[\frac{mV}{K}\right], \tag{2.143}$$

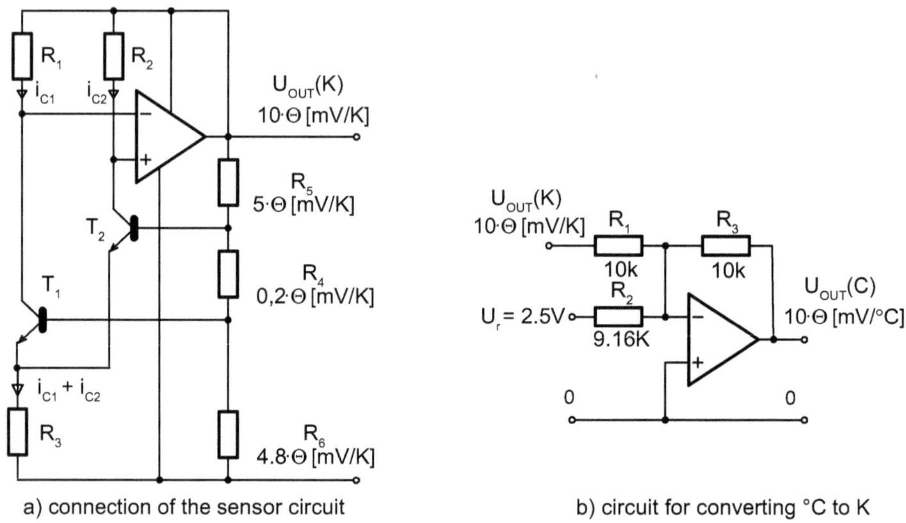

Figure 2.104 Temperature sensor with BE junctions of bipolar transistors.

where $k = 1.38 \cdot 10^{-23}$ is the Boltzmann constant, $q_e = 1.56 \cdot 10^{-19}$ C is the electron charge, and U_T is the thermal voltage, which at 27 °C (300 K) has a value of 26 mV.

In practice, a connection with two transistors T_1, T_2 with different collector currents is used (Figure 2.104(a)).

Because for the U_{BE} voltage of bipolar transistors it applies

$$U_{\mathrm{BE1}} = U_T \cdot \ln\frac{I_{C1}}{I_{s1}}, \qquad U_{\mathrm{BE2}} = U_T \cdot \ln\frac{I_{C2}}{I_{s2}}, \tag{2.144 a, b}$$

where I_{C1}, I_{C2} are their collector currents and I_{s1}, I_2 are their saturation currents that apply to the voltage across resistor R_4

$$U_{R4} = U_{\mathrm{BE2}} - U_{\mathrm{BE1}} = U_T \cdot \ln\frac{I_{C2}}{I_{s2}} \cdot \frac{I_{s1}}{I_{C1}}. \tag{2.145}$$

With identical saturation currents and the ratio of collector currents $I_{C2}/I_{C1} = 10$, the voltage on resistor R_4 is

$$U_{R4} = \frac{k \cdot \Theta}{q_e} \ln 10 = 0.2 \cdot \Theta \left[\frac{\mathrm{mV}}{\mathrm{K}}\right]. \tag{2.146}$$

A voltage drop of $4.8 \cdot \Theta$ [mV/K] will then occur on resistor $R_6 = 24 \cdot R_4$ and a voltage drop of $4.8 \cdot \Theta$ [mV/K] on resistor $R_5 = 25 \cdot R_4$. The output

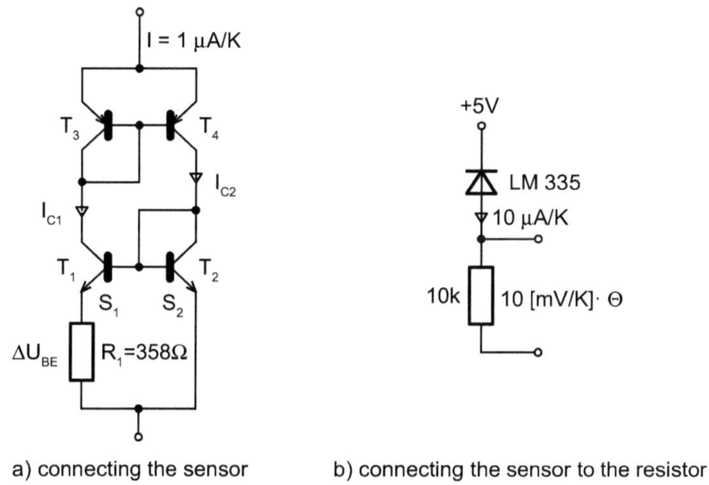

Figure 2.105 Temperature sensor with current output.

voltage of the circuit is thus directly proportional to the temperature of the sensor in K

$$U_{OUT} = 10 \cdot \Theta \left[\frac{\text{mV}}{\text{K}} \right]. \tag{2.147}$$

For example, at 0 °C (273 K), the output voltage of the circuit is $U_{OUT} = 2.73$ V.

The circuit in Figure 2.104(b) is used to convert the voltage in mV/°C to the voltage mV/K.

At ratio

$$\frac{2.73}{R_1} = \frac{U_r}{R_2} \tag{2.148}$$

and the resistance values of the resistors $R_1 = R_3 = 10$ kΩ, $R_2 = 9.16$ kΩ and the reference voltage $U_r = 2.5$ V, the output voltage of the circuit is directly proportional to the measured temperature in °C

$$U_{OUT} = 10 \cdot \theta \left[\text{mV} / ^{\circ}\text{C} \right]. \tag{2.149}$$

Temperature sensors with current output work on a similar principle (Figure 2.105).

Transistors T_1, T_2 with different emitter surfaces S_1, S_2 are fed by identical collector currents $I_{C1} = I_{C2} = I_C$ from the current divider formed by transistors T_3, T_4. The voltage across resistor R_1 is proportional to the absolute temperature Θ

$$\Delta U_{\mathrm{BE}} = U_T \ln\frac{S_1}{S_2} = \frac{k}{q_e}\ln\frac{S_1}{S_2}. \tag{2.150}$$

The resulting current of the current source is then determined by the sum of the collector currents

$$I_C = I_{C1} + I_{C2} = 2 \cdot I_C = 2\frac{\Delta U_{\mathrm{BE}}}{R_1}. \tag{2.151}$$

With the area ratios of the emitters $S_1/S_2 = 8$ and the resistance of the resistor $R_1 = 358\ \Omega$, the circuit generates a current

$$I = \Theta[\mu\mathrm{A} / \mathrm{K}]. \tag{2.152}$$

If, for example, we connect this circuit in series with a resistor R = 10 kΩ, then the voltage drop across it has a temperature dependence of 10 mV/K.

Typical temperature sensors with voltage output are, for example, TMP35 and TMP36 circuits with a sensitivity of 10 mV/°C. The AD590 and LM335 circuits have sensitivities of 1 and 10 μ/AK, respectively. Both groups of sensors have an accuracy of up to 0.5% in the temperature range of –55 to 125 °C. The TMP03 and TMP04 circuits additionally contain *delta sigma modulators*, whose output pulse signal alternations determine the measured temperature in °C and K

$$\theta[°\mathrm{C}] = 235 - 400\frac{T_2}{T_1}, \qquad \Theta[\mathrm{K}] = 455 - 720\frac{T_2}{T_1}, \tag{2.153 a, b}$$

where T_1 and T_2 are the pulse lengths with higher and lower TTL levels.

The circuits of the AD7410 series, additionally equipped with 16-bit A/D converters, make it possible to measure the temperature in the range of –40 to 105 °C with an accuracy of up to 0.5 °C.

2.8.4 Thermocouple temperature sensors

They use *thermocouples* to measure temperature, which are made up of a pair of different metals or their alloys (Figure 2.106).

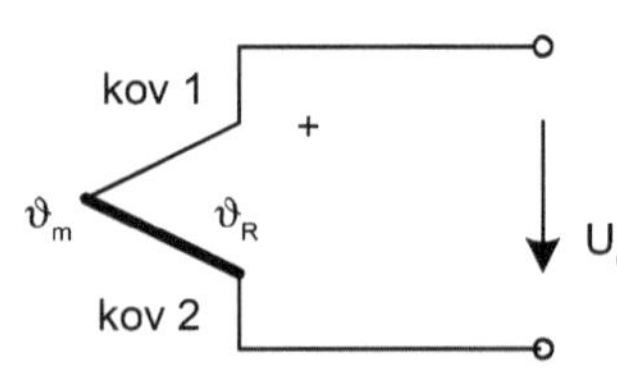

Figure 2.106 Thermocouple.

At different temperatures of the measuring and reference end of the thermocouple, a

Table 2.6 Parameters of typical thermocouples.

Type	Kov 1	Kov 2	k_t (mV/°C)	Temperature range (°C)
T	Cu	Cu-Ni	39.2	–200 to 400
J	Fe	Cu-Ni	50.2	–200 to 700
E	Cr-Ni	Cu-Ni	6.9	–200 to 1000
K	Cr-Ni	Al-Ni	40.5	–200 to 1300

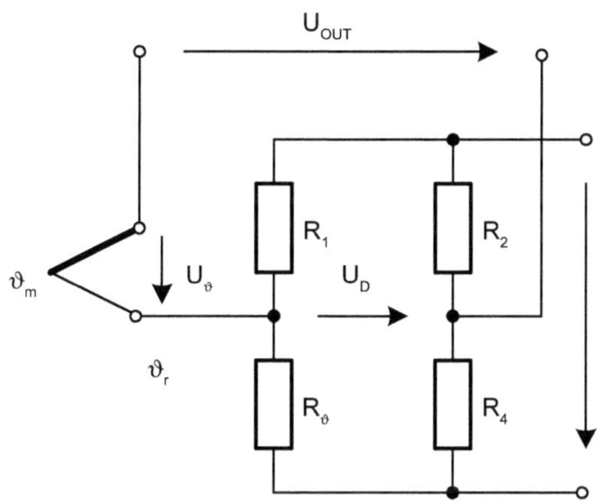

Figure 2.107 Kompenzační krabice.

thermoelectric voltage proportional to the difference in these temperatures is generated in the metal contact

$$U_{\vartheta} = k_t\left(\vartheta_m - \vartheta_r\right), \tag{2.154}$$

where k_t is the average voltage constant of the thermocouple in the given temperature range.

Table 2.6 shows the parameters of typical thermocouples.

To eliminate temperature measurement errors caused by temperature fluctuations of the reference end of the thermocouple, compensation boxes were previously used, which are currently being replaced by isothermal terminal blocks.

The *compensation box* consists of a resistance bridge with temperature-independent resistors R_1, R_2, and R_4 and a temperature-dependent resistor R_{ϑ}, which is powered by the reference voltage U_r (Figure 2.107).

The condition for the elimination of changes in the temperature of the reference end of the thermocouple is determined by the equality of the temperature changes of the voltage on the diagonal of the bridge and the voltage of the thermocouple

$$\frac{dU_D}{d\vartheta} = \frac{dU_{\vartheta}}{d\vartheta}. \tag{2.155}$$

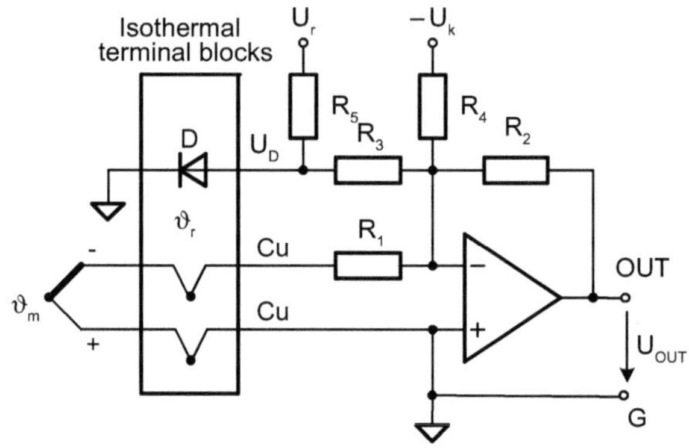

Figure 2.108 Thermocouple amplifier with isothermal terminal block.

Tension on the diagonal of the bridge

$$U_D = \left(\frac{R_1}{R_1 + R_\vartheta} - \frac{R_2}{R_2 + R_4} \right) U_r \tag{2.156}$$

has a temperature change

$$\frac{dU_D}{d\vartheta} = -\frac{R_1}{\left(R_1 + R_\vartheta\right)^2} \frac{dR_\vartheta}{d\vartheta} U_r, \tag{2.157}$$

from which the required temperature change of the resistor R_ϑ can be determined

$$\frac{dR_\vartheta}{d\vartheta} = -\frac{\left(R_1 + R_\vartheta\right)^2}{U_r \cdot R_1} \frac{dU_D}{d\vartheta}. \tag{2.158}$$

The resistances of the resistors are determined by the equation for the reference temperature $\vartheta_r = 0$ °C

$$R_2 \cdot R_\vartheta\left(0^\circ\mathrm{C}\right) = R_1 \cdot R_4. \tag{2.159}$$

At this temperature, the voltage across the diagonal of the bridge is $U_D = 0$.

A thermocouple amplifier with isothermal terminal block temperature measurement is shown in Figure 2.108.

The transmission of the amplifier, determining the change of its output voltage with temperature, is given by the ratio of resistors R_2/R_1 and the average temperature coefficient k_t of the thermocouple

$$\frac{R_2}{R_1} = \frac{\Delta U_{OUT} \;/\; \Delta\vartheta}{k_t}. \tag{2.160}$$

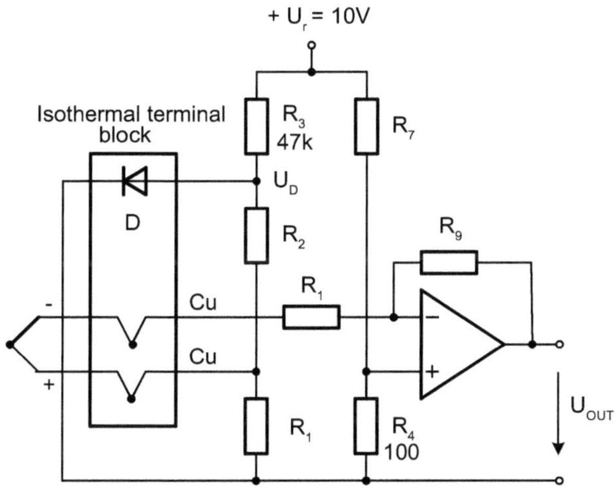

Figure 2.109 Bridge connection of the thermocouple voltage amplifier.

The temperature dependence of the voltage of the diode U_D with the temperature coefficient $\alpha \cdot U_D = dU_D/d\vartheta = -2.2$ mV/°C compensates under the assumption

$$\frac{\alpha_{U_D}}{k_t} = \frac{R_3}{R_1} \tag{2.161}$$

effect of temperature change of the isothermal terminal block.

The condition of zero output voltage of the amplifier at temperature $\vartheta_m = 0$ °C is fulfilled when the currents are equal

$$\frac{U_k}{R_4} = \frac{U_D(0°\mathrm{C})}{R_3}. \tag{2.162}$$

If, for example, the output voltage of the amplifier is to change in the range of 0–1 V when the temperature of the measuring end of the *K*-type thermocouple changes from 0 to 100 °C, then when choosing resistance $R_1 = 1\text{k}\Omega$, it applies to resistances $R_2 = 255$ kΩ and $R_3 = 56$ kΩ. At the compensation voltage $U_k = -5$ V and the voltage drop on the diode U_D (0 °C) = 0.689 V, $R_4 = 406$ kΩ. Resistor $R_5 = 4.2$ kΩ determines that at the voltage $U_r = 5$ V and a temperature of 27 °C, the current flowing through the diode is 1 mA.

The *bridge connection* of the thermocouple voltage amplifier, requiring only one supply voltage, is shown in Figure 2.109.

Table 2.7 shows the resistances of the resistors for amplifiers of the most used thermocouples, whose output voltage has a temperature change of 10 mV/°C.

Table 2.7 The nominal resistances of the resistors of the amplifier in Figure 2.109.

Type	R_1	R_2	R_7	R_9
K	110 Ω	5.76 kΩ	102.5 kΩ	269 kΩ
J	100 Ω	4.02 kΩ	80.7 kΩ	200 kΩ
S	100 Ω	20.5 kΩ	392 kΩ	1.07 MΩ

Typical monolithic thermocouple signal processing circuits with an internal isothermal terminal block are the AD594 and AD596 circuits for J-type thermocouples and the AD595 and AD597 circuits for *K*-type thermocouples. Temperature measurement accuracy of up to 0.5% can be achieved with these circuits.

Modern circuits for processing thermocouple signals are, for example, the AD8496 and AD8497 circuits, which are designed for *J*- and *K*-type thermocouples and whose output voltage change is 5 mV/°C. The circuits have temperature compensation for temperature fluctuations of the reference ends of the thermocouples.

Universal circuits for processing temperature sensor signals are, for example, circuits of the LTC 298x series, enabling two-wire and four-wire connection of resistance and semiconductor temperature sensors, including thermocouples. In addition to 24-bit delta sigma A/D converters, these circuits also contain an internal EPROM memory, enabling digital linearization of the conversion characteristics of temperature sensors with an accuracy of up to 0.1 °C.

2.8.5 Non-contact temperature sensors

For non-contact temperature measurement, *pyrometers* are used, which work on the principle of measuring the radiation energy that is emitted by any object whose temperature is higher than zero absolute temperature, i.e., –273 °C. The radiated energy of an absolutely black body with an area of 1 m^2 for a period of 1 s is determined by the *Stefan–Boltzmann law*

$$E = \varepsilon \cdot \sigma \cdot \Theta^4, \tag{2.163}$$

where ε is the emissivity of the body whose temperature we are measuring, Θ is its absolute temperature and $\sigma = 5.67 \cdot 10^{-8}\ \mathrm{W \cdot m^{-2} \cdot K^{-4}}$ and *Stefan–Boltzmann´s constant.*

where ε is the emissivity of the body whose temperature we measure, Θ is its absolute temperature, and $\sigma = 5.67 \cdot 10^{-8}\ \mathrm{W \cdot m^{-2} \cdot K^{-4}}$ is the *Stefan–Boltzmann constant.*

The *emissivity* of the body depends not only on the temperature but also on the color and type of surface of the body whose temperature is measured.

Table 2.8 Emissivities of the most common materials.

Material	Emissivity	Material	Emissivity
Bronze, oxidized (200 °C)	0.61	Copper, oxidized (130 °C)	0.76
Wood (70 °C)	0.94	Lead (40 °C)	0.43
Aluminum, polished (100 °C)	0.09	Plast: PE, PP, PVC (20 °C)	0.94
Aluminum, heavily oxidized (93 °C)	0.21	Glass (90 °C)	0.83
Chrome (40 °C)	0.08	Zinc, oxidized (400 °C)	0.11
Chrome, polished (150 °C)	0.06	Cast iron (100 °C)	0.80
Copper, polished (40 °C)	0.03	Steel (20 °C)	0.77

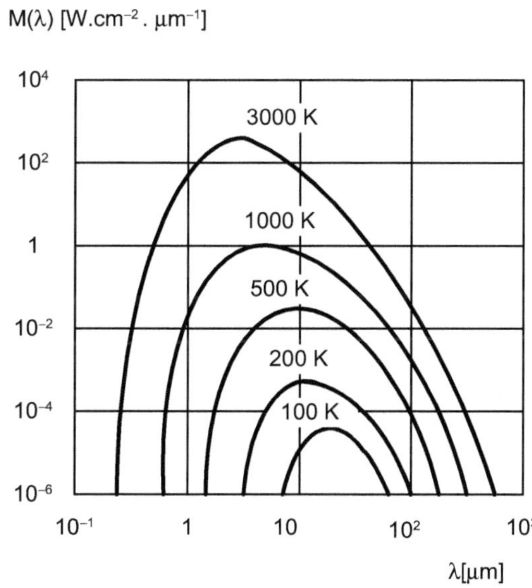

Figure 2.110 Spectral density of radiation intensity of an absolute black body.

An absolutely black body has an emissivity of 1. Other materials always have an emissivity lower than 1. Table 2.8 shows the emissivities of the most common materials.

The dependence of the *spectral density of the radiation intensity* of an absolute black body on its wavelength is defined by *Planck's law*

$$L(\lambda) = c_1 \cdot \lambda^{-5}\left[\exp\left(\frac{c_2}{\lambda \cdot \Theta} - 1\right)\right]^{-1}, \qquad (2.164)$$

where $c_1 = 3.74 \cdot 10^{-16}\,\mathrm{W \cdot m^2}$, $c_2 = 1.44 \cdot 10^{-2}\,\mathrm{m \cdot K}$ are *Planck's constants*, λ is the wavelength of radiation, and Θ is the absolute temperature of the *black body* (Figure 2.110).

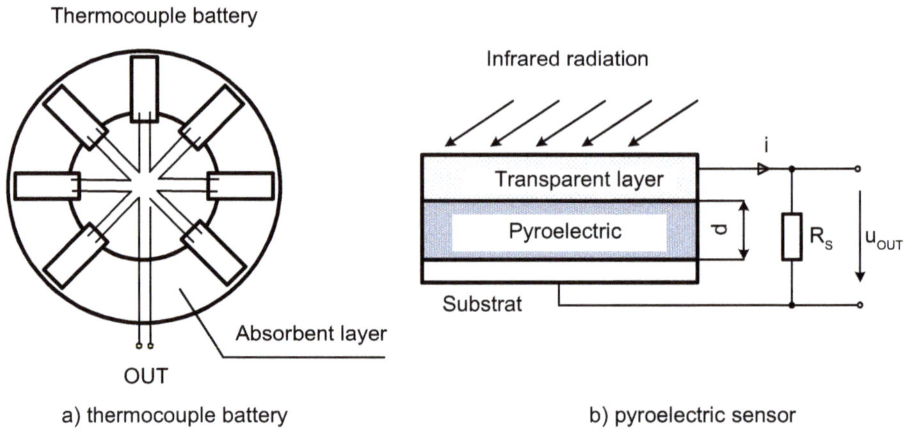

Figure 2.111 Total pyrometer sensors.

From the point of view of the range of wavelengths, we distinguish between total and band pyrometers.

Broadband pyrometers operate in the wavelength range of 2.5–20 μm. These pyrometers use either thermopile batteries (*Thermopile*), consisting of tapes approximately 30-μm thick, resistance temperature sensors, consisting of MgO, MnO, NO, TiO_2, and $Tl_2SeAs_2Te_3$ oxides, which are deposited on a ceramic item, or *pyroelectric sensors* based on $PbTiO_3$ and $PbZrO_3$, using spontaneous polarization when their temperature changes (Figure 2.111(a)).

The *pyroelectric sensor* consists of two plane-parallel electrodes, while the front electrode of the sensor is transparent to infrared radiation.

When the sensor is irradiated with thermal radiation, an electric charge q is induced between its electrodes, the size of which is proportional to the area of the sensor S, the average radiation intensity I_e, and the temperature of the measured object Θ

$$q = k_p \cdot S \cdot I_e \cdot \Theta, \tag{2.165}$$

where k_p is the sensor's pyrometric constant (Figure 2.111(b)).

This charge causes the inter-electrode voltage of the sensor

$$u_{\mathrm{OUT}}(t) = \frac{q}{C_s} = \frac{q \cdot d}{\varepsilon \cdot D}, \tag{2.166}$$

where C_s is the inter-electrode capacitance of the sensor, given by the pyroelectric permittivity ε, its area S, and thickness d.

To calibrate total pyrometers, temperature calibrators are used, containing a black heated cell, the emissivity of which is close to one. When introducing the temperature Θ_0, at which the measured body emits thermal energy with the same intensity of thermal radiation as the measured object with the actual temperature Θ_s, the temperature of the measured body can be expressed by the equation based on the *Stefan–Boltzmann law*

$$\Theta_s = \Theta_0 \cdot \sqrt[4]{\frac{1}{\varepsilon_s}}, \tag{2.167}$$

where ε_s is the emissivity of the body whose temperature we are measuring.

Monochromatic pyrometers measure body temperature based on the intensity of thermal radiation emission in a narrow band of wavelengths, ranging from tens of nm to units of μm. For the total radiation energy of such frequency-limited radiation in the range of wavelengths λ_1–λ_2 then applies

$$L(\lambda_1, \lambda_2, \varepsilon_\lambda) = \int_{\lambda_1}^{\lambda_2} \varepsilon_\lambda \cdot L_\lambda \cdot d\lambda, \tag{2.168}$$

where ε_λ is the spectral emissivity of the body and L_λ is the spectral radiation density.

Due to the difficulty of determining these quantities, the *effective emissivity* value ε_{ef} is introduced during the calibration of these pyrometers, and a similar relation (2.168) applies to the temperature of the measured body, as for total pyrometers.

To measure the radiation intensity, these pyrometers use *quantum radiation sensors* consisting of photodiodes connected in the photodiode mode. The sensitivity of these photodiodes depends not only on the composition of their semiconductor materials and temperature but also on the wavelength of the radiation. The most commonly used photodetector materials are *silicon* (Si), *gallium arsenide* (GaAs), *indium antimonide* (InSb), *indium arsenide* (InAs), and others. These materials absorb light radiation in the range of, e.g., 250–1100 nm for silicon and 800 nm to 2 μm for GaAs.

Dual-band pyrometers determine body temperature based on two intensities of thermal radiation emission with different wavelengths λ_1 and λ_2. These pyrometers are calibrated at the so-called temperature of the spectral composition Θ_p, corresponding to the temperature of the black body, at which the ratio of emitted energy for both wavelengths is the same as the ratio of emitted energies of the measured body at its actual temperature Θ_s.

Based on *Planck's law*, the temperature of the measured body can be expressed by the equation

$$\frac{1}{\Theta_s} = \frac{1}{\Theta_p} - \frac{1}{c_2} \cdot \frac{\lambda_1 \cdot \lambda_2}{\lambda_1 - \lambda_2} \cdot \ln \frac{\varepsilon_{\lambda 1}}{\varepsilon_{\lambda 2}}, \tag{2.169}$$

where $\varepsilon_{\lambda 1}$ and $\varepsilon_{\lambda 2}$ are the emissivities of the measured body at wavelengths λ_1 and λ_2.

With these pyrometers, temperatures in the range of –50 to 3500 °C can be measured with an accuracy of up to a few °C.

To investigate temperature images, e.g., mechanical parts of electrical machines and their windings, *thermal cameras* are used, consisting of an optical system with a flat image sensor with a matrix of photodiodes. These sensors use CCD (charge coupled devices) structures. To transfer the charges from the photosensitive capacitors of these sensors, the interline or frame method is described in detail in Section 2.1.4 (Figure 2.44).

In the focal point of the optical system of these thermal cameras, usually formed by germanium lenses due to their permeability in the given waveband, a flat matrix FPA (*focal plane array*) is placed, e.g., with a resolution of 480 × 640 pixels, enabling to measure the distribution of the temperature field in the range of temperatures 20–160 °C with an accuracy of a few °C. In order to suppress noise, the FPA matrices of higher-quality thermal cameras are cooled by Peltier cells to a temperature of up to –30 °C.

2.8.6 Thermal energy sensors

Thermal energy sensors work on the principle of measuring the thermal output transmitted by a heat-carrying medium, which is most often warm water or steam. The heat output is determined by the product of the mass flow of the heat-carrying medium Q_m with the difference in its heat contents i_1, i_2 at the input and output of the measuring system

$$P_q = Q_m \left(i_1 - i_2\right). \tag{2.170}$$

The heat content of the heat-carrying medium is determined by the product of its specific heat capacity c_p and its temperature ϑ

$$i = c_p \cdot \vartheta. \tag{2.171}$$

If we express the *mass flow rate* Q_m by the product of the *volume flow rate* Q_v with the specific density c_p of the heat-carrying medium, then for the heat output

$$P_q = Q_v \left(\rho_1 \cdot c_{p1} \cdot \vartheta_1 - \rho_2 \cdot c_{p2} \cdot \vartheta_2\right), \tag{2.172}$$

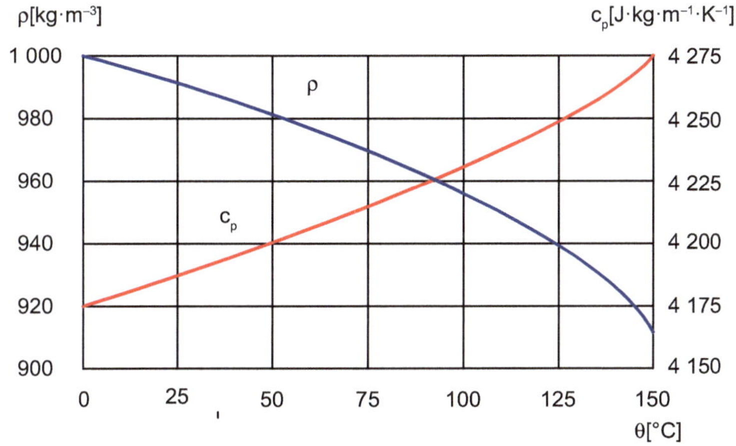

Figure 2.112 Temperature dependences of density ρ and specific heat capacity c_p of water.

where c_{p1}, c_{p2} are heat capacities and ρ_1, ρ_2 are densities of the heat-carrying medium at temperatures ϑ_1, ϑ_2.

If the heat-carrying medium is *water*, in the operating temperature range of 20–150 °C, the product of its density and specific heat capacity can be considered practically independent of its temperature (Figure 2.112).

This product can then be replaced by the thermal coefficient k_ϑ, using which the transferred heat output can be expressed by the approximate relation

$$P_q \approx Q_v \cdot k_\vartheta \left(\vartheta_1 - \vartheta_2\right). \tag{2.173}$$

The *thermal energy* consumed in the heat consumer is then determined by the time integration of the thermal output for the period $\Delta T = t_2 - t_1$

$$Q_q = \int_{t_1}^{t_2} P_q(t) \cdot dt. \tag{2.174}$$

The measuring system of the thermal power and thermal energy meter with a volume flow meter is shown in Figure 2.113. The system consists of temperature sensors S_1, S_2, measuring the temperature of the supplied and discharged medium, the difference between the signals of which is multiplied by the signal of the volume flow meter in a multiplier. The thermal energy is then determined by the time integration of this signal, which determines the transmitted thermal power.

These thermal energy sensors use resistance or semiconductor temperature sensors to measure temperatures, and turbine, ultrasonic, or thermal mass flow sensors to measure volume flow.

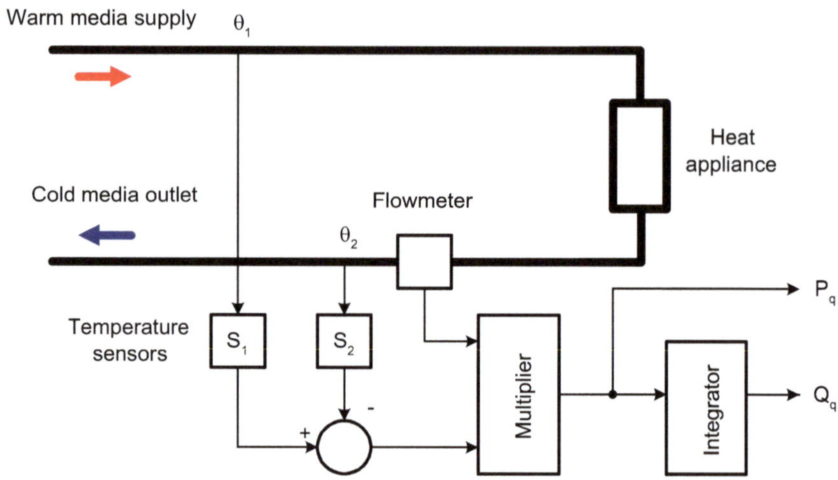

Figure 2.113 Block diagram of heat output and energy sensor.

Industrially produced heat output sensors with ultrasonic mass temperature sensors achieve a measurement accuracy of several percentages in the temperature range of 20–150 °C and a flow rate of 1–50 m^3 of hot water.

2.9 Sensors of Electric and Magnetic Quantities

The most commonly used sensors of electric and magnetic quantities include DC and AC voltage and current sensors, active and reactive power sensors, and charge sensors and magnetic field intensity sensors.

2.9.1 Electrical voltage sensors

They are designed to measure direct and alternating voltages, including the determination of their mean and effective values. Resistive or capacitive voltage dividers or voltage transformers are used to reduce high DC and AC voltages to values suitable for their processing. Asymmetric and differential amplifiers are used to amplify voltage signals.

The *resistive voltage divider* has, assuming that the internal resistance of the voltage source R_i is significantly smaller than one of the resistors of the divider $R_i << R_1, R_2$, the nominal dividing ratio (Figure 2.114(a))

$$\frac{u_2}{u_1} = \frac{R_2}{R_1 + R_2}. \tag{2.175}$$

When sensing AC voltages at higher frequencies, when resistors with high nominal resistance values are used, their parasitic capacities C_{p1}, C_{p2} are

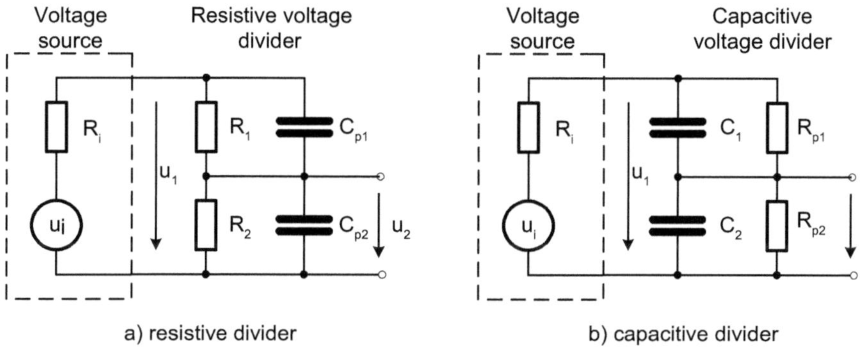

Figure 2.114 Voltage dividers.

also applied, which affect the frequency transfer of the dividing ratio of the divider, for which

$$\frac{u_2}{u_1} = \frac{R_2\left(1 + j \cdot \omega \cdot \tau_1\right)}{R_1\left(1 + j \cdot \omega \cdot \tau_2\right) + R_2\left(1 + j \cdot \omega \cdot \tau_1\right)}, \tag{2.176}$$

where $R_1 \cdot C_1$ and $R_2 \cdot C_2$ are the time constants of resistors R_1 and R_2.

Therefore, for undistorted transmission of AC voltages, the divider must be frequency compensated, which requires that the time constants of both its resistors be identical

$$R_1 \cdot C_1 = R_2 \cdot C_2$$

Resistive voltage dividers are manufactured with division ratios in the range of 1:10 to 1:10,000 with nominal resistance values in the range of units MΩ to hundreds of MΩ for voltage ranges of hundreds of V to hundreds of kV with an accuracy of 0.1%–1%.

To reduce high AC voltages, capacitive voltage dividers, formed by capacitors with a low loss factor, are used (Figure 2.114(b)).

For the ideal division ratio of a capacitive divider

$$\frac{u_2}{u_1} = \frac{C_1}{C_1 + C_2}. \tag{2.177}$$

This dividing ratio is also influenced by the leakage resistances R_{p1}, R_{p2} of the capacitors. Therefore, capacitors with low dielectric losses, e.g., with ceramic, polypropylene, or vacuum dielectrics, are used to produce capacitive dividers with a range of several units to tens of kV. HV capacitive dividers use capacitors whose dielectrics are made of sulfur fluoride SF6, which, although it has excellent insulating and self-extinguishing properties, is the

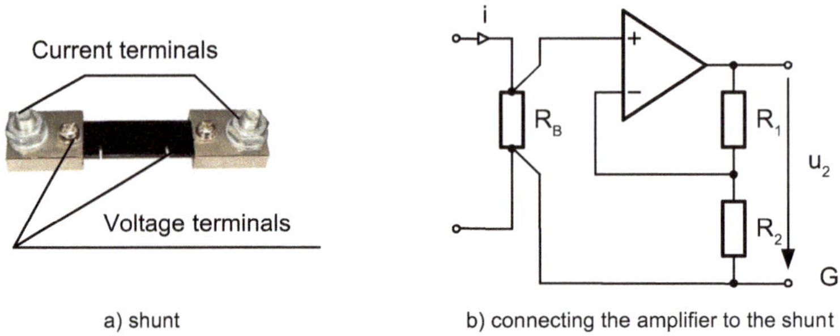

Figure 2.115 Measurement of shunt currents.

cause of the greenhouse effect. This dielectric is replaced by liquid insulators, e.g., mineral and silicone oils.

2.9.2 Electric current sensors

To measure large currents, shunts are used, whose current terminals are connected to the measured circuit and between whose voltage terminals the voltage drop caused by the passage of the measured current is measured (Figure 2.115).

For the output voltage of the connected non-inverting amplifier, the voltage applies

$$u_2 = i \cdot R_B \left(1 + \frac{R_2}{R_1}\right). \tag{2.178}$$

The shunts are manufactured in the range of 10 mΩ to 100 Ω with an accuracy of 0.5%–0.05% and allow to measure currents in the range of hundreds of mA to hundreds of A.

Current amplifiers with a voltage output are used to sense small currents, in the order of units to tens of mA, whose ideal transmission is (Figure 2.116)

$$u_2 = i \cdot R_1, \tag{2.179}$$

measuring extremely small currents instead of ordinary operational amplifiers, *electrometric amplifiers* are used, whose input quiescent currents I_0 reach units of pA (see Table 2.9).

2.9.3 Voltage and current transformers

Voltage and current measuring transformers and current probes are used for non-contact sensing of alternating voltages and currents.

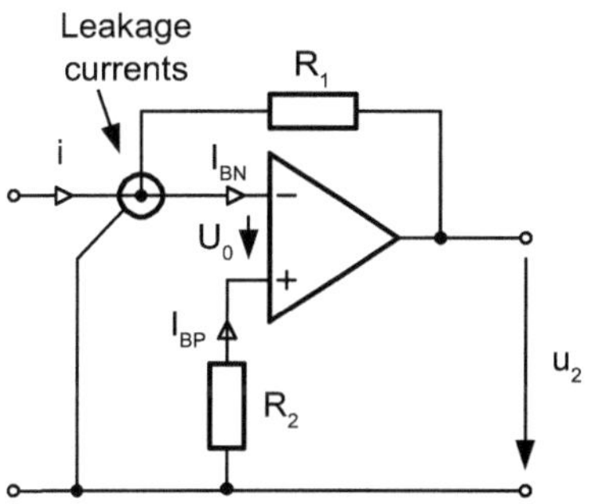

Figure 2.116 Current amplifier with voltage output.

Table 2.9 Parameters of typical electrometric amplifiers.

Type	A_D	*CMRR*	U_0	I_0	u_n	i_n	*SR*	f_T
AD795	120 dB	130 dB	0.5 mV	1 pA	60 nV/√Hz	10 fA/√Hz	1 V/μs	1 MHz
MAX420	140 dB	140 dB	1 μV	2 pA	85 nV/√Hz	10 fA/√Hz	0.5 V/μs	1 MHz
LT1050	127 dB	130 dB	1 μV	1 pA	100 nV/√Hz	10 fA/√Hz	4 V/μs	3 MHz

Voltage measuring transformers are transformers whose secondary winding is practically not loaded, and therefore voltage drops caused by their internal impedance are not applied.

Transformers used in the energy industry are manufactured with nominal input voltages of 6–110 kV, while the nominal output voltage is 100 V. For instrument applications, voltage transformers with significantly lower nominal output voltages, e.g. several V, are used (Figure 2.117).

Current measuring transformers are transformers whose secondary winding is usually connected to a current amplifier with a voltage output (Figure 2.118).

The sensed current i_1 creates an induction flux in the core of the transformer, which in its secondary winding with the number of turns N_2 generates a current i_2

$$i_2 = \frac{i_1}{N_2}, \tag{2.180}$$

which is converted to voltage in the current amplifier

$$u_2 = R \cdot i_1 = R\frac{i_1}{N_2}. \tag{2.181}$$

Current measuring transformers used in the energy sector are produced in the range of nominal input currents from a few A to hundreds of A with a nominal output current of 5 A, or 1 A. Current transformers with significantly lower nominal output currents, e.g. several units to tens of mA, are used for instrument applications.

Figure 2.117 Voltage and current measuring transformers.

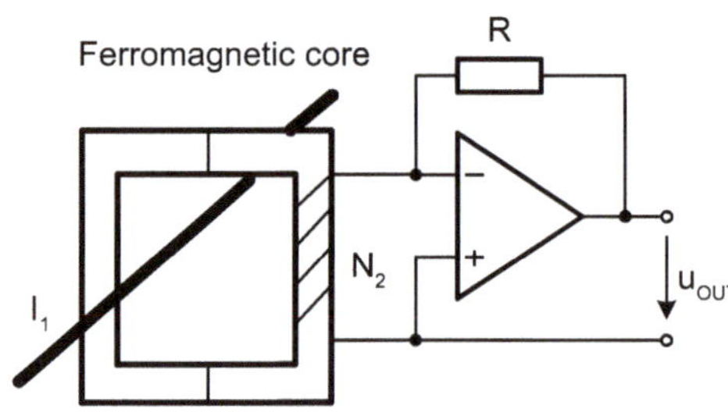

Figure 2.118 Connecting a current transformer to a current to voltage amplifier.

Both types of transformers are produced in accuracy classes of 0.1%, 0.2%, 0.5%, 1%, and 3% and their frequency span is in the range of hundreds of Hz. Their advantage is high insulation strength and robustness.

For non-contact measurement of large alternating currents, current probes are used, working on the *Rogowski coil* principle, which is wound on a toroidal non-ferromagnetic frame (Figure 2.119).

A voltage is induced in the coil by the passage of the measured alternating current I_1 of frequency f

$$u_2 = M \cdot \omega \cdot i_1 = \frac{\mu_v \cdot N \cdot S}{l_s} \cdot 2 \cdot \pi \cdot f \cdot i_1 \qquad (2.182)$$

where M is the mutual inductance between the conductors through which the measured current flows, μ_v is the air permeability, which is approximately equal to the vacuum permeability, N is the number of turns of the coil, S is the area of its winding, and $l_s = \pi \cdot (D_1 + D_2)/2$ is the mean length of one turn of the coil winding.

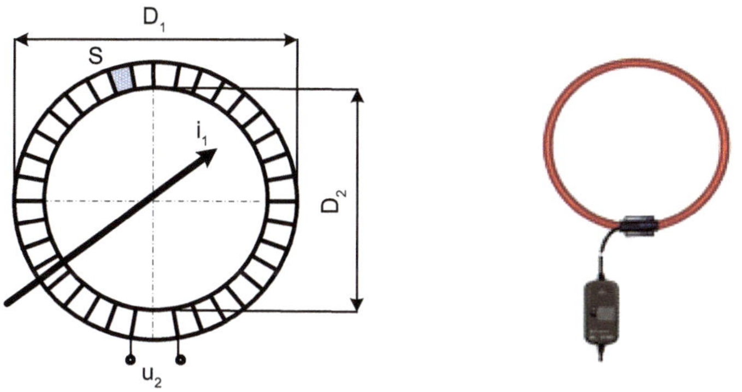

a) diagram of a toroidal coil b) execution of a toroidal coil

Figure 2.119 Rogowski's toroidal coil.

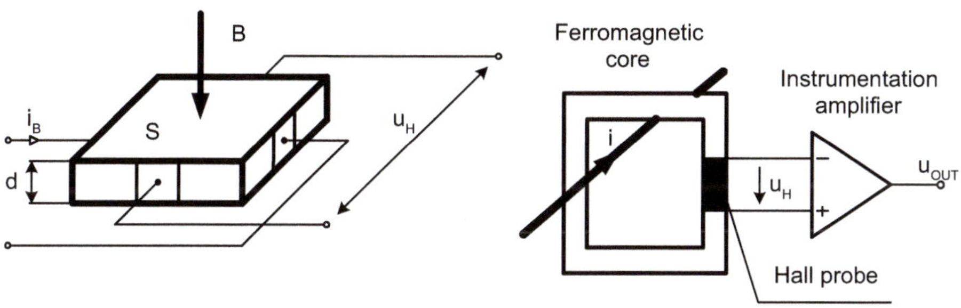

a) Hall probe b) current transformer with Hall probe

Figure 2.120 Current transformer probe with Hall probe.

The Rogowski coil enables the measurement of currents in the range of up to several kA in the frequency range of tens of kHz with an accuracy of up to 1%. Their advantage is the impossibility of overexciting them, and the disadvantage is considerable sensitivity to alternating interfering magnetic fields.

2.9.4 Current sensors with Hall probe

The *current probe* for measuring alternating and direct current has a *Hall probe* inserted in the air gap of the ferromagnetic core, whose output voltage U_H is directly proportional to the vector product of the magnetic induction B and the excitation current I_B (Figure 2.120).

$$u_H = k_H \cdot i_B \cdot \frac{B}{d} = k_H \cdot i_B \cdot \frac{i_1}{d \cdot S}, \qquad (2.183)$$

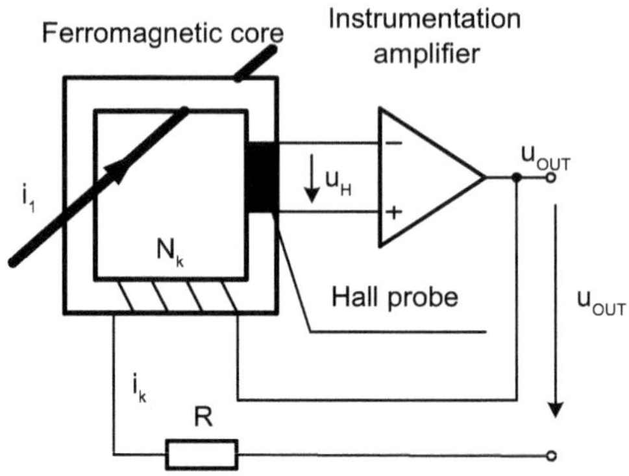

Figure 2.121 Feedback connection of a current transformer with a Hall probe.

where k_H is the Hall constant, d is the thickness of the Hall probe, and S is its area, perpendicular to the magnetic induction vector.

To eliminate the non-linear course of the magnetization characteristic of the ferromagnetic material of the transformer core, these probes use a feedback connection of a current transformer, which contains a compensating winding with the number of turns N_k, through which the compensating current i_k flows (Figure 2.121).

In order not to apply the non-linearity of the magnetization characteristic of the ferromagnetic core of the transformer, the condition for the compensation current must be met

$$i_k = \frac{u_{\text{OUT}}}{R} = \frac{i_1}{N_k}. \tag{2.184}$$

In this case, a zero induction flux always flows through the ferromagnetic core, and therefore the non-linearity of the course of its magnetization characteristic does not apply. Thus, current measurement accuracy of up to 0.05% can be achieved in the range of hundreds of mA to hundreds of A in the frequency range of up to 20 kHz.

2.9.5 Electrical power and energy sensors

These sensors are designed to measure electrical active, reactive, and apparent power, as well as active and reactive electrical energy.

The *active power* of alternating current is given by the mean value of its instantaneous power

$$P = \frac{1}{T_p}\int_0^{T_p} p(t)\cdot dt = \frac{1}{T_p}\int_0^{T_p} u(t)\cdot i(t)\cdot dt\,[W], \tag{2.185}$$

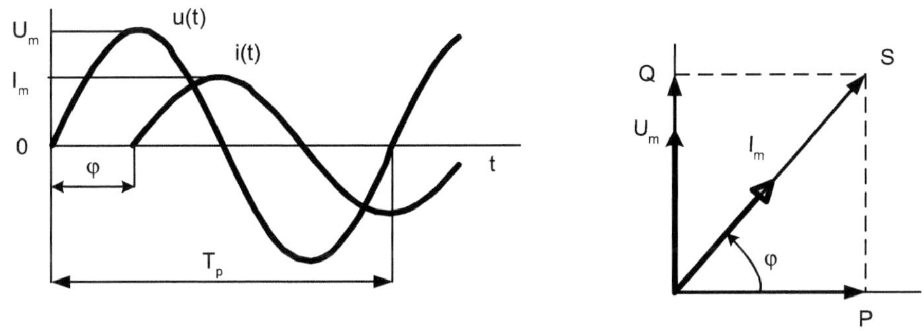

Figure 2.122 Time and vector representation of voltage, current and power waveforms.

where T_p is the period of the voltage and current time courses (Figure 2.122).

The *energy* of alternating current is determined by the mean value of the instantaneous power

$$E = (t_2 - t_1)\int_{t_1}^{t_2} p(t) \cdot dt = \int_{t_1}^{t_2} u(t) \cdot i(t) \cdot dt\,[Ws], \tag{2.186}$$

where $t_2 - t_1$ is the time for which we determine the energy.

For a sinusoidal waveform of voltage and current with amplitudes U_m and I_m and a phase shift φ between them

$$u(t) = U_m \cdot \sin\omega t, \qquad i(t) = U_m \cdot \sin(\omega t + \varphi), \tag{2.187 a, b}$$

active power can be defined as the product of their effective values $U = U_m/\sqrt{2}$ and $I = I_m/\sqrt{2}$

$$P = U \cdot I \cdot \cos\varphi = \frac{U_m \cdot I_m}{2}\cos\varphi[W], \tag{2.188}$$

where cosφ is the *power factor* of the single-phase system.

Reactive power is then defined by the equation

$$Q = U \cdot I \cdot \sin\varphi[\mathrm{VAR}]. \tag{2.189}$$

Based on the knowledge of active and reactive power, *active* and *reactive energy* can be defined

$$E_P = \frac{1}{t_2 - t_1}\int_{t_1}^{t_2}[p(t) \cdot \cos\varphi]dt = \frac{1}{t_2 - t_1}\int_{t_1}^{t_2}[u(t) \cdot i(t) \cdot \cos\varphi]dt\,[\mathrm{Ws}], \tag{2.190}$$

$$E_Q = \frac{1}{t_2 - t_1}\int_{t_1}^{t_2}\left[p(t)\cdot \sin\varphi\right]dt = \frac{1}{t_2 - t_1}\int_{t_1}^{t_2}\left[u(t)\cdot i(t)\cdot \sin\varphi\right]dt\ \left[\text{VARs}\right]. \tag{2.191}$$

The *apparent power* of alternating current is determined by the squared sum of active and reactive power

$$S = U \cdot I = \sqrt{P^2 + Q^2}\ \left[\text{VA}\right]. \tag{2.192}$$

If the voltage and current waveforms are distorted due to higher harmonic components, then the active power of the non-harmonic voltage and current waveforms is defined by the sum of the active powers of their spectral components

$$P_n = \sum_{k=0}^{\infty} U_k \cdot I_k \cdot \cos\varphi_k\ \left[\text{W}\right], \tag{2.193}$$

where $U_0 \cdot I_0 = P_0$ is the *power* of the *DC component* of voltage and current, U_k, I_k are the effective values of the *k*th spectral components of voltage and current, and φ_k are the phase shifts between them. Similarly, the reactive power of non-harmonic waveforms of voltages and currents is defined

$$Q_n = \sum_{k=0}^{\infty} U_k \cdot I_k \cdot \sin\varphi_k\ \left[\text{VAR}\right]. \tag{2.194}$$

For non-harmonic time courses of voltages and currents, the relationship between active, reactive, and apparent power is determined by the equation

$$S_n = \sqrt{P^2 + Q^2 + D^2}\ \left[\text{VA}\right], \tag{2.195}$$

where D_n is the *deformation power*, caused by the existence of higher order harmonic components of voltage and current.

An example of power measurement in a single-phase network using voltage and current transformers is shown in Figure 2.123.

The alternating current source generates a current through the primary coil of the current transformer to the load *Z*. The voltage on the load is transformed by the voltage transformer to the input voltage of the power meter.

If the conversion constants of the voltage and current transformer are $p_u = U_1/U_2$ and $p_i = I_1/I_2$, then for the active and reactive power on the load

$$P = p_u \cdot p_i \cdot U_1 \cdot I_1 \cdot \cos\varphi, \qquad Q = p_u \cdot p_i \cdot U_1 \cdot I_1 \cdot \sin\varphi. \tag{2.196 a, b}$$

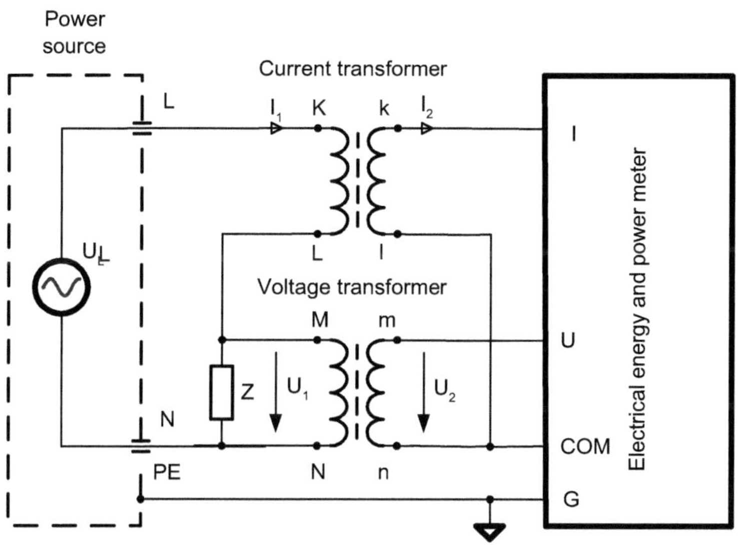

Figure 2.123 Connection to measure power and energy in a single phase network with voltage and current transformers.

The common terminal COM is connected to the secondary windings of the transformers and thus forms the signal ground of the meter to which the measured current and voltage are related. For safety reasons, the protective conductor PE of the power source is connected to the protective terminal G of the meter.

To multiply the signals, modulation multipliers or multipliers with a *Hall probe* are used in power sensors. The *modulation multiplier* is formed by a delta sigma modulator with an integration amplifier Z_1, a comparator K, and an amplitude modulator formed by a switched inverting amplifier Z_2 (Figure 2.124).

The *delta sigma modulator* works on the principle of equalizing the charge on the integration capacitor C_1 so that its average value is zero in the steady state. With identical resistances of resistors $R_1 = R_2 = R_3$ and zero input voltage $u_1 = 0$, the output voltage of the integrator has a symmetrical sawtooth waveform, i.e., the times T_1 and T_2 are identical. At non-zero input voltage u_1 of the modulator, it applies to the ratio of times

$$\frac{T_2 - T_1}{T_2 - T_1} = \frac{u_1}{u_2}. \quad (2.197)$$

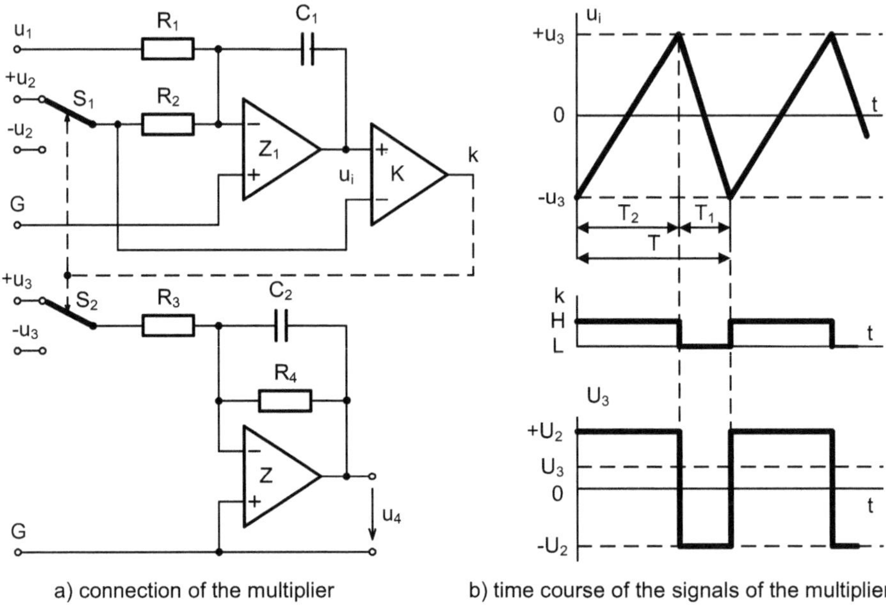

Figure 2.124 Modulating multiplier with delta sigma and modulator.

This means that for its positive voltage $u_1 > 0$, the time is $T_2 > T_1$ and for its negative voltage $u_1 < 0$, the time is $T_2 < T_1$. The signal modulated in this way controls the switch S_2, which, when the resistances of the resistors $R_4 = R_5$ are equal, generates a voltage modulator at the output, the mean value of which is directly proportional to the product of the input voltages of the multiplier

$$u_{4s} = u_1 \cdot u_3. \tag{2.198}$$

Multipliers with a *Hall probe* use the fact that the Hall voltage U_H, sensed on their sensing contacts, is directly proportional to the product of the excitation current i_b and the induction B, whose vector is perpendicular to the active surface of the Hall probe of size S (Figure 2.125).

Through the first voltage amplifier Z_1 with current output, the magnetizing winding of the ferromagnetic core with the number of turns N, in the gap of which the Hall probe is placed, is excited. The excitation current of the magnetizing winding of the core i_1 induces an induction B in its gap, which is given by the ratio of the inductance Φ to the cross section of the core S. Through the second voltage amplifier Z_2 with a current output, the Hall probe is excited, which has a voltage at its output

$$u_H = k_H \cdot \frac{B}{d} \cdot i_B = k_H \cdot \frac{\Theta}{S \cdot d} \cdot i_B = k_H \cdot \frac{N \cdot i_1}{S \cdot d} \cdot i_B, \tag{2.199}$$

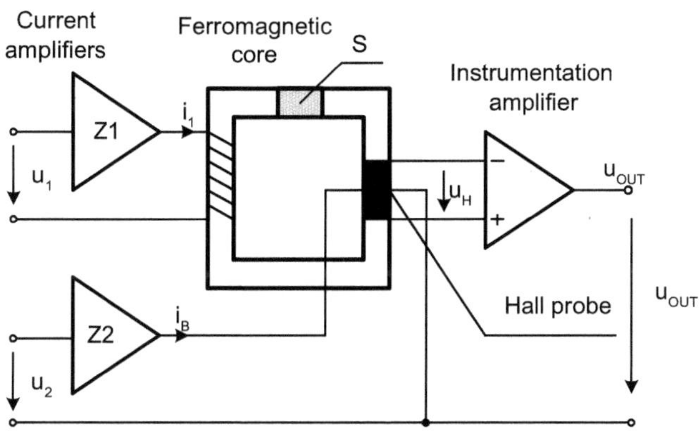

Figure 2.125 Multiplier with Hall probe.

where k_H is the Hall constant and d is the thickness of the Hall probe.

In modern electrical power and energy sensors, voltage and current signals are first digitized into corresponding digital signals, which are then digitally multiplied and integrated.

2.9.6 Electric charge sensors

A plate *dynamic capacitor* is used to sense the electrostatic charge on the surface of non-conductive objects. It consists of one fixed electrode, usually in the shape of an annulus, and a second movable circular electrode, the movement of which is controlled by a piezoelectric actuator (Figure 2.126).

As a result of the time change of the capacity $C(t)$ of the dynamic capacitor, a voltage is induced between its electrodes by the action of an electrostatic field of intensity E

$$u(t) = \frac{Q}{C(t)}, \tag{2.200}$$

which is directly proportional to the size of the detected surface charge Q. The voltage modulated in this way is amplified by an *electrometric amplifier* with a high input resistance and rectified by a *synchronous detector*, the use of which significantly eliminates the influence of external interfering signals on the accuracy of charge determination.

Inductive and capacitive probes are used to detect *partial discharges* in the insulation systems of HV electrical machines.

A *simple inductive probe* consists of an air coil into which voltage pulses are induced due to changes in the magnetic field, which after integration determine the charge of the partial discharges. If a conductor is located

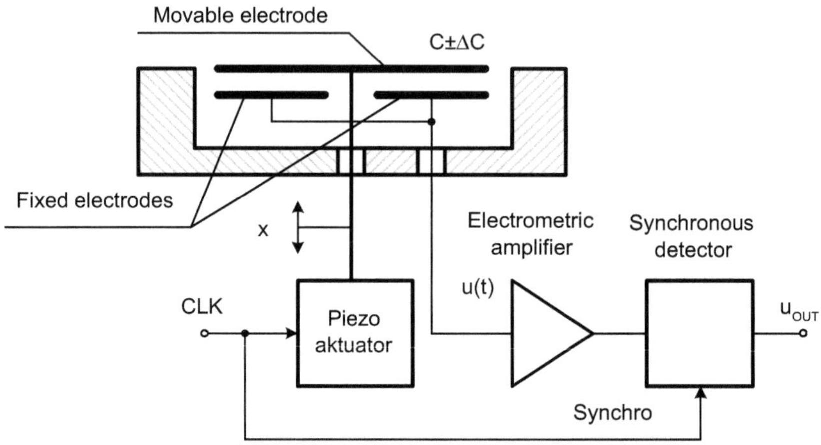

Figure 2.126 The principle of operation of a dynamic capacitor.

at a distance *r* from the coil through which a time-varying current *i*(*t*) flows, then this current causes a temporal change in the magnetic field intensity $dH = di(t)/r$ at a distance *r* from the conductor. A voltage is then generated in a coil with *N* turns and an area *S*, *perpendicular* to the phasor of the intensity of the magnetic field voltage

$$u(t) = N \cdot \frac{d\Phi(t)}{dt} = \frac{N \cdot S \cdot \mu_0}{r} \cdot \frac{di(t)}{dt}, \tag{2.201}$$

where $d\Phi/dt = \mu \cdot S \cdot dH/dt$ is the time variation of the magnetic flux impinging on the coil, and μ is the air permeability, which is approximately equal to the vacuum permeability $\mu_o = 4\pi \cdot 10^{-7}$ [H/m] (Figure 2.127).

If the integrating element with a time constant $R_1 \cdot C_1$ is connected to the coil, then there is a voltage at its output

$$u_{1i}(t) = \frac{1}{R_1 \cdot C_1} \cdot \int u(t) \cdot dt = \frac{N \cdot S \cdot \mu_0}{r} \cdot i(t). \tag{2.202}$$

If this voltage is integrated once more with the time constant $R_2 \cdot C_2$, then the output voltage of the second integrator is directly proportional to the charge of the pulses that pass through the conductor

$$u_{2i}(t) = \frac{1}{R_2 \cdot C_2} \cdot \int u_{1i}(t) \cdot dt = \frac{1}{R_1 \cdot C_1 \cdot R_2 \cdot C_2} \cdot \frac{N \cdot S \cdot \mu_0}{r} \cdot q. \tag{2.203}$$

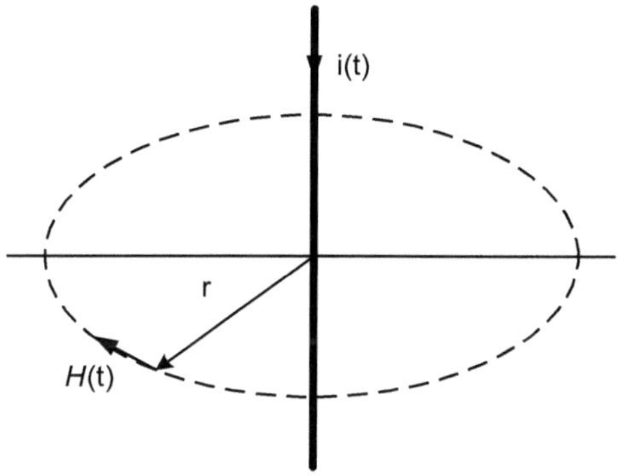

Figure 2.127 A simple inductive probe.

The prerequisite for the reliability of charge determination by this method is that the time constants of both integrators are always several orders of magnitude greater than the time constant corresponding to the lower limit frequency f_d of charge measurement, i.e., that $R_1C_1 >> 1/2 \cdot \pi \cdot f_d$ and $R_2C_2 >> 1/2 \cdot \pi \cdot f_d$. The upper limit frequency of the probe is limited by its resonance frequency $f_o = 1/2 \cdot \pi(L_cC_c)^{-1/2}$, where L_c is the inductance and C_c is the self-capacitance of the coil.

The disadvantage of this probe is its low sensitivity, which decreases with increasing distance from the point of origin of charges, e.g., partial discharges in insulators.

Contact probes are therefore used to diagnose the stator windings of HV rotating electrical machines, in which the rotor of the machine is removed, which consists of a *ferrite core* with a wound coil. These probes are placed above the groove of the machine's stator winding and, together with its magnetic circuit, form a *current transformer*, in whose winding current pulses are induced, caused by partial discharges in the machine's winding (Figure 2.128).

Differential electromagnetic probes are mainly used to localize the occurrence of partial discharges in the stator grooves of HV electrical rotating machines. The probes consist of two inductive probes, each of which is located at one end of the stator slot and whose sensing windings are electrically connected to each other (Figure 2.129).

Under the influence of the electromagnetic field, which is caused by the currents of partial discharges, impulses from the source of the discharges spread in both directions. If the partial discharge current pulse passes through both probes in the same direction, the output voltages of the probes are of opposite polarity and therefore read off, meaning that the location of the partial

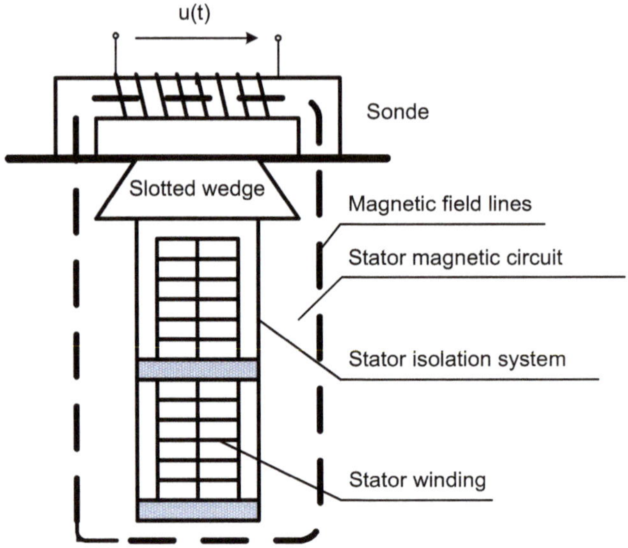

Figure 2.128 Inductive probe of the winding of an electric machine.

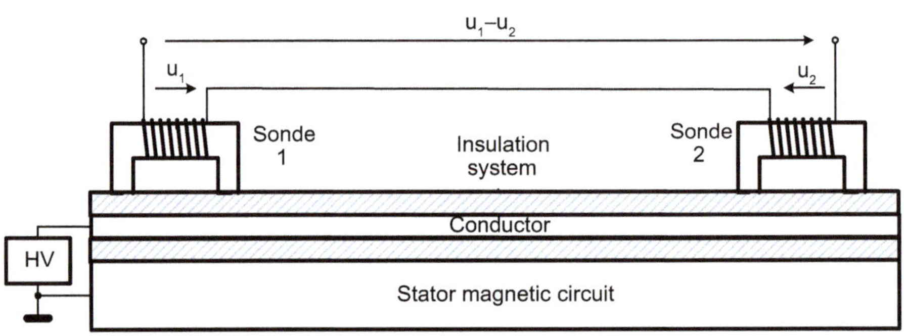

Figure 2.129 Measurement of charges of partial discharges with a differential electromagnetic probe.

discharges lies outside the part that is bounded by the probes. If the occurrence is between the probes, the probe signals have the same polarities and their induced voltages are added. This connection also eliminates the effect of external interference and the effect of discharge activity in neighboring grooves.

A *simple capacitance probe* consists of a circular electrode, which is insulated and attached to the probe housing, in the front part of which there is a cover made of non-conductive material, most often made of quartz glass (Figure 2.130).

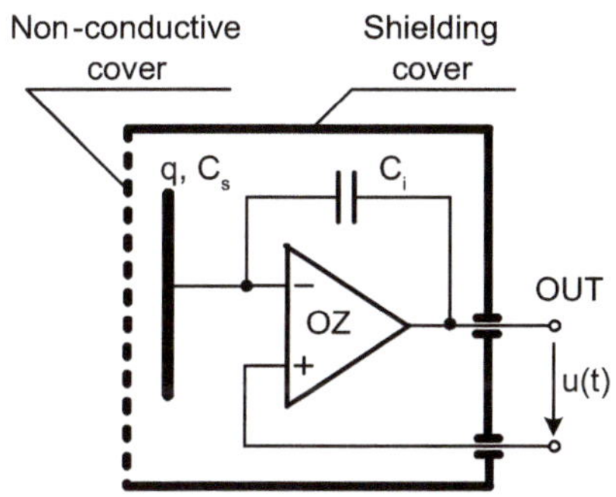

Figure 2.130 Simple capacitance probe.

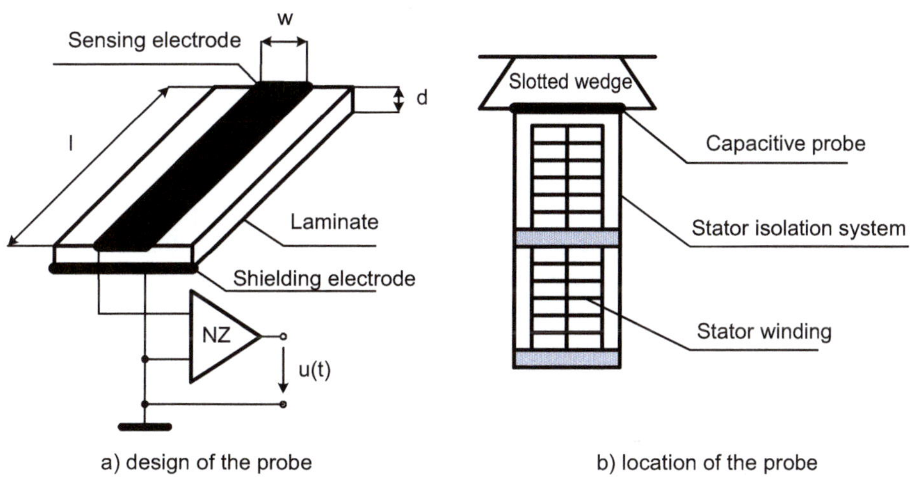

Figure 2.131 Slot capacitance probe.

Due to the action of an *alternating electrostatic field* in the vicinity of the sensing capacitor C_s, caused by partial discharges, a charge q is induced on the electrode of the sensing capacitor, which is converted into a voltage by a *charge amplifier*

$$u(t) = \frac{q}{C_i}, \tag{2.204}$$

where C_i is the integration capacitance of the charge amplifier.

Capacitance groove probe consists of an insulating laminate board on which a shielding copper layer is placed on one side and a sensing electrode in the form of a narrow copper strip is placed on the other side of the board. The probe is inserted into the stator grooves of the machine under the groove wedges. Figure 2.131 shows the design of the groove probe and its location in the stator groove of a rotating electric machine.

It applies to the capacitance between the shielding and sensing electrode of the probe

$$C = \varepsilon_0 \cdot \varepsilon_r \cdot \frac{w \cdot l}{d}, \tag{2.205}$$

where $\varepsilon_0 = 8.854 \cdot 10^{-12}$ [F·m^{-1}] is the permittivity of the vacuum, $\varepsilon_r \approx 2$ is the relative permittivity of the laminate, d is its thickness, and w and l are the width and length of the sensing electrode.

If a partial discharge q occurs near the capacitive probe, then a voltage will arise between the electrodes of the probe which, after integration with the charge amplifier, is proportional to the charge of this discharge

$$u(t) = k \cdot \frac{q}{C}, \tag{2.206}$$

where $k < 1$ is the charge transfer coefficient from the place of its origin to the dielectric of the capacitor.

The *coaxial capacitance probe*, suitable for detecting partial discharges in HV cables, consists of a coaxial capacitor, for which the capacitance between the inner conductor and the inner sheath C_1 and the capacitance between the inner and outer shield sheath C_2 apply

$$C_1 = \frac{2 \cdot \pi \cdot \epsilon_1}{\ln \frac{d_2}{d_1}}, \qquad C_2 = \frac{2 \cdot \pi \cdot \epsilon_2}{\ln \frac{d_3}{d_2}}, \tag{2.207 a, b}$$

where d_1 is the diameter of the inner conductor, d_2 and d_3 are the inner diameters of the inner and outer sheath of the probe, and ε_1 and ε_2 are the permittivity of the environment between the inner conductor and the inner sheath, and between the inner and outer shielding sheath (Figure 2.132).

If a partial discharge q occurs in the dielectric between the inner conductor and the inner sheath, then this charge causes a voltage between the two sheaths which, after integration by a *charge amplifier*, is proportional to the charge of this discharge

$$u(t) = \frac{C_1}{C_1 + C_2} \cdot q. \tag{2.208}$$

Capacitance probes enable accurate localization of the source of partial discharges, especially in HV cables. Their disadvantage is the impossibility of determining the source of partial discharges of insulation systems that are electrically shielded, because then the electric field radiated into the surroundings cannot be measured.

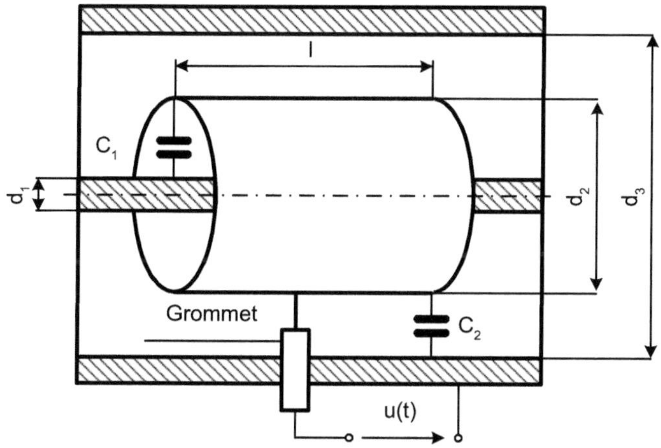

Figure 2.132 Coaxial capacitance probe.

2.9.7 Intensity sensors and electromagnetic field induction

In addition to inductive probes and *Hall probes*, sensors with a *ferromagnetic probe* and magnetoresistive sensors are used to measure the intensity and induction of the electromagnetic field.

Extremely sensitive measurements of magnetic induction are made possible by sensors working on the principle of *nuclear magnetic resonance* and sensors with *superconducting elements*. These sensors make it possible to measure the induction of magnetic fields in the range of 10^{-14} to 10^{-6} T with a spectral noise component of 3 fT/√Hz.

Sensors with a ferromagnetic probe FG (*flux gate*) consist of a closed core of magnetically soft ferromagnetic material, e.g., permalloy NiFe, with high initial permeability and minimal coercive force and hysteresis. The core of the sensor is wrapped with a magnetizing and sensing winding. When the core is excited by a magnetizing current of a harmonic course, a voltage with an inharmonic time course is induced in the sensing winding, the course of which corresponds to the course of the magnetization curve of the core

$$B(t) = f\left[(H(t))\right], \tag{2.209}$$

where $B(t)$ is the time change of the induction in the core, and $H(t)$ is the time change of the magnetic field intensity determined by the magnetizing current $i_b(t)$, the number of turns of the magnetizing winding N_b, and the mean length of the magnetizing flux of the core l_s

$$H(t) = \frac{N_b}{l_s} \cdot i_b(t). \tag{2.210}$$

The course of the magnetization curve of the core causes only the occurrence of odd harmonic components of the induced voltage in the sensing winding. This winding is made up of two identical sections connected against each other (Figure 2.133).

If no external magnetic field acts on the core in the direction of the core axis, voltages of the same magnitude but opposite phases are induced in the pickup coils. Therefore, their total voltage is ideally zero. Due to the action of the external electromagnetic field in the direction of the sensor core, the differential voltage on the sections of the sensing coils is non-zero, and the amplitude of its second harmonic component with respect to the frequency of the excitation current is directly proportional to the intensity of this external field.

This alternating voltage is rectified by a synchronous detector with a high dynamic margin, since the detector rectifies relatively small voltages with a high component of alternating voltage with the frequency of the excitation current. Ferromagnetic probes make it possible to measure the induction of magnetic fields in the range of 10^{-10} to 10^{-2} T.

Magnetoresistive sensors use the *anisotropic magnetoresistance* of doped semiconductor materials, which consists in different shifts in the energy levels of electrons with positive and negative spins. When a magnetic field is applied to these semiconductors, their resistivity decreases in the direction of the magnetic field vector.

In practice, *ferromagnetic magnetoresistors* are most often used, which are formed by a thin layer of meander-shaped permalloy, which is placed on a ceramic substrate (Figure 2.134).

The resistance dependence of the magnetoresistor is defined by the equation

$$R = R_{\min} + \left(R_{\max} - R_{\min}\right) \cdot \left[1 - \left(\frac{H}{H_s}\right)^2\right], \tag{2.211}$$

where $R_{\min}$ and $R_{\max}$ are the minimum and maximum values of the sensor, and H_s is the intensity of the electromagnetic field, corresponding to the saturation of the sensor.

To increase the sensitivity of the sensor, a permalloy layer is placed between other permalloy layers, which act as a concentrator of the magnetic field lines of the sensed magnetic field. With this modification, a magnetic induction measurement sensitivity of up to 10^{-10} T can be achieved.

2.10 Sensors of Liquid Properties

The most common types of these sensors are conductivity and liquid acidity sensors. These sensors use their physicochemical properties.

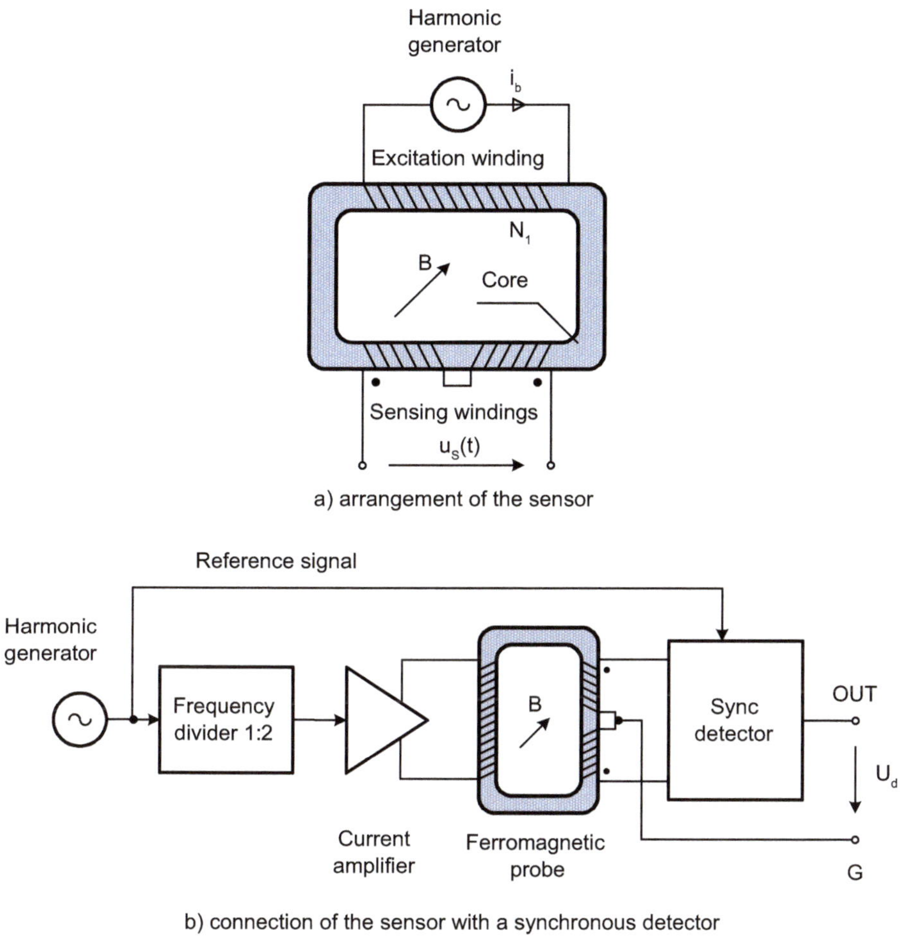

Figure 2.133 Ferromagnetic probe.

2.10.1 Liquid conductivity sensors

Electrode sensors with planar, cylindrical, or rod electrodes are used to determine the electrical conductivity (conductivity) of liquids. These electrodes are usually made of either platinum or stainless steel with a layer of titanium carbide TiC or graphite.

For these sensors, the *conductivity* of the solutions is determined by measuring the resistance between these electrodes, and alternating current is usually used for this measurement in order to prevent the polarization of the solution, which creates an ionic double layer of oppositely charged ions on

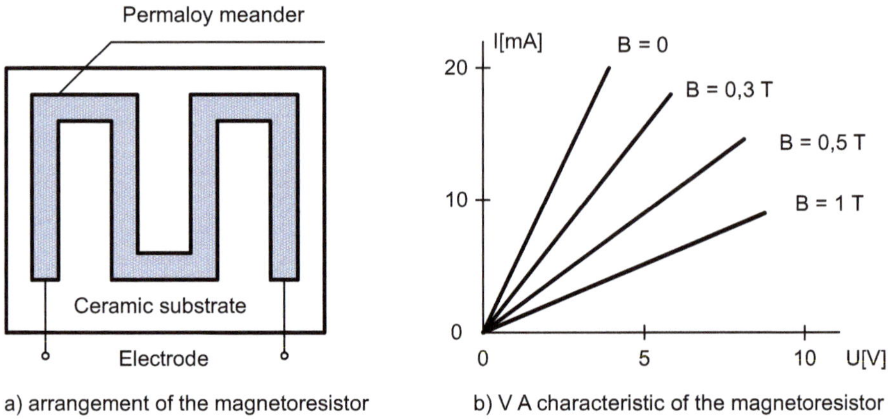

Figure 2.134 Magnetoresistor.

the electrodes. Polarization manifests itself especially in highly conductive solutions and when using a low-frequency alternating current supply.

The replacement diagram of the electrode sensor in Figure 2.135 contains the electrolyte resistance R_x, the geometric capacitance of the sensor electrodes C_g, the polarization capacitance C_p, the polarization resistance R_p, and the capacitance of the electrode leads C_K, including the capacitance of the connection cable.

For example, at a frequency of the excitation voltage of 5 kHz, the polarization capacity ranges from 10 to 100 μF, the polarization resistance ranges from 10 mΩ to 10 Ω, and the geometric capacity of the sensor, depending not only on its dimensions but also on the permittivity of the liquid, ranges from 10 up to 100 pF.

In order to ensure the homogeneity of the intensity of the electric field H between the planar electrodes, a *four-electrode connection* is used, consisting of external electrodes connected to the alternating current source I_k and internal sensing electrodes between which the voltage U_k is measured (Figure 2.136).

For the *specific conductivity* of liquids, given by their conductivity G_k per unit length, it holds

$$\gamma = \frac{G_k}{d_2} = \frac{I_k}{d_2 \cdot U_k}, \qquad (2.212)$$

where d_2 is the distance between the sensing electrodes of the sensor.

Induction sensors consisting of a non-magnetic tube through which the liquid flows and two toroidal transformers, one of which functions as a

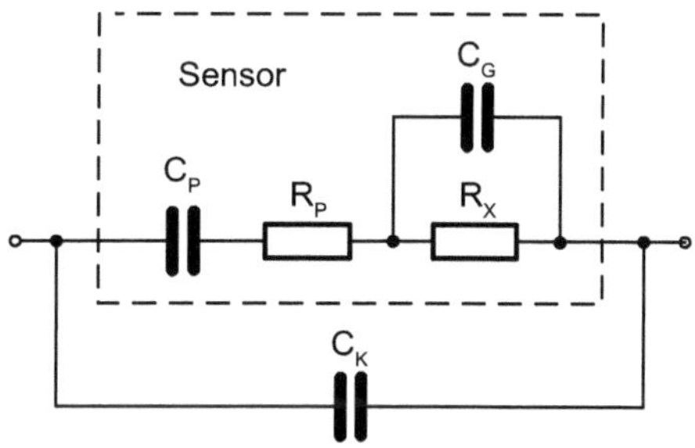

Figure 2.135 Liquid conductivity sensor replacement diagram.

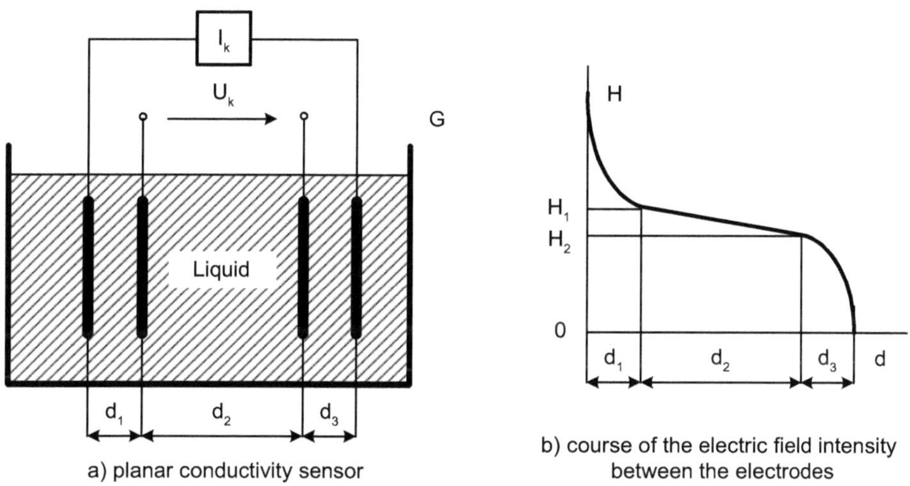

Figure 2.136 Four electrode connection of a planar liquid conductivity sensor.

voltage exciter and the other as a current sensing transformer, are used for non-contact measurement of the conductivity of liquids (Figure 2.137).

The excitation current i_1 induces an induction current i_k through a coil with i_1 turns, which in a conductive liquid with conductivity G_k induces a current i_k, which in the magnetic circuit of the sensing toroid induces an induction current i_2, in whose winding with the number of turns N_3, a current i_2 is induced. The system created in this way represents two transformers, of which the secondary winding of the excitation transformer and the primary winding of the sensing transformer have the same unit number of turns. A current amplifier with a feedback resistor R_z is connected to the output of the sensing current transformer, for which the output voltage is

$$u_2 = \frac{N_1}{N_3} \cdot \frac{u_1}{G_k} \cdot R_z. \tag{2.213}$$

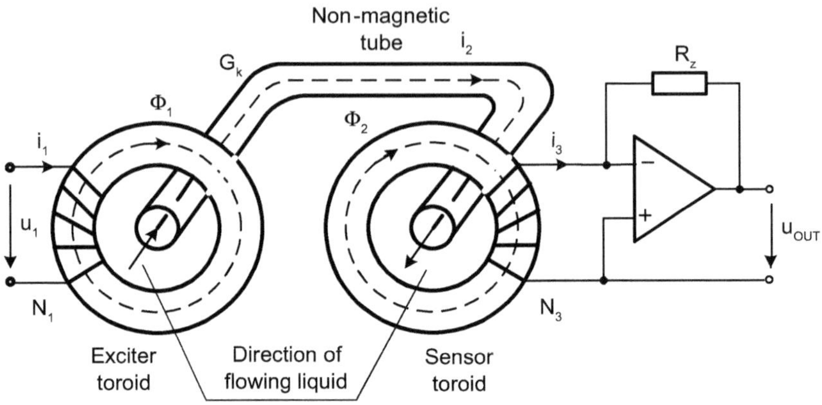

Figure 2.137 Non contact inductive sensor of conductivity of liquids.

An alternating current with frequencies in the range of hundreds of Hz to several kHz is used to excite these sensors. These sensors are especially suitable for continuous conductivity measurement of aggressive and toxic liquids.

2.10.2 Liquid acidity sensors

The principle of operation of these sensors is based on electrochemical reactions occurring on electrodes placed in the liquids whose acidity we measure. At these electrodes, either *anodic oxidation* occurs, i.e., the acceptance of electrons, or *cathodic reduction*, i.e., they give them up.

When an electrode is immersed in an electrolyte solution containing ions of the metal from which the electrode is made, an electric potential is created between it and the solution, the size of which depends on the type of material from which the electrode is made, the concentration of ions in the solution and its temperature. This potential is defined by the *Nernst equation*

$$V = V_0 \pm \frac{R \cdot \Theta}{n \cdot F} \ln \alpha_i, \tag{2.214}$$

where V_0 is the standard potential at ion activity 1, $R = 8.314\ \mathrm{J{\cdot}K^{-1}{\cdot}mol^{-1}}$ is the gas constant, Θ is the temperature in K, n is the valence (oxidation number) of the ions, $F = 9.64787 \cdot 10^4\ \mathrm{C{\cdot}mol^{-1}}$ is the *Faraday constant*, and a_i is the ion activity, with the positive sign corresponding to positively charged cations and the negative sign to negatively charged anions.

The given equation applies to all ions in the solution. The hydrogen cation ion H+ and its counterpart the *hydrogen anion* H^- are found in all aqueous solutions, which is why these ions have a prominent position among other liquid ions. For this reason, the acidity of liquids is expressed by the number of *hydrogen ions* H^+ in units of PH (*power of hydrogen*), which are defined as the negative logarithm of the concentration of these ions

$$\mathrm{PH} = -\log_{10}\left[\mathrm{H}^+\right]. \tag{2.215}$$

Since the concentration of hydrogen ions in neutral distilled water is 10^{-7} mol^{-1}, it is assigned the degree of acidity PH = 7. Solutions with a pH lower than 7 are called acidic, and solutions with a pH higher than 7 are called basic.

In the *potentiometric measurement* of the acidity of solutions, two electrodes are used, of which the measuring electrode is usually made of porous glass, permeable to hydrogen ions of the measured solution, and the reference electrode, the upper part of which is filled with a solution with a precisely defined PH value, the so-called *puffer*. The lower part of the electrode is filled with a saturated solution of potassium chloride KCl, the potential of which is independent of the acidity of the measured solution (Figure 2.138).

Platinum wires are inserted inside both electrodes, between which a voltage is created, proportional to the acidity of the solution

$$U = k - 0.059 \cdot \mathrm{PH}, \tag{2.216}$$

where the constant k can be determined from the voltage between the electrodes immersed in distilled water with an acidity of PH = 7.

In practice, to measure the acidity of solutions, combined electrodes are used, located in one housing, in which a resistance temperature sensor with a Pt sensor is usually located, enabling compensation of the temperature dependence of the inter-electrode voltage. For example, a temperature change of 10 °C causes a change in the measured acidity value of 0.15 PH.

In addition to these glass electrodes, ion-selective electrodes are also used to measure the acidity of solutions, the membranes of which allow, e.g., ions Na^+, Ca^{2+}, K^+, Cl^-, NH_4^+, and NO_3^+.

Electrometric amplifiers with high input resistance, ranging from hundreds of MΩ to units of TΩ, are used to amplify the inter-electrode voltages. Figure 2.139 shows an example of connecting an electrometric amplifier with an operational amplifier AD795, whose input impedance is 1 TΩ.

An amplifier with a nominal gain of $(R + R_T)/R_T$ amplifies the inter-electrode voltage of the pH probe with a change of 50 mV/pH, the temperature coefficient of which is $3.3\cdot10^{-3}$/°C. When choosing resistor R = 19.6 kΩ with zero temperature coefficient and resistor R_T = 1 kΩ with temperature resistance coefficient $3.5\cdot10^{-3}$/°C, the output voltage of the amplifier changes in the range of 1 V/pH with zero temperature change. The leakage resistance of

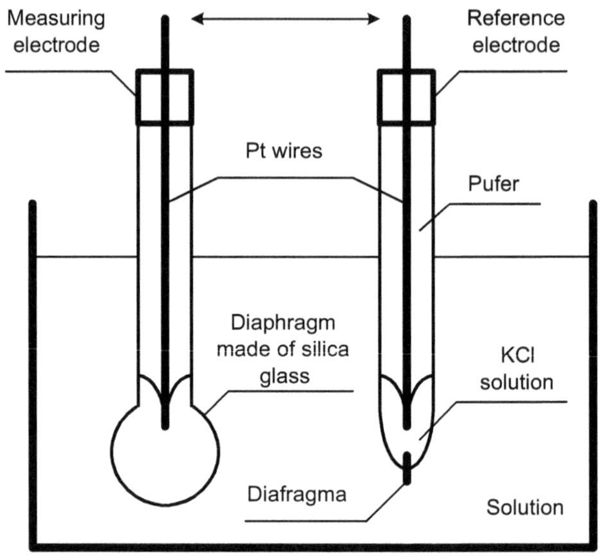

Figure 2.138 Potentiometric measurement of PH.

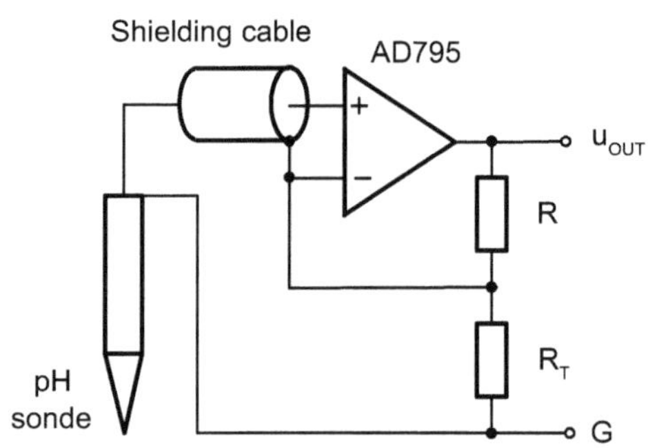

Figure 2.139 Electrometric pH probe amplifier with temperature compensation.

the supply cable is eliminated by the equipotential connection of the amplifier, in which the inner conductor and the sheath of the cable are practically at the same potential, given by the input differential residual voltage of the operational amplifier, the typical value of which is 1 mV.

In addition to glass probes, ISFET (*ion sensitive field effect transistor*) semiconductor solution acidity sensors are also used in technical practice, which are made up of FET transistors with *S* (*source*), *D* (*drain*), and *G* (*gate*) electrodes (Figure 2.140).

By choosing a sensitive gate layer, the desired selectivity for a certain ion of the analyzed solution is achieved. For example, ISFET sensors with

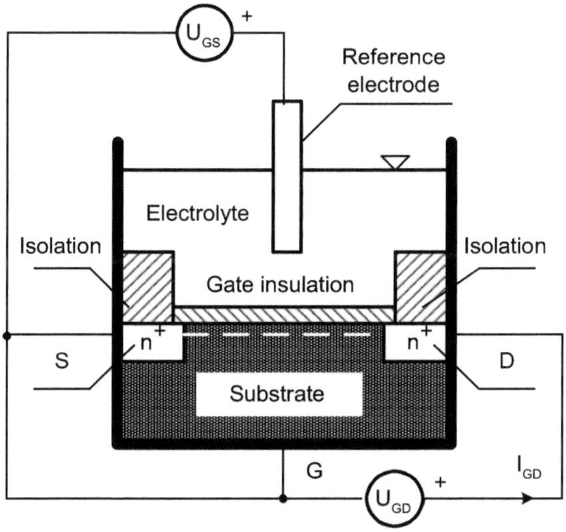

Figure 2.140 ISFET electrochemical sensor.

ion-sensitive layers of silicon oxide SiO_2, silicon nitride Si_3N_4, and tantalum oxide Ta_2O_5 are used to determine the acidity of solutions.

By applying a selectively permeable membrane to the sensitive layer of the gate, allowing only the selected type of ions of the analyzed solution to pass through, an increase in the sensitivity and selectivity of the sensor can be achieved. The gate voltage U_{GD}, which determines the acidity of the solution, is determined from the gate current, I_{DS}, passing between the sensor substrate and its electrode *D*. Compared to potentiometric pH sensors, ISFET sensors allow for miniaturization but fall short of their accuracy and durability.

2.11 Gas Property Sensors

The most frequently analyzed gases and their compounds include oxygen O_2, carbon dioxide CO_2, carbon monoxide CO, methane CH_4, and hydrogen sulfide H_2S. Their physical, chemical, and optical properties are used to detect these gases.

2.11.1 Oxygen concentration sensors

Capacitive, magnetic, thermomagnetic, potentiometric, amperometric, and electrochemical sensors are used to determine the *oxygen concentration*.

Capacitive oxygen concentration sensors use its high permittivity, the relative value of which reaches 80 at a temperature of 20 °C and a pressure

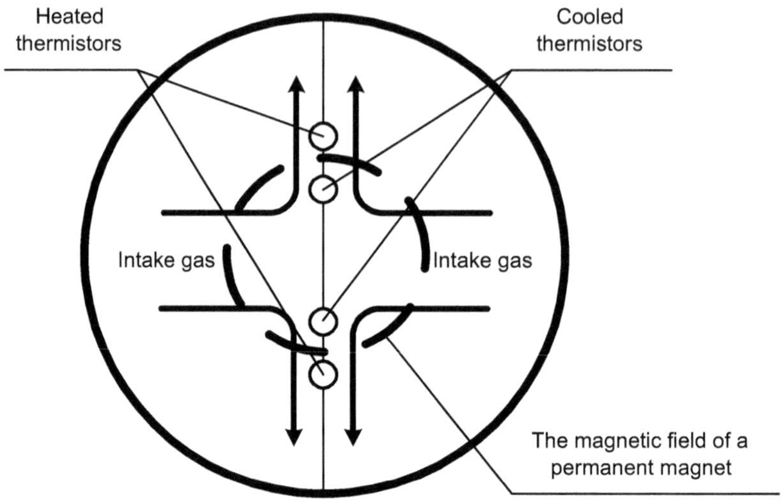

Figure 2.141 Thermomagnetic oxygen concentration sensor.

of 101 kPa. In the simplest version, these sensors are formed by plate capacitors, and their dielectric is the analyzed gas. Due to their simplicity and low price, these sensors are mainly used to measure oxygen concentration in living spaces. The continuous measurement of the oxygen content uses the coaxial design of the capacitors, between the electrodes of which the analyzed gas flows.

Magnetic oxygen concentration sensors use its higher permeability than other common diamagnetic gases to determine its concentration in a gaseous environment. With these sensors, the analyzed gas flows through a coil, the change in inductance of which is proportional to the oxygen concentration in the gas.

Thermomagnetic oxygen concentration sensors use the growth of its permeability with the square of its temperature. The sensor consists of a chamber through which the measured mixture of gases flows. The chamber temperature is maintained at a constant temperature. The principle of operation of the sensor is based on the combination of the gradient of magnetic susceptibility (relative magnetic conductivity) of oxygen in the field of a permanent magnet and the inversely oriented thermal gradient of two pairs of thermistors (Figure 2.141).

The temperature gradient and the resulting gas flow cause two pairs of thermistors, one of which is heated. The magnetic susceptibility gradient occurs due to the change in oxygen temperature. At the sensor, the change in resistance of the cooled thermistor is measured as a function of the gas

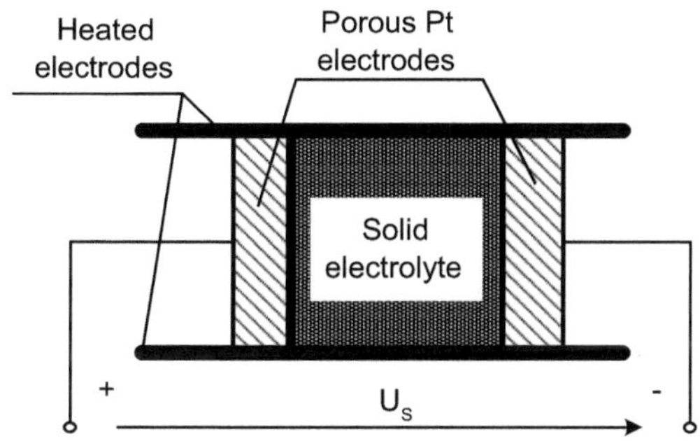

Figure 2.142 Potentiometric oxygen concentration sensor.

flow rate. The measuring circuit is a bridge, where one branch consists of heated thermistors connected in series and the other branch consists of cooled thermistors. The measurement results of both methods can be affected by interference with other paramagnetic gases, e.g., nitrogen dioxide NO_2 and nitric oxide NO.

Potentiometric oxygen content sensors use a solid electrolyte consisting of carbon zirconium ZrO_2 doped with yttrium oxide Y_2O_3 to determine its concentration. This electrolyte is heated by heating electrodes to a temperature in the range of 650–950 °C (Figure 2.142).

Due to the chemical reaction of the electrolyte placed between the two platinum sensing electrodes, the *Nernst equation* for the inter-electrode potential applies

$$V_S = \frac{R \cdot \Theta}{4 \cdot F} \ln\left(\frac{p_M}{p_S}\right), \tag{2.217}$$

where $R = 8.314\ \text{J}\cdot\text{K}^{-1}\cdot\text{mol}^{-1}$ is the *gas constant,* Θ is the temperature in K, $F = 9.64787\cdot 10^4\ \text{C}\cdot\text{mol}^{-1}$ is the *Faraday constant,* p_M is the partial pressure of oxygen in the analyzed gas, and p_S is the partial pressure of oxygen in air. If both partial pressures are equal, their ratio can be replaced by the ratio of oxygen concentrations in the analyzed gas and air c_M/c_S.

This measurement principle is also used to measure the oxygen concentration in the exhaust gases of combustion engines. For these *Lambda sensors*, the designation of the symbol λ, used to denote the ratio of the oxygen concentration c_M supplied to the combustion space, to the oxygen concentration c_{MOPT} required for optimal combustion of the mixture.

Amperometric oxygen concentration sensors work on a similar principle to potentiometric sensors, with the difference that the oxygen concentration is determined from the nature of the VA characteristic of the sensor.

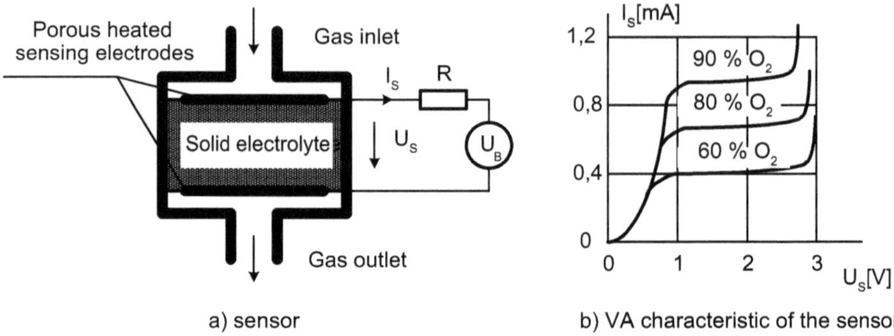

Figure 2.143 Amperometric oxygen concentration sensor.

The sensor again consists of a solid solution of ZrO_2, heated to a temperature of approximately 500 °C, during which, due to the chemical reaction of the electrolyte with oxygen, ions are released, the frequency of which determines the current flowing between the sensing electrodes (Figure 2.143).

Compared to potentiometric sensors, these sensors allow measuring the oxygen concentration in a larger range, in the order of tens of %.

2.11.2 Thermocatalytic sensors of gaseous hydrocarbons

Thermocatalytic sensors (*Pellistors*) use heat measurements caused by their combustion to detect the concentration of flammable gas compounds and their mixtures. The sensor consists of a sintered metal combustion chamber through which the analyzed gas diffuses. A heating resistor R_p, made of platinum wire, is placed in the chamber. This is heated to a temperature of approximately 500 °C by the flow of current.

The heating resistor is also used as a resistance temperature sensor. To suppress the effect of ambient temperature, humidity, and thermal conductivity of the environment on the determination of the concentration of flammable gas compounds, the heating resistor is connected to a resistance bridge, the other resistors of which are located outside the combustion chamber (Figure 2.144).

2.11.3 Semiconductor gas concentration sensors

Semiconductor gas concentration sensors work either on the oxidation principle or use changes in the parameters of transistors controlled by the FET (*field effect transistors*).

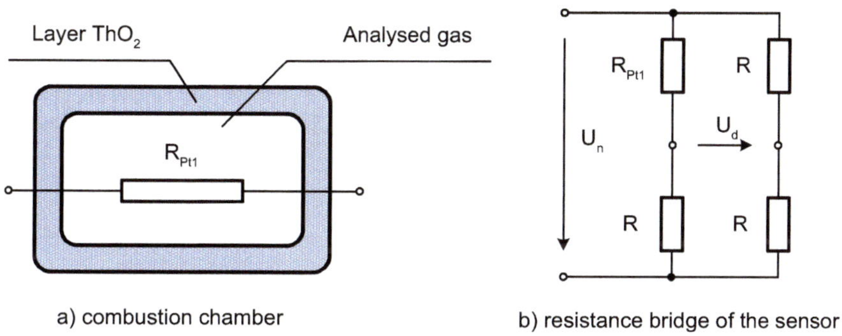

Figure 2.144 Thermocatalytic sensor of gaseous hydrocarbon concentration.

Oxidation sensors use either surface or volume detection of the amount of released ions, which is caused by the absorption of the detected gas into the doped semiconductor layer. *Reducing gases*, such as H_2, CO, CH_4, H_2S, isobutene, and ethanol, which are electron donors, cause an increase in the conductivity of the surface layer of the sensor. *Oxidizing gases,* e.g., O_2, NO_2, and Cl_2, on the other hand, cause a drop in the conductivity of the layer. The appropriate choice of semiconductor and temperature changes the size of the *Fermi energy level*, thereby achieving different sensitivities to the desired type of gas or its concentration. To increase the selectivity of these sensors, the doped semiconductor layer is usually provided with an additional filter layer (Figure 2.145).

If the sensor resistance is R_s, then for the output voltage of the inverting operational amplifier,

$$U_{\mathrm{OUT}} = \frac{R}{R_s} U_n. \tag{2.218}$$

Oxidation sensors with surface detection of ions use sensitive layers of SnO_2 and ZnO type N semiconductors. The heating temperature of these semiconductors ranges from 200 to 450 °C.

Oxidation sensors with volume detection of ions are most often made of oxides TiO_3, $BaTiO_3$, and $SrTiO_3$ and the heating temperature reaches up to 900 °C. For both types of sensors, the sensor substrate consists of aluminum and silicon oxides Al_2O_3, SiO_2, ceramics, or ferrite.

Sensors with transistors controlled by an electric field, referred to as CHEMFET (*chemically sensitive field effect transistors*), use the structure of these transistors, formed by metal–insulator–semiconductor transitions, while the insulator in these transitions is usually made of oxides SiO_2, TIO_2, or ZnO. The conductivity of the channel formed between its electrodes G

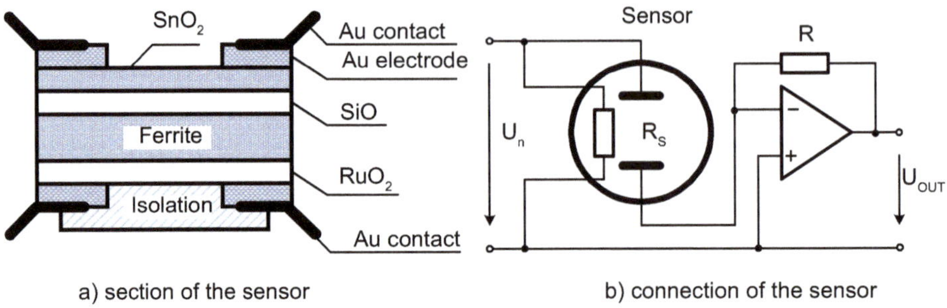

Figure 2.145 Oxidizing gas sensor with surface gas detection.

(*gate*) and S (*substrate*) is influenced by the intensity of the electric field in these transistors. Unlike classic FET transistors, where this potential difference determines the electrical properties of the gate–substrate transition, the electrical properties of the gate of CHEMFET sensors are determined by the absorption of analyzed gases into this substrate.

Most often, the gates are made of palladium Pd, which is sensitive to hydrogen and other gases from which hydrogen can be split off. Therefore, other gases, such as H_2S and NH_3, can also be analyzed with these sensors. Since catalytic reactions can also take place on the palladium layer, oxidizing gases such as O_2, CO, and CO_2 can also be analyzed when using it.

Figure 2.146 shows a CHEMFET sensor of hydrogen H_2, using the change in capacitance C_{GS} between the gate and substrate of the sensor at its potential difference U_{GS}.

2.11.4 Optical gas concentration sensors

These sensors use different *transmittances of light radiation*, which depend not only on the type of gas, its pressure, and temperature but also on the wavelength of the radiation. When the radiation passes through the analyzed gas, it is attenuated at specific wavelengths which correspond to the composition of the gas. This phenomenon is caused by the formation of a permanent dipole moment in gases composed of more than two different atoms. Figure 2.147 shows the relative spectral transmittance of typical hydrocarbons.

It is clear from the above spectrogram that, for example, hydrocarbons such as methane CH_4, ethane C_2H_6, propane C_3H_8, and butane C_4H_{10} show maximum transmittance at wavelengths of infrared radiation in the range of 3–4 μm, carbon dioxide CO_2 at a wavelength of 4.25 μm, carbon monoxide CO at a wavelength of 4.7 μm, freon SF_6 at a wavelength of 8.35 μm, and ammonia NH_3 at a wavelength of 10.45 μm.

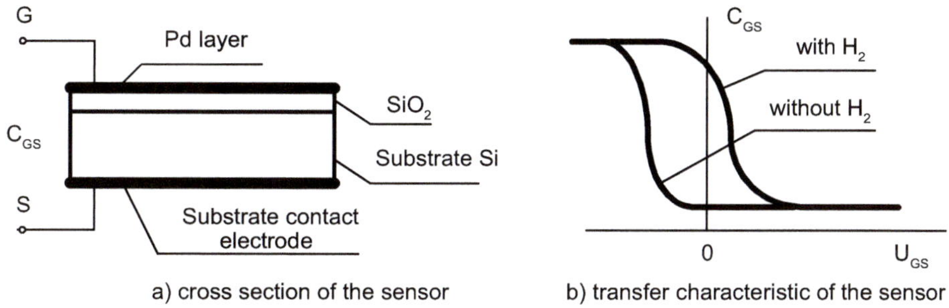

Figure 2.146 CHEMFET hydrogen sensor.

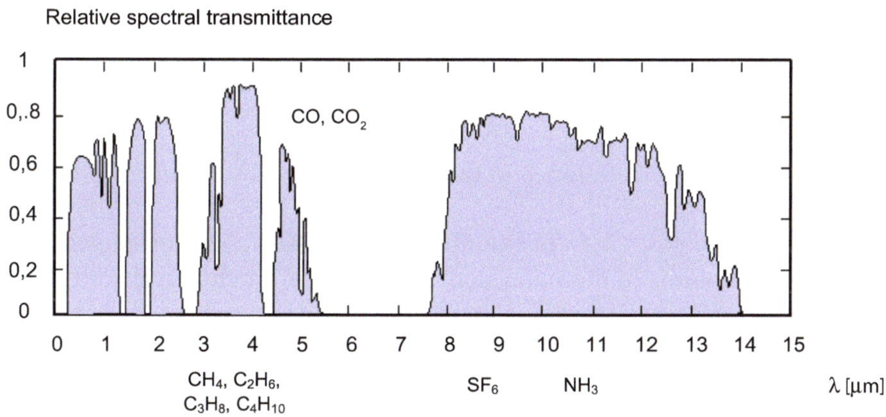

Figure 2.147 Relative spectral transmittance of typical hydrocarbons and freon.

From the point of view of the used range of *infrared radiation* wavelengths in the wavelength range of a few μm to tens of μm, we distinguish between scattering and non-scattering gas concentration sensors.

Dispersive infrared detectors (DIRs) are formed by a source of infrared radiation, which is usually generated by a broadband infrared diode, the light flux of which is divided by a monochromator into several frequency bands in which the analyzed gas is detected (Figure 2.148).

Common monochromators are, for example, monochromatic filters, glass prisms, or gratings. *Infrared red detectors* (IRDs) are used to detect the components of the infrared spectrum divided in this way. Based on changes in the spectral transmittance of the analyzed gases, their composition and concentration is then determined.

By simultaneous measurement of the analyzed gas in the measuring chamber and the chamber with the reference gas, it is possible to remove

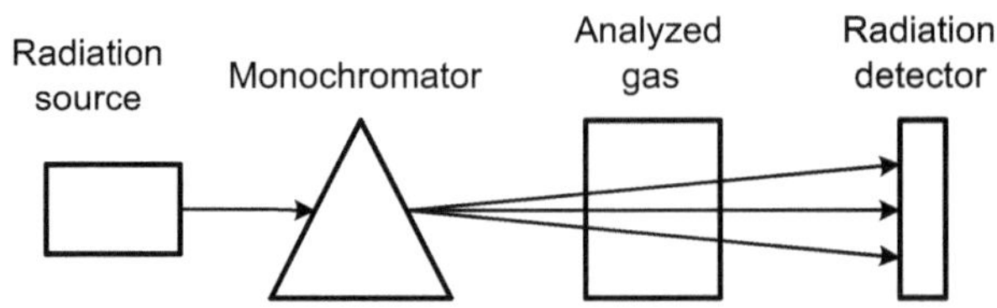

Figure 2.148 Schematic diagram of the dispersion sensor DIR.

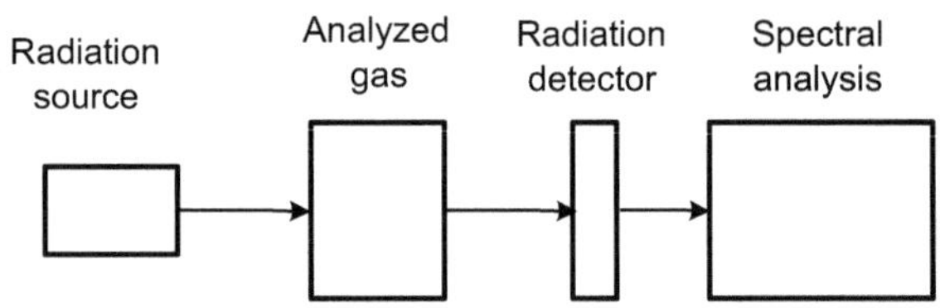

Figure 2.149 Schematic diagram of a non-dispersive NDIR sensor.

the influence of the difference in radiation intensity by means of differential amplification.

Reference gases are, for example, H_2, O_2, and N_2, as their molecules are composed of atoms of the same element and therefore do not absorb infrared radiation.

Non-dispersive infrared detector (NDIR) sensors also consist of a source of infrared radiation, the light flux of which passing through the analyzed gas is detected by a broadband infrared detector and subsequently processed by spectral analysis methods (Figure 2.149).

Photo-ionized detectors (PIDs) use the energy of ultraviolet radiation to ionize their molecules to detect gases. The energy supplied by the photon to the molecule must be high enough to allow the release of an electron from the valence layer of the gas atom, and thus the formation of an ion and a free electron. The resulting charged particles create a measurable voltage on the electrodes. The current generated in this way is directly proportional to the ionized gas concentration (Figure 2.150).

By supplying ionization energy in the form of ultraviolet radiation, all substances with an ionization potential lower than the radiation energy are ionized. Depending on the radiation energy, groups of substances are detected, e.g., volatile organic compounds. These compounds can be divided into groups according to the required ionization energy. For example, benzene has a cutoff potential of 9.5 eV, ammonia, ethanol, and acetone 9.5–10.6 eV, acetylene, formaldehyde, and methanol 10.6–11.7 eV, argon 8.4 eV, krypton 10.0 eV, and xenon 11.7 eV.

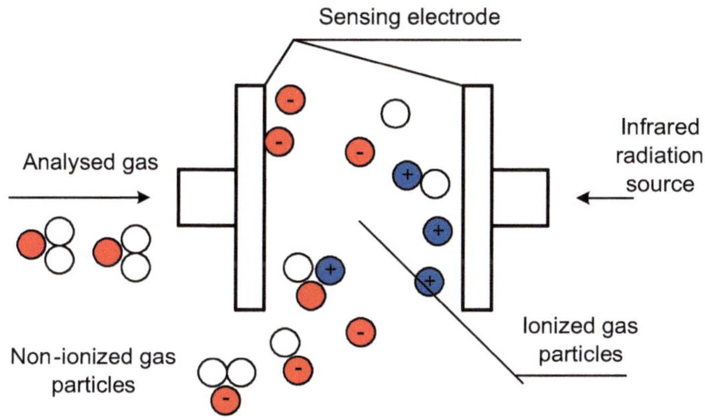

Figure 2.150 Principle diagram of a photoionization detector.

Photo-ionized detectors are usually calibrated for a specific concentration of calibration gas, e.g., isobutylene. For other gases of the sensor calibrated in this way, correction constants for concentration recalculation are set by the manufacturer.

Photoluminescence detectors use radiant recombination of excited charge carriers to detect gases. This occurs after the absorption of a photon of excitation radiation. In the case of using ordinary porous silicon, whose band gap energy reaches a value of 1.12 eV at a temperature of 27 °C, the quantum yield of luminescence is practically negligible. Therefore, for these detectors, etched *crystalline silicon* is used, which enables the creation of *nanostructured formations*, during which the band structure changes from indirect to direct and, as a result, the band gap widens. This modification increases the number of atoms near the surface. Irregularities of the surface and associated defects cause the emergence of new energy levels in the forbidden zone. These then participate in the recombination of excited electrons and holes.

For these sensors, the source of excitation radiation is usually an ultraviolet LED diode with a wavelength of 375 nm, which causes an increasing frequency of extinction with increasing concentration of the analyzed gas. The rate of quenching of the photoluminescence of the analyzed gases also depends on the dielectric constant and the pressure of the saturated vapors. The quenching *time depends* on the degree of oxidation of the porous silicon and on the type and concentration of the analyzed gas. Higher alcohols have the greatest influence on the quenching time, followed by toluene, and non-polar substances have the least influence. The quenching time increases with increasing oxidation of the silicon surface.

2.11.5 Integrated arrays of chemical sensors

By integrating several different chemical sensors, it is possible to realize the so-called *electronic nose*, which is suitable as a detection tool for resolution, usually for volatile organic compounds. This sensor system mimics the action of an animal's nose, where inhaled air with odors flows onto mucus-covered olfactory receptors, where molecules of compounds in the gaseous state are captured. Receptor proteins stimulate enzymes and, with the help of the transfer substance CAMP (*cyclic adenosine monophosphate*), create a transfer channel for impulses to the brain. The neural network compares the input signal with already known odors. An electronic nose uses a similar procedure and components to an animal nose, i.e., sampling the gas, detecting it with an array of sensors, and processing the sensor signals. By comparing it with a library of known odors and evaluating them using artificial intelligence tools, it is possible to determine not only the composition, but also the mass representation of the individual components of the detected gases.

Receptors of chemical transducers are most often formed on the basis of polymers. The sensor array must exhibit significant cross-sensitivity. Therefore, individual sensors cannot have the same sensitivity and selectivity as the same gas molecules. The resulting signal composed of the response of all sensors is compared to a library of known signals. The electronic nose is completely dependent on the repeatability and reproducibility of measurements. A library of known molecules (atoms) of substances, their usual mixtures, or a typical complex of chemical reaction products forms a necessary software accessory of a hardware solution. It is usually necessary to create a library for a new sensor array by measuring known odors (gases).

2.11.6 Gas humidity sensors

Resistive and *capacitive sensors* are used to sense the humidity of gases. Both types of sensors use changes in the impedance of the sorption, i.e., absorbing material, caused by the equilibrium state of adsorption, i.e., the accumulation of gas particles (atoms, molecules) on the surface of the sorption material due to the effect of intersurface attractive forces, and desorption, i.e., the release of water molecules from the water vapor of gases.

Resistive gas humidity sensors are made up of a supporting aluminum electrode, on which a thin layer of Al_2O_3 with a thickness of several μm is applied, which is covered with an extremely thin gold layer with a thickness of several tenths of a μm, allowing water vapor molecules to pass through.

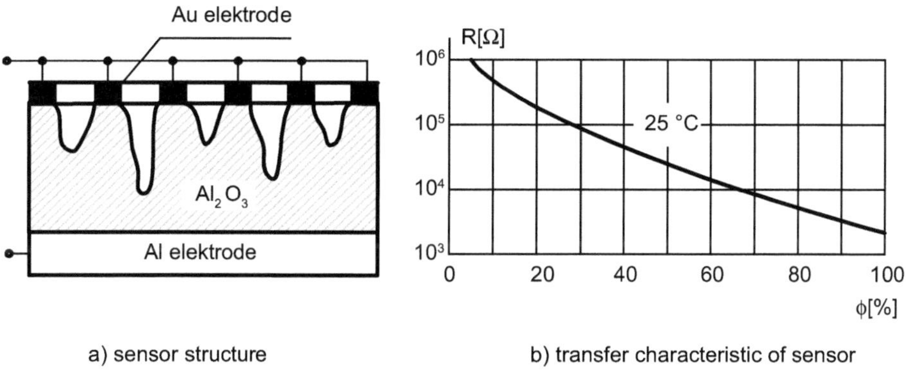

Figure 2.151 Resistive humidity sensor.

Figure 2.151 shows the structure of a resistive humidity sensor and the dependence of its resistance on the relative humidity of the air.

Capacitive humidity sensors are usually formed by a capacitor placed on a glass substrate, on which a layer of hygroscopic polymer with a thickness of approximately 1 μm is evaporated with Cr Ni electrodes in a meander shape. This layer is covered by an extremely thin layer of gold with a thickness of several tenths of μm, which allows water vapor molecules to pass through. The capacity of the sensor is similarly determined by the sorption balance with the environment whose humidity is being measured, as with resistive humidity sensors. These sensors are also manufactured using thin film technology. Figure 2.152 shows the structure of a capacitive humidity sensor and the dependence of its capacity on the relative humidity of the air.

2.12 Radiation Sensors

Gas, scintillation and semiconductor detectors are used to detect the intensity of ionizing radiation.

2.12.1 Ionization detectors

These detectors work on the principle of *gas ionization*, created by the action of ionizing radiation. Gas detectors consist of two insulated electrodes, which are placed in an environment with inert gases, e.g., argon mixed with nitrogen or methane (Figure 2.153).

After direct voltage is connected to the electrodes of the detector, the gas is ionized due to ionizing radiation. After galvanic separation of the ions created in this way by the HV coupling capacitor C_v, the charges are

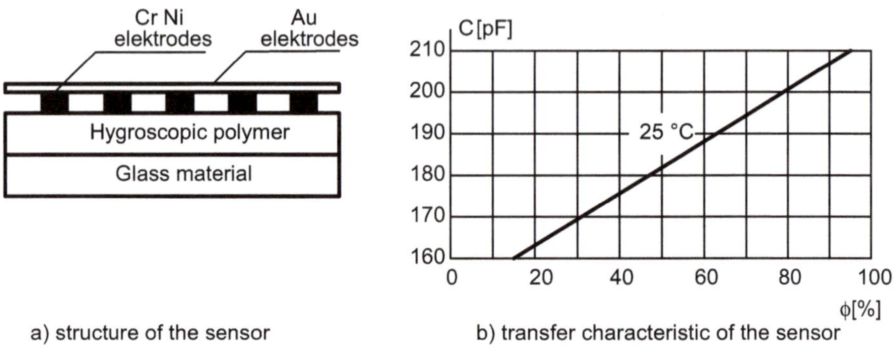

Figure 2.152 Capacitive humidity sensor.

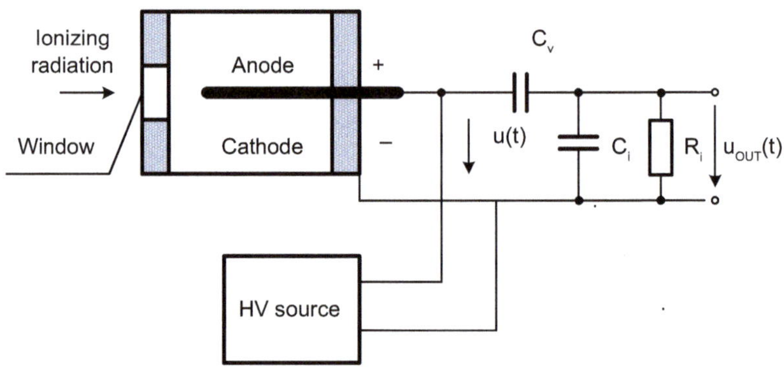

Figure 2.153 Gas detector of ionizing radiation.

integrated by an integration cell with a time constant $\tau_i = R_i \cdot C_i$, which must be significantly greater than the charge pulse collection time.

The dependence of the mean value of the *ionization curren*t, which is proportional to the output voltage of the integration cell, on the intensity of the electric field in the detector for α and β radiation, shows four areas (Figure 2.154).

In region 1, in which there is no reverse recombination of ions, the ionization current is smaller than it corresponds to the electric charges created by the ionization of gas molecules. This region, called *Ohm's law* region, is not used to measure ionizing radiation.

Region 2, referred to as the *saturated current region,* is typical for the operation of ionization chambers. In this region, there is practically no recombination of ions between the electrodes, and the mean value of the ionization current is practically independent of the intensity of the electric

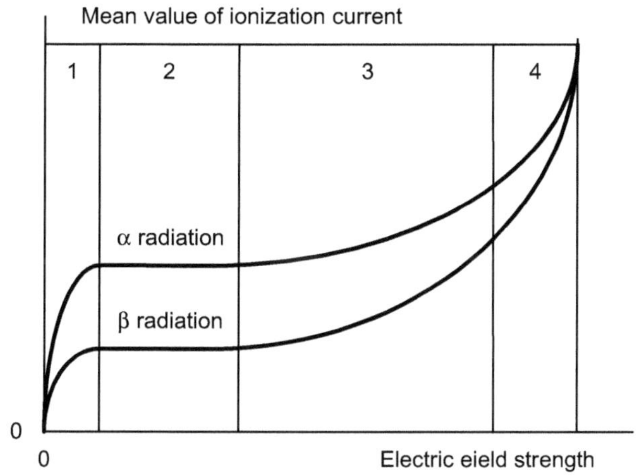

Figure 2.154 Areas of operation of gas radiation detectors.

field in the detector. In addition to integration of charges, during which their average value is measured, the frequency of occurrence of current pulses can also be evaluated, which makes it possible to carry out spectrometric measurements of radiation.

When measuring neutron radiation, in which the neutron energy is not able to ionize the gas filling of the detectors, the cathodes of the detectors are covered with isotopes of uranium U235 and U238, plutonium Pu239, thallium Th232, and beryllium B10. The disadvantage of these detectors, which is the sensitivity to γ radiation, is eliminated by using two ionization chambers of identical dimensions, one of which is covered with these isotopes, while the charges from the anodes of these chambers are subtracted.

Region 3, in which the *impact ionization* of gas molecules occurs, is typical for the use of proportional detectors. For these detectors, the amplitudes of the current pulses depend on the type and energy of the particles, the intensity of the electric field, the type, pressure, and temperature of the inert gas of the detectors. The average frequency of occurrence of individual types of particles can thus be determined by amplitude discrimination of pulses (Figure 2.155).

Area 4, in which the so-called *Geiger threshold* is exceeded, avalanche ionization of the gas filling of the detectors does not depend on the type and energy of the ionizing parts, and the dependences for both types of α radiation and β radiation practically merge. For these detectors, the so-called *dead time* is defined, given by the pulse wear time, during which the detector is unable to detect the following current pulses.

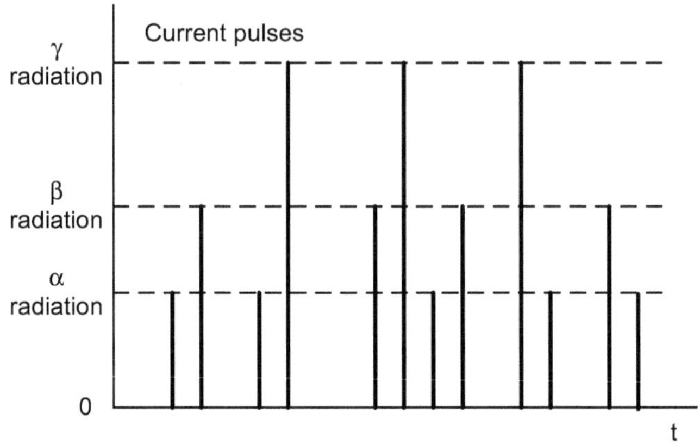

Figure 2.155 Current pulses of proportional radiation detectors.

2.12.2 Scintillation detectors

These detectors transform absorbed radiation energy into emitted photon energy in the visible or ultraviolet region of electromagnetic radiation. *Scintillators* in these detectors are usually quartz crystals activated by heavy metals, e.g., lead and beryllium, which form a scintillation detection unit with a light-conducting fiber and a photomultiplier (Figure 2.156).

The function of the photomultiplier is conditioned by an *external photoelectric phenomenon*, in which a beam of electrons is emitted from the photocathode of the multiplier, which is amplified on the dynodes D_1–D_4, and the electron current obtained in this way is sensed at the anode of the photomultiplier. Its amplifying effect is conditioned by the fact that the incident electrons on the dynodes excite more than one secondary electron. The active layer of photocathodes is usually formed by a vaporized layer of Sb-Cs, Cs-K-Sb, and Cs-Rb-Sb compounds, the maximum sensitivities of which are for radiation wavelengths in the range of 350–500 nm.

2.12.3 Semiconductor radiation sensors

They consist of large-area silicon or germanium photodiodes, and the width of the band gap of the used semiconductor determines the energy required to create an electron–hole pair in their semiconductor structures. This energy ranges from 2.96 to 3.66 eV. Therefore, these detectors are significantly more sensitive compared to gas and scintillation detectors. Their disadvantage is their significant dependence of sensitivity on temperature. That is why germanium photodetectors in particular are cooled by built-in *Peltier cells*.

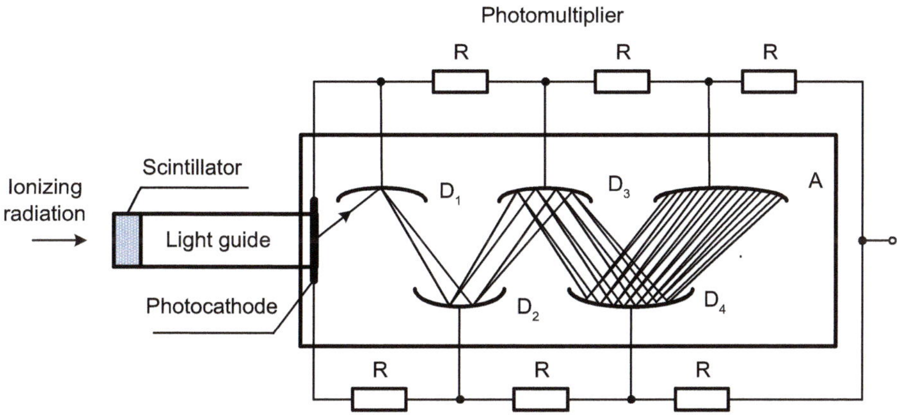

Figure 2.156 Scintillation detection unit with photomultiplier and optical light guide.

To detect ionizing radiation, *semiconductor detectors* with a surface barrier, which are equipped with a layer of Au-Si-N alloy, implemented PN detectors, suitable for spectrometry of low-energy photons, and planar PIN detectors are most often used.

The *planar PIN detector* contains an intrinsic layer made of high-purity germanium, which is used to equalize the number of excited donors and acceptors of lithium alloys. When radiation hits this layer, free electrons are excited in the semiconductor, the intensity of which determines the current pulses of the PIN diode connected in the closing direction (Figure 2.157).

$$u_{\mathrm{OUT}} = \frac{1}{C_i}\int i_d(t)\cdot dt, \tag{2.219}$$

where C_i is the capacitance of the feedback capacitor of the charge amplifier.

To evaluate these current pulses, a *charge amplifier* is used, whose output voltage determines the size of these charges.

Since there is a negligible voltage between the anode and cathode of the diodes, given by the input residual voltage U_0 of the operational amplifier, the leakage resistance R_d and the capacitance C_d in the closing direction, including the leakage resistances and the capacities of their inputs, are practically not applied.

In practical applications, therefore, a resistor R_z with a high resistance, of the order of units to tens of MΩ, is connected in parallel to the integrating capacitor, which ensures that the output of the operational amplifier does not go into a saturation state due to its input current at the inverting input, which would significantly delay its time response.

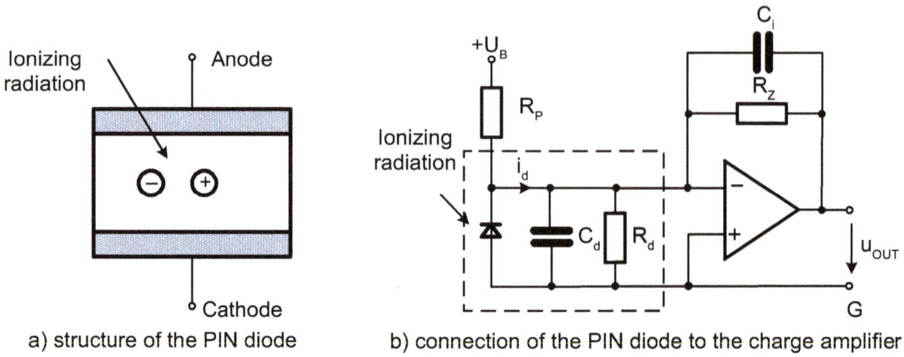

Figure 2.157 Semiconductor ionizing radiation detector with PIN diode.

References

[1] Ramon, P. A., Webster, J. *Sensors and Signal Conditioning*, 2nd Edition, Wiley 2002, ISBN 978-1118-58593-1.

[2] Ripka, P., Tipek, A. *Modern sensors handbook.* London 2007. Instrumentation and measurement series (ISTE). ISBN 978-1-905-20966-8.

[3] Nawrocki, W. *Measurement systems and sensors.* Second edition, Artec House 2016. ISBN 1-58053-945-9.

[4] Fraden, J. *Handbook of Modern Sensor, Physics, Design, Application.* Springer Nature, Fifth Edition, Berlin 2016, ISBN 978-3-319-19302-1.

[5] Kanagachidam, B. *Senosors and Protolos for Industry 4.0. Industrial Application of Marker Tech.* APress, Calcata 2023, ISBN 978-1-48429-0-064.

[6] Webster, J., Eren, H. *Measurement, instrumentation and sensors handbook: Spatial, mechanical, thermal, and radiation measurement.* Second edition. Boca Raton: CRC Press, 2014. ISBN 978-14398-4888-3.

[7] Yongpan, L., Youn-Long, L., Chong-Min, K., Hiroto, Y. *Smart Sensors and Systems. Technology Advancement an Application Demonstrations.* Springer, Berlin 2020. ISBN 978-3030-4223-32.

[8] Kester, W. *Practical Design Techniques for Sensor Signal Conditioning.* Analog Devices, Norwood 1999, ISBN 0-916-550-20-6.

[9] Soloman, S. *Sensors handbook.* Mc Graw Hill, Second edition. New York 2010. ISBN 978-0-07-160571-7.

3

Personal Computers in Measuring Systems

Computers in any form (personal, industrial, single board) are an integral part of all numerically controlled measuring systems. Their use enables not only the control of measurement systems but also the transfer and processing of measured data, including their presentation. In addition, their standard hardware interfaces, e.g., RS-232, USB, and LAN, allow easy connection of measuring units even in relatively complex and sophisticated measuring systems.

3.1 Computer Architecture

A standard personal computer is made up of a motherboard, on which, in addition to the microprocessor, there are also so-called chipsets, which form one or two integrated circuits. These are designed to work together and are usually delivered as a single product. Manufacturers of such chipsets are usually independent of motherboard manufacturers. The chipset provides communication between the processor, buses, slots, controllers, and other components on the motherboard. The most famous chip producers include NVIDIA, AMD, VIA Technologies, SiS, and Intel. Figures 3.1 and 3.2 show a typical personal computer motherboard and its simplified block diagram.

The basis of every personal computer is the CPU (*central processing units*), which is clocked by the clock pulse generator CLK and controlled by the BIOS (*basic input-output*) operating system, stored in the fast fixed semiconductor memory FLASH ROM. To transfer data from this memory to the processor, the LPC (*low pin count*) bus was used in older computers, which is replaced by the SPI (serial peripheral interface) bus in modern computers. This bus is also used to connect keyboards and computer positioning devices (mice and joysticks) that do not require high data transfer speed requirements. The fast parallel bus FSB (*fast serial bus*) is used to transfer data from the processor to the slot drivers.

Other connectors (*slots*) are used to connect internal solid-state memories, external memory disks, graphics and sound cards, and other peripherals, e.g., drivers of the RS-232 serial interface, universal serial interface USB

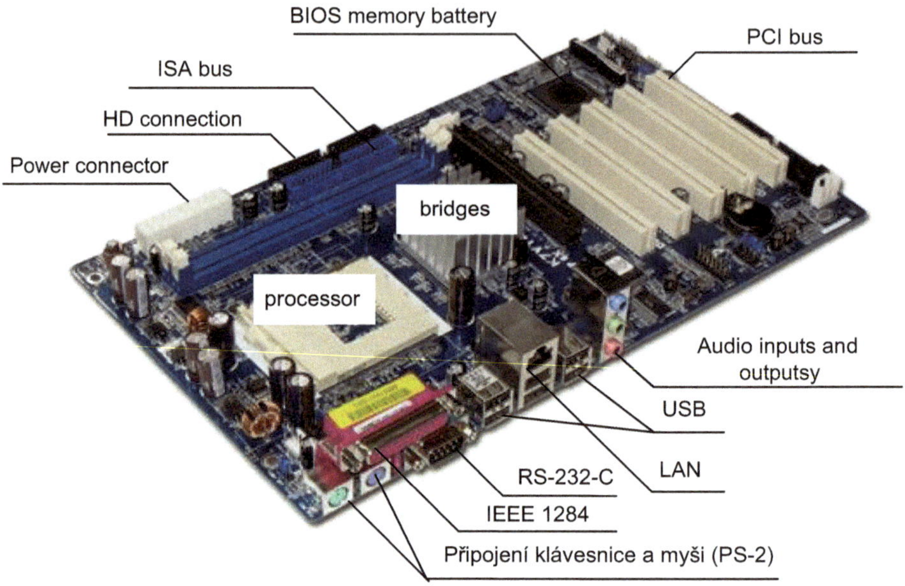

Figure 3.1 Motherboard of a typical personal computer.

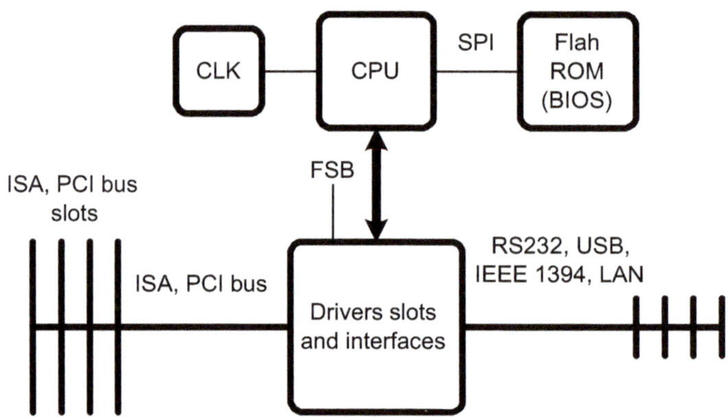

Figure 3.2 Motherboard of a typical personal computer.

(*universal serial bus*), IEEE-1394 interface (*Fire Wire*), and LAN (*local area network*) computer networks.

In older personal computers, the ISA (*industrial standard architecture*) bus was used to drive peripherals that did not require fast data transfer, which was able to transfer 16- or 32-bit data at a transfer rate of up to 256 Mb/s. In modern computers, the PCI (peripheral component interconnect) bus is used for data transfer, which enables a transfer speed of up to 528 Mb/s when transferring 32-bit or 64-bit data. This bus is used, for example, to connect plug-in cards intended, for example, to control measurement systems based on the IEE-488 protocol and DAQ (*data acquisition systems*) data collection systems with low data transfer speed requirements.

The *PCI Express* bus with serial data transfer is used to drive peripherals requiring fast data transfer. The bus designation is not quite correct because it is a transfer between which data is transferred without device addressing. Serial data transmission brings significant advantages over parallel transmission, because when increasing the transmission frequency in parallel data transmission, due to the existence of parasitic capacities and inductances between the bus conductors, their transmission properties are undesirably influenced by each other and thus problems with the synchronization of data transmission (*clock skew*) occur.

Several independent ISA and PCI buses can be placed on the computer motherboard, or their industrial variants.

3.1.1 Personal computer processors

Older personal computers used the so-called single-core processors SCP (*single computer processor*), in which data operations were performed sequentially, and therefore did not allow high operating speeds to be achieved. These so-called single-task systems, using, e.g., the classic MS DOS operating system, allowed the execution of at most one program, which was inconvenient for the users of these computers, as the work in progress had to be saved first, the program currently running had to be terminated, and another program started in which performed the required action, and then the system could revert back to the previously used program and saved work.

The advantage of this operating system was its simplicity, and therefore it was not demanding not only on programming but also on the system resources of computers, which at that time were slow and did not require a high capacity of operating memory.

In modern personal computers, multi-core processors (MCPs) are used to increase the operating speed of processors, which work on the principle of time sharing and the principle of parallel data processing.

The principle of time sharing (*multitasking*) consists in the alternating method of processing computational tasks for certain periods of time, the length of which is quantized in multiples of predetermined time intervals. According to the method of allocating and removing time quanta to individual tasks, we distinguish between cooperative and non-cooperative multitasking.

Cooperative multitasking requires active participation in running tasks. Each task is obliged to pass control back to the operating system often enough, which can start another task, which after a while voluntarily gives up the processor again, etc. The advantage of this method of calculation is the simpler implementation of the operating system. Its main disadvantage is that a wrongly programmed task does not return to the control of the operating system, thus causing it to stop completely. Typical users of cooperative multitasking are, for example, the *Windows* 95 and *Windows* 98 operating systems.

Non-cooperative multitasking consists in the priority of allocating and removing the processing processor of individual tasks at regular time intervals (typically approximately 100× to 1000× per second.) With the help of a timer, an interruption of the currently running program (process) will be triggered, the current situation will be evaluated (which tasks request about processor allocation), and either the task that was interrupted by the system or another task that is interested in processor allocation is allowed to run again. However, even in preemptive multitasking, a task can voluntarily request a context switch and give up the rest of its quantum. In this case, the task so-called "sleeps" and the process either goes into a state of rest (Sleep Mod) or the system goes into a so-called slow input–output operation (*Low Speed Input Output Mod*), such as reading data from the hard disk.

This system uses, for example, the *Windows* NT operating system, *Linux*, and some other less well-known operating systems. The advantage of this method of time sharing is that the computer does not "freeze," because even if the task cycles, the operating system takes control of the given task with the help of a timer and allocates the processor to another task. The disadvantage of this solution is the more complex implementation of the operating system and the necessity of hardware support in the computer processor in the so-called *privileged mode*.

Parallel data processing (*multithreading*) is based on the simultaneous running of several data file processing processes, called threads. This method of data processing uses shared operating memory. Although its use reduces the time overhead of the operating system when switching between individual processes, which is necessary to ensure the correct operation of the system, it increases the risk of concurrent requests for its access to data (*race condition*).

Current multi-core processors consist of up to 16 processors and are capable of handling up to 32 processes independently. They usually also have built-in fast *CASH memory* with a capacity of up to 65 MB and a graphics processor. The clock frequency of these processors reaches up to 4.5 GHz. To

Figure 3.3 Multi-core processor and its cooling system.

cool them, either additional fans are used, or water cooling with a radiator is used in case of greater power losses (Figure 3.3).

3.1.2 Semiconductor memories of personal computers

To write and store the processed data, personal computers use fast memories with a random selection of RAM (*random access memory*), in which the data is stored through the states of bistable flip-flops with CMOS transistors, which have minimal consumption in the idle state, in the order of tens of µW/bit.

The first types of personal computer RAM data memories were static memories with a random selection of SRAM (*static random access memory*), whose capacity was up to 256 MB and whose data throughput was in the range of several tens of Mb/s.

Since 1998, these memories have been replaced by SDRAM (*synchronous data random access memory*) with a capacity of up to 512 MB and a transfer speed from 66 to 133 Mb/s. These memories were usually supplied with a voltage of 3.3 V and to ensure the direction of their insertion into the memory slot of the computer's motherboard, they were provided with two notches.

Modern dynamic DDR (*double data rate*) memories enable, in comparison with SDRAM memories, also referred to as SDR (single data rate) memories, to achieve a higher transfer rate due to stricter control of the timing of electrical data and clock signals. These memories are produced with a capacity of up to 8 GB and their transfer speed ranges from 3.2 to 57.6 Gb/s. The supply voltage of these memories ranges from 1.2 to 2.5 V. For reasons of non-interchangeability, these memories are equipped with one notch.

Figure 3.4 Kingston DDR400 memory.

Table 3.1 Parameters of typical DDR memories.

Memory type	Clock frequency	Transfer speed	Data access time	Power voltage	Launch year
DDR-400	100 MHz	3.2 Gb/s	10 ns	2.5 V	1998
DDR2-800	200 MHz	6.4 Gb/s	12 ns	1.8 V	2003
DDR3-1600	200 MHz	1.8 Gb/s	15 ns	1.5 V	2007
DDR-3200	400 MHz	25.6 Gb/s	25 ns	1.2 V	2014
DDR-6400	400 MHz	51.2 Gb/s	36 ns	1.1 V	2018
DDR-7200	450 MHz	57.6 Gb/s	36 ns	1.1 V	2020

Typical representatives of these memories are, for example, memories of the Kingston DDR series (Figure 3.4).

Table 3.1 shows the parameters of typical DDR memories.

3.1.3 Memory disks of personal computers

Memory disks of personal computers are used to store, read, and delete data. Currently, solid-state, hard disc drive, and solid-state hybrid memory drives are used.

A disk HDD (*hard disk drive*) stores data on a disc covered with a magnetic layer. Data is written to and read from this disk using write and read heads that move over the surface of the magnetic disk. The number of these heads ranges from 2 to 256. The capacity of the disk depends on its diameter and the number of recording tracks, the number of which varies from 300 to 3000. The speed of writing and reading data of these disks depends on the number of revolutions of the magnetic disk. Laptop drives have a 2.5" (inch) disc diameter and 5400 rpm. Desktop drives are typically 3.5" and 7200 rpm.

The advantage of HDD hard disks is the possibility of achieving extremely high capacity, low price, and long data retention time. If the mechanical parts of the disk are damaged, the data disk can be removed and moved to an undamaged mechanical disk system. The disadvantages of HDD disks compared to solid-state SSD disks are longer response times when writing and reading data, mechanical wear, greater noise, and consumption.

a) hidden HDD disk b) exposed HDD disk

Figure 3.5 Hard HDD disk.

Figure 3.5 shows a hard disk including a data disk with a *Western Digital* recording head.

A *solid*-state disc SSD stores data in a similar way to a USB flash drive, with the difference that an SSD has a larger capacity and is faster. The disk is therefore undemanding to mechanical damage, is completely silent, and has significantly lower consumption compared to an HDD disk. The main advantage of SSD disks is their speed of writing and reading data. This is particularly evident when the operating system is started. Theoretical reading and writing speeds are 10× to 100× higher than with a classic HDD disk.

SSD drives are produced in various designs. SSDs with the size of a classic 2.5" HDD or M.2 disks are most often used in laptops. The difference can also be in the controller. The most commonly used SATA III controller reaches transfer speeds of up to 600 Mb/s. The newer NVMe standard is designed specifically for SSD disks and can make full use of its capabilities, allowing transfer speeds of up to 3.5 Gb/s.

The *lifetime* of SSD disks is determined by the limited number of writes to the memory cell. While the number of rewrites is not physically limited with a magnetic disk, with an SSD, the insulator between the memory cells is broken with each rewrite, and the cells are then unable to maintain an electric charge after a certain period of time. Depending on the technology used, each cell has a lifespan of only a few thousand write cycles. SLC cells have a lifespan of approximately 100,000 transcripts, EMLC cells 30,000, MLC cells 10,000, and TLC only about 1000–5000 transcripts.

To increase the life of the disks, such algorithms are used so that the disk wears out evenly and the cells in which the most frequently overwritten data are not damaged first. SSDs also have free memory available, which is used to replace non-functional cells. It is stated that, for example, disks with a capacity of 120 GB have a usable capacity of 95% of their original value after

Figure 3.6 Solid state SSD disk.

30,000 hours of operation. Figure 3.6 shows a solid-state SSD from Kingston with a capacity of 2 TB.

The SSHD (*solid-state hybrid drive*) hybrid disk combines the advantages of HDD and SSD disks. This disk consists of a smaller SSD disk with a capacity of several GB and a magnetic HDD disk, which makes up the larger part of the disk capacity. The SSHD disk has a built-in memory controller and divides and writes data by itself according to how often it is used. After installing the operating system, its loading (boot) speed is quite slow. After several of its uploads, the disk already evaluates the system files as frequently used, moves them to the SSD part, and thus the next system startup time is already close to a classic SSD disk. Their advantage is a very good ratio of their capacity, speed, and price. The disadvantage is the possibility of mechanical damage, higher consumption, and noise.

3.2 Personal Computer Bus

The *personal computer bus* is used to transfer data and control functions between its parts, while the data transfer is defined by a standardized protocol that must be accepted by all functional parts of the computer that are connected to the bus.

From the point of view of the method of data transmission, we distinguish between serial and parallel buses. A serial bus consists of at least two wires, along which data and addresses are transmitted by serial protocol in serial order. These buses are used to transfer data with lower requirements for transfer speed. Parallel buses are made up of separate addressing, control, and data conductors, and data can be transmitted in these buses either in full parallel form or can be transmitted sequentially in serial–parallel form.

3.2.1 ISA bus

The ISA bus (*industrial standard architecture*) was the oldest parallel computer bus, introduced in 1981 by IBM. This bus is based on 16-bit Intel 8088 microprocessors, used in PC/XT computers, Intel 80 286 processors, used in PC/AT computers and 32-bit Intel 80 386 processors, and Intel 80 387 coprocessors, used in Pentium computers.

The *standard ISA bus* consisted of a 16-bit data bus and a 24-bit address bus, supplemented by a control bus and power supply. The bus makes it possible to use a multi-level interrupt system of data transfer with different service priority, which gives priority to data transfer with a higher priority. This priority is declared by the IRQ command. The bus also enables direct access to the DMA (direct memory access) memory, which, like a multi-level interrupt system, has a different priority for data access. The DRQ command can thus be used to transfer data over the bus independently of the processor's activity. Acknowledgment of the DMA transfer request is generated by the DACK command. The T/C (*terminal count*) signal is used to confirm the transmission of the selected data block.

Since the DMA controller works with a 16-bit address, the maximum capacity of transferred data is 64 kB. Another 4-bit page register is used to determine the DMA channel, the direction of data transfer from the memory to the application plug-in board register, the starting address of the computer memory block, and the number of transferred bytes.

Because personal computers of the IBM class have a so-called open architecture, it is possible to expand the use of these computers with plug-in boards inserted into the ISA bus, e.g., through the IEEE-488 (IEC-625) protocol, or directly use DAQ (data acquisition system) single-purpose boards. In this case, signals IRQ2–IRQ7, IRQ9–IRQ12, and IRQ14 and IRQ15 are used to interrupt data transfer, and signals DRQ0–DRQ3 for 8-bit channels and DRQ5–DRQ7 for 16-bit channels are used to determine DMA channels.

Although the processors used in IBM series computers allow addressing up to 65,536 devices (ports), only 1024 ports are available on the standard ISA bus. The first 256 ports with addresses 0000 to OFFH are reserved for I/O (*input/output*) devices on the computer's system board. Another 768 addresses 0100 to 3FFH are for plug-in boards, with addresses 300H to 31FH being for prototype boards.

The ISA bus allows data transfer in either 8-bit or 16-bit word format with a clock frequency of 8 MHz. When transferring a 16-bit word to an even address device, the word is transferred at once in one write read cycle. When transferring this word to an odd address device, the lower 8 bits are transferred first and then the higher 8 bits.

Two types of dual-row ISA connectors are used on IBM-class personal computer motherboards. The longer type contains 2 × 32 positions, and the shorter type contains 2 × 16 positions.

Figure 3.7 Single-board computer PC104.

In addition to IBM class computers, the ISA bus was also used in the *industrial standard* PC/104, which was designed by Ampro in 1987 and standardized in 1992 under the designation IEEE P 996.1. The PC/104 system is designed to minimize space and power consumption. This system usually consists of a single-board microcomputer with a processor, chipset, BIOS operating system memory, and SIMM dynamic data memory, including connectors for connecting power, a keyboard, and I/O circuits. In this system, the ISA bus is used to connect several boards placed one above the other with standard dimensions of 90 mm × 96 mm, on which additional PBM application modules (*piggy back modules*) can be placed, e.g., modules for controlling simpler control and measurement systems. This system uses a shorter type of two-row connectors with 2 × 16 contacts. Figure 3.7 shows a single-board computer in the PC/104 standard.

3.2.2 PCI bus

The PCI (*peripheral component interconnect*) bus was introduced in the mid-1990s. This processor-independent bus has been further improved. It is currently the most widely used personal computer bus. Similar to the industrial versions of computers with the ISA 96/AT96 or PC/104 bus, the industrial version of the IPCI (*industrial PCI*) and compact PCI buses is used with a more massive construction, adapted for use in standard 19-inch instrument cabinets. Figure 3.8 shows the PCI bus on a personal computer board.

Since this bus is *processor-independent*, it is connected to the system bus of the processor via a PCI bridge, which enables its use in computers other than IBM computers, such as Apple Macintosh or Power PC computers. The bridge enables voltage and impedance matching of transmitted signals.

Figure 3.8 PCI bus on a personal computer board.

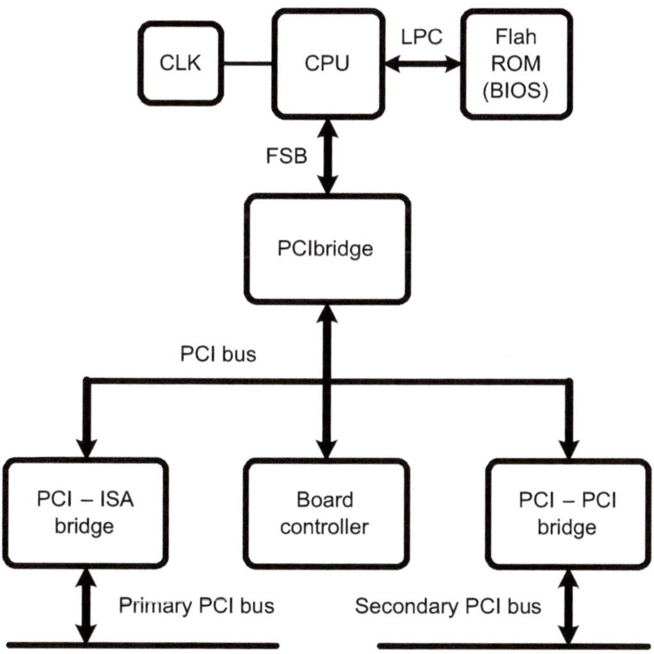

Figure 3.9 Block diagram of a personal computer with extended PCI buses.

Special integrated circuits of *Intel* series 430xx, 440xx, and 450xx or PCI to PCI bridge and PCI to card bus controller circuits from *Texas Instruments* are used to implement bridges. Typically, a maximum of 10 devices can be connected to the primary PCI bus, e.g. plug-in boards to connect sensor signals to physical quantities (Figure 3.9).

The bus clock signal is derived from the CLK clock signal of the processor. Based on the V2.1 recommendation, the frequency of the PCI bus clock signal can range from 25 to 33 MHz, or double the frequency of 66 MHz can be used.

The PCI bus consists of *data and address wires*, which are multiplexed with a basic width of 32 bits and an extended version of 64 bits to save the number of wires. When transferring 32-bit words, a transfer speed of up to 132 Mb/s can be achieved. When transferring 64-bit words, a transfer speed of up to 528 Mb/s can be achieved. The control signals of the PCI bus are the RST and IDSEL signals for initialization and detection of the connected device, the REQ, GNT, REQ64, and ACK64 signals for determining the connection priorities (*arbitration*) of the selected bus speed, and the FRAME, DEVSEL, IRDY, TRDY, STOP, and LOCK signals for communication. The C/BE0 and C/BE7 signals determine the validity of the data, the PERR and SERR signals indicate errors in their transmission, and the INTxx signals are interrupt signals. The test signals are the TCK, TDI, TDO, TMS, and TRST signals. Additional PCI bus auxiliary signals are the PAR, PAR64, PRSNT, JTAG, SBO, and SDONE signals.

In order to distinguish the supply voltages of these buses, keys are placed on the plug-in boards at the place of their contacts 12, 13, 50, and 51, which make it impossible to insert them into the PCI bus slot by mistake. Signals on PRSTN 1 and PRSTN 2 are used to identify the presence and power consumption of plug-in boards. The permitted power consumptions of individual plug-in boards are 7.5, 15, and 25 W. Table 3.2 identifies the presence and wattage of plug-in boards in PCI-based computers.

3.2.3 PCI-Express bus

Referred to as PCI-E or 3GIO, this bus was created to replace the older PCI, PCI-X, and AGP standards. The bus designation is not quite correct, as these are two-point connections between which data is transferred without the need for device addressing. The PCI-Express bus uses serial–parallel data transfer, because this data transfer makes it possible to increase the clock frequency of the transfer compared to parallel transfer, and thereby increase the transfer speed of the bus, because with only parallel data transfer, unwanted mutual influence of wires occurs at high frequencies and the related creation problems with synchronization (*clock skew*), Figure 3.10.

The original design of the PCI-Express bus was brought by the AWG (*Arapaho Work Group*), consisting only of *Intel* employees. The final version 1.0 standard from 2004 was already the result of collaboration between

Table 3.2 Identification of the presence of the board and its power consumption in PCI buses.

PRSTN 1	PRSTN 2	Identification of the presence of the board and its power consumption
X	x	Plate absent
L	x	Board with 25 W power input inserted
X	L	Board with 15 W power input inserted
L	L	Board with 7.5 W power input inserted

Note: x means the bus is in neutral due to open collector circuits.

Figure 3.10 Types of PCI buses on a personal computer board.

several hardware corporations, including *Dell, IBM*, and *Hewlett-Packard*. In 2004, the PCI-Express bus became the new standard for personal computers. In 2007, most graphics cards for PCI-Express were already sold in *PCI-Express 1.x* variants, whose transmission speeds are listed in Table 3.3.

The *PCI-Express 2.0 bus*, launched on the market in 2007, made it possible to increase the transfer speed of one serial line during one-way transfer from 250 to 500 Mb/s compared to the previous PCI-Express 1.0 bus. For example, two-way data transfer over this bus could be up to 16 Gb/s.

The *PCI-Express 3.0 bus*, launched on the market in 2010, allowed a double increase in transfer speed compared to the PCI-Express 2.0 bus, and in addition, instead of 8b/10b, it used a method of changing the sequence

Table 3.3 Transfer speeds of the PCI-Express1.x bus.

Bus type	One-way transmission	Two-way transmission
PCI-Express 1.1	250 Mb/s	500 Mb/s
PCI-Express 1.4	1 Gb/s	2 Gb/s
PCI-Express 1.8	2 Gb/s	4 Gb/s
PCI-Express 1.16	4 Gb/s	8 Gb/s

Figure 3.11 SAS controller plug-in board with PCI-Express bus.

of bits (*scrambling*) to encode data using a known binary polynomial. As a result of this encoding change, the data can be recovered using an inverse polynomial. In addition, instead of 8b/10b coding, 128b/130b coding is used, which reduced the overhead during data transfer from the original 20% for the PCI-Express 2.0 bus to 1.5%. The maximum transfer speed of this bus in the *PCI-Express 3.16* design reaches up to 32 Gb/s for two-way data transfer. Figure 3.11 shows the plug-in board of the SOS controller, controlled via the PCI-Express bus.

The *PCI-Express 4.x bus*, introduced in 2011, makes it possible to double the transfer speed again compared to the PCI-Express 3.x bus to a maximum value of 64 Mb/s for bidirectional data transfer. In addition, the transmission reliability is increased with this bus. The fastest variant of this bus, *PCI-Express 3.16*, enables bidirectional data transfer to reach a transfer speed of up to 64 Mb/s.

The *PCI-Express 5.x bus*, which was launched in 2021, can also use optical cables instead of metal wires for data transmission, thereby significantly eliminating the influence of interfering signals during data transmission even in highly noisy environments. The fastest version of this bus, PCI-Express 5.16, enables bidirectional data transfer to achieve a transfer speed of up to 128 Mb/s.

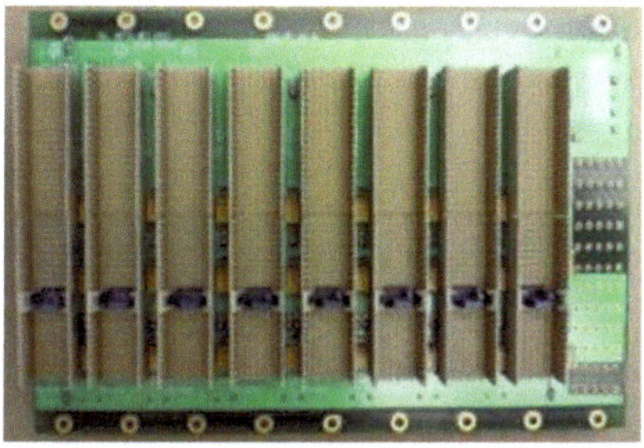

Figure 3.12 Compact PCI bus connector array.

3.2.4 Industrial modifications of PCI buses

Similar to the ISA bus, the PCI bus has its industrial modifications. There are currently two different PCI industry bus standards.

The *compact PCI bus* was designed in 1958 by the PICMG (*Personal Industrial Computer Manufacturers Group*) from the USA. This bus is a passive variant of the PCI bus, and therefore its clock frequency is limited to 33 MHz. Due to the standard PCI bus, double the number of PCI devices can be connected to this bus without using a PCI/PCI bridge. This bus uses five-row connectors in a 5 × 47 arrangement for 64-bit boards, with mechanical keys placed on the 13th and 15th contacts to prevent the board from being inserted into the bus by mistake. For 32-bit boards, connectors are used in the arrangement of 5 × 25 with the same location of the mechanical key (Figure 3.12).

The *compact PCI industrial bus* also allows the use of plug-in boards of the VME bus system. For this purpose, the standardized interface compact PCI/VME (*PCI Mezzanine Card Specification*) is used. This interface uses connectors according to the VME (*Versa Module Europa Bus*) specification, which are located on the top of the VME module.

The PCI EX bus (*PCI Extension for Instrumentation*) was created in 1995 based on the initiative of the *National Instruments* company. This bus uses Eurocard connectors. The connector pin assignment is standardized by PICMG US and PICMG Europe for this bus.

Unlike the original *Eurocard boards*, which used a 0.1" (2.54 mm) pin pitch, the compact PCI system uses metric connectors with a 2 mm pin pitch, designed according to IEC 1076. The 3U plug-in boards have a 110-pin connector (J1), which carries the 32-bit PCI bus signals, and an optional 110-pin connector (J2) that allows the upper 32 bits to define the I/O interface. 6U plug-in boards have identical J1 and J2 connectors, which are always used for 64-bit PCI buses.

Compact PCI plug-in boards can be designed for +3.3 or +5 V supply. Boards with +3.3 V supply are marked with a yellow key, and boards with

Figure 3.13 Typical personal computer interfaces.

+5 V supply are marked with a blue key. Plug-in boards compatible with both supply voltages do not have a code key.

The *industrial PCI bus* was created in 1996 in the BRD (*Bundes Republic Deutschland*). This passive bus has a maximum of five slots, of which the first slot is for the system slot, and the other four slots are for application slots. The clock frequency of this bus can be up to 66 MHz. The control signals of this bus are supplemented with some additional signals compared to the classic PCI bus. These are, for example, the SLEEP signals, enabling the disconnection of the plug-in board from the system board, and the RSTIN signal, enabling manual, or program system reset. This bus uses five-row connectors in the arrangement of 5 × 48 contacts, which are divided into three parts of 5 × 12, 5 × 24, and 5 × 12 contacts.

3.3 Personal Computer Interface

Typical interfaces of older personal computers were the RS-232-C serial interface and the IEEE 1284 (*Centronics*) parallel interface. These interfaces were gradually replaced in the 1980s by better and faster RS-485, IEEE-1394 (*Fire Wire*), USB (*universal serial bus*), and LAN *(local area network)* interfaces.

Figure 3.13 shows typical personal computer interfaces with these interfaces, complete with connectors for connecting a keyboard and mouse via the PS2 serial interface, and audio connectors for connecting a microphone and speakers.

3.3.1 RS-232-C interface

The *RS-232-C serial interface*, standardized in 1969 by the EIA (*Electronic Industries Association*), was originally intended to connect DTE (*data*

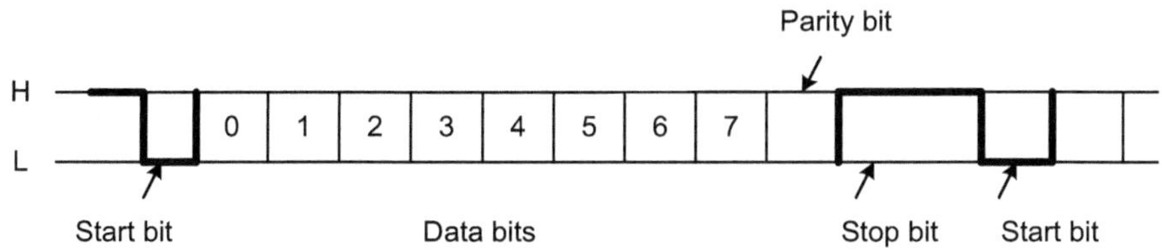

Figure 3.14 Data transfer format for RS-232-C interface.

terminal equipment) with DCE (*data communication equipment*). Over time, these interfaces began to be used to control and transfer data from many other devices, including electronic digital measuring devices. Despite its shortcomings, which consist of a two-point connection, low transmission speed, and low resistance to interference, we encounter this interface in simpler and cheaper measuring devices.

Data through this serial interface is transmitted in the form of a sequence of 8 data bits, starting with the *start bit* and ending with the *stop bit.* Logic level L corresponds to a voltage in the range of +3 to +15 V, and level H corresponds to a voltage in the range of −3 to −15 V. Since the drive and receive voltage signals of this interface are related to the neutral conductor, parasitic leakage currents are applied during their transmission resistances and capacitances of conductors. As a result, this significantly limits the transmission speed of this interface. With the recommended line resistive load of this interface, which is in the range of 3–7 kΩ and its drain capacitance less than 2.5 nF, a transfer rate of up to 20 kB/s can be achieved with a line length of tens to hundreds of m. Format data transfer for this interface is shown in Figure 3.14.

Asynchronous data transfer with this interface requires identical settings for their transfer format and transfer rate. Each message begins with a start bit, followed by 5–8 data bits, a parity bit, and a stop bit. The transmission speed is selected from the range of 110, 300, 600, 1200, 2400, 4800, 9600, and 19,200 bit/s. The RS-232-C standard defines 20 signals, which it assigns to positions on the 25-pin *Canon* connector (*D Shell Connector*), but for most applications in instrumentation, a connector with only nine contacts is used (Figure 3.15). Table 3.4 shows the pin designations of the nine-pin RS-232-C connector.

In practice, two methods of communication between devices are used to connect devices.

Modem communication uses all seven interface signals according to Figure 3.16.

In most applications of the RS-232-C interface in measuring technology, only two DTE devices (usually a computer and a measuring device, or a

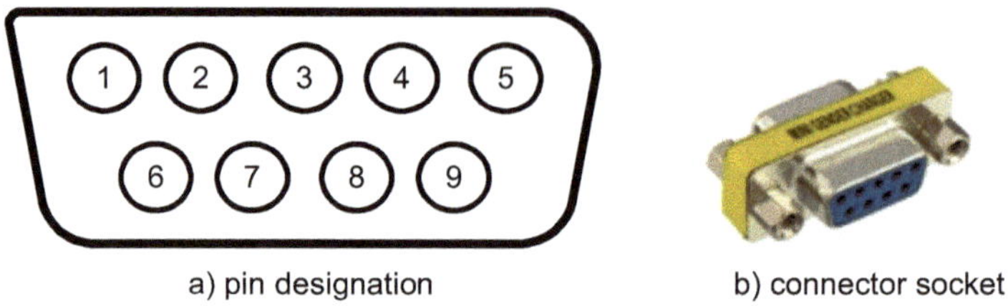

Figure 3.15 Nine-pin RS-232-C connector.

Table 3.4 Assignment of RS-232-C interface signals in a nine-pin connector.

Pin	Signal name	Signal	Significance of the signal
1	Data Carrier Detected	DCD	Signal detection
2	Received Data	RxD	Received data
3	Transmitted	TxD	Broadcast data
4	Data Terminal Ready	DTR	Receiver and transmitter functionality
5	Signal Ground	SG	Signal ground
6	Data Set Read	DSR	Functionality and readiness to communicate
7	Request To Send	RTS	Readiness to receive data
8	Clear To Send	CTS	Readiness to receive and transmit data
9	Ring Indicator	RI	Ringtone

measuring device and a printer) are connected at a distance of units, or tens of m. In these cases, a so-called *modemless connection* is used, in which the transmission channel is replaced by a specially connected NMC cable (*null modem cable*). In addition, device manufacturers often use only some signals and assign them different functional meanings.

Examples of three- and five-wire connection of two devices without a modem are shown in Figure 3.17.

With a *three-wire connection*, only data wires TxD, RxD, and ground wire SG are used. In this case, the data transfer is controlled only by software (*software handshaking*), e.g., by the *Xon/Xoff protocol*, in which the device that receives the data controls the communication by sending special control characters. If the device is ready to receive data, it sends the Xon signal (ASCII 19), and the other device responds by interrupting data transmission. This activity is repeated until the message is transmitted without errors. With a five-wire connection, RTS and CTS signal wires are also used, which simplifies and speeds up data transfer.

Full modem connection (FMC) uses all data and signal wires for data transmission, while two connection variants are used (Figure 3.18).

As a rule, *ASCII encoding* is used for data encoding, in which characters are sent individually in a seven-bit format, usually with parity, as well as

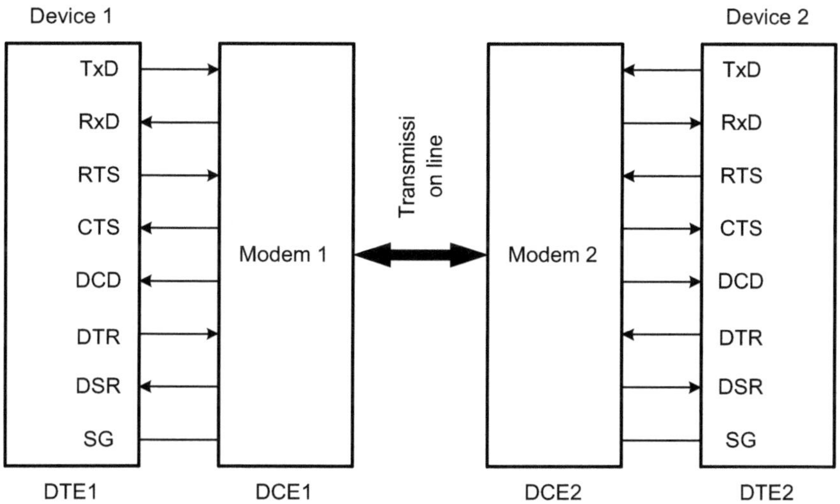

Figure 3.16 Connecting devices with standard RS-232-C modem communication.

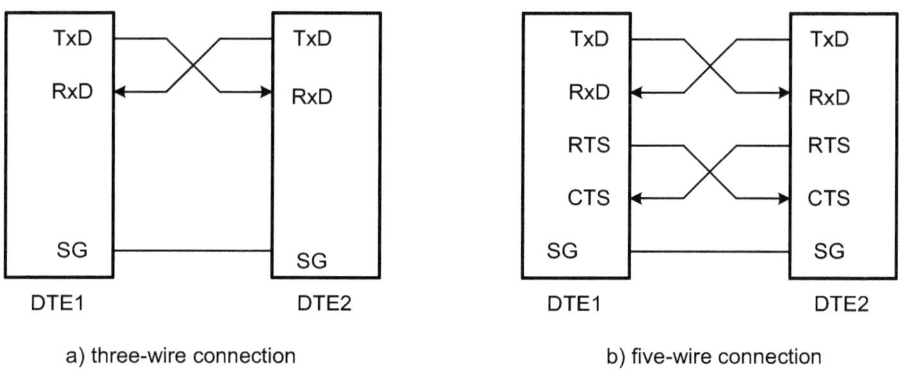

Figure 3.17 Modemless RS-232-C connections.

binary encoding, in which the data bytes are sent unmodified in an eight-bit format without parity, or HEX *hexadecimal encoding*, in which the characters are encoded in a seven-bit format with parity. Coding in the XMODEM protocol, in which data packets are sent in 128 binary coded bytes with check parity at the end of the transmission, is mainly used in telecommunications.

The programming languages C, PASCAL, and BASIC are used to program the RS-232-C interface, which allow direct access to the registers of the communication circuit.

To connect the units via the RS-232C interface, shielded multicore cables are used in less demanding applications. In more demanding

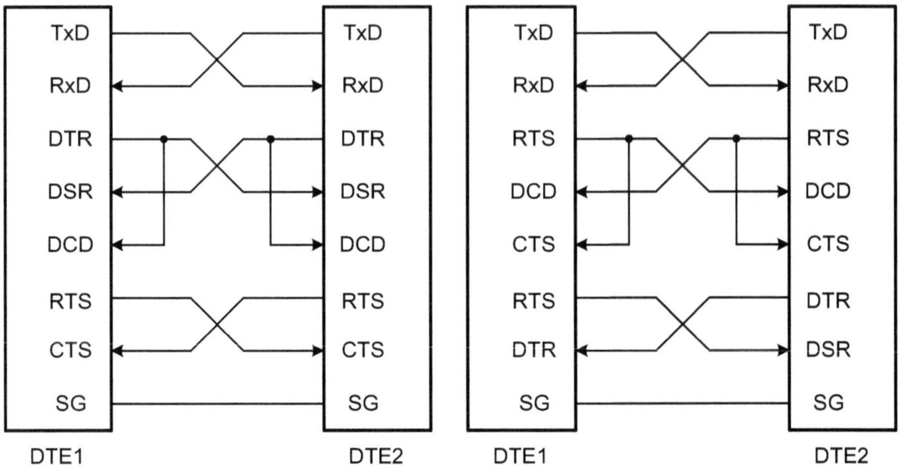

Figure 3.18 Two variants of a full RS-232C modemless connection.

applications with *twisted pairs* of wires, which, in order to suppress the occurrence of reflections, at their ends are provided with damping resistors of 120 Ω, the value of which is the same as the wave impedance of the twisted pairs of conductors. Typical two-sided drivers and receivers for the RS-232 interface are, for example, the MAX 202E double driver, or quadruple exciter MAX 206 E.

3.3.2 IEEE-1394 interface

The impetus for the creation of this interface, also referred to as *Fire Wire* or *iLink*, or *Digital Link* was *Apple's* effort to develop an interface to replace the SCSI (*Small Computer System Interface*) parallel interface that was used to connect external magnetic hard drives, DVD (*digital versatile disc*) optical carriers, and scanners to personal computers. The development of the IEEE-1394 interface was completed in 1995 and subsequently standardized in 2008 by the IEEE Std. 1394-2008.

This serial interface is used not only for fast data transfer between peripherals of personal computers with 32-bit buses but also in centralized measuring systems, consisting of, e.g., VXI or PXI frames. The data transfer in these systems is used by, e.g., *National Instruments* as a standard. Although the IEEE-1394 interface is in principle intended for the connection of only two devices (peer-to-peer), e.g., a computer with an external memory disk, this interface also allows the creation of tree structures of a data network (Figure 3.19).

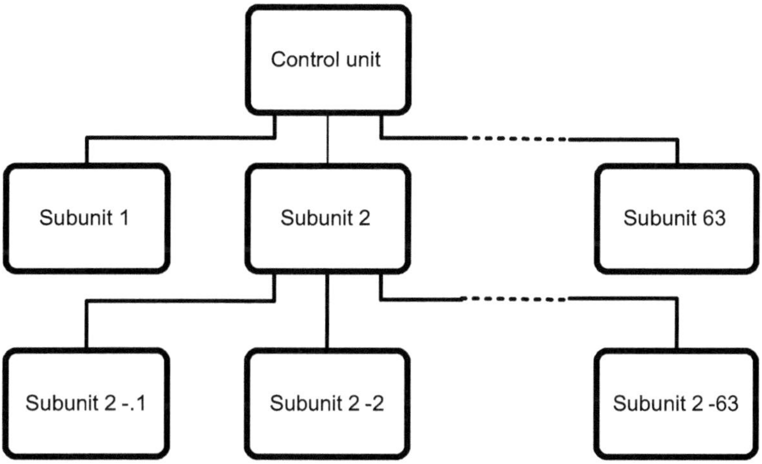

Figure 3.19 Tree structure of an IEEE-1394 bus network.

The basic element of this network is its control unit *rooter*, to which subsidiary units are connected via network *branch*. Up to 63 units can be connected to this bus, while it is possible to connect another 63 units to each secondary unit. The control program of the network enables its automatic reconfiguration when these secondary units are disconnected and connected. In addition, the IEEE-1394 protocol allows the transfer of data between individual sub-units without the involvement of the control unit.

The original version of the IEEE-1394-1995 interface, referred to as *FireWire 400*, transmits data in HP (*half duplex*) mode, in which data is transmitted at different transmission rates for transmission and reception. This version of the interface allows data transfer at speeds of 98.304, 196.608, or 393.216 Mb/s, which is 12.2888 Mb/s, 24.576 Mb/s, and 49.152 Mb/s. These bit rates are referred to as S100, S200, and S400. The maximum cable length for this interface is limited to 4.5 m, but it is possible to connect up to 16 cables using active repeaters using so-called external or internal hubs. In this way, a maximum cable length of up to 72 m can be achieved.

To connect the units, this interface variant uses a special cable with two pairs of shielded twisted conductors TPA+, TPA–, and TPB+, TPB–, supplemented by a +V supply wire and a ground wire SG, representing the signal ground, while the inner shielding of the wires and the outer shielded jacket of the cable is grounded (Figure 3.20).

In the *half-duplex data transmission*, one pair of twisted wires TPA+, TPA– is used for data transmission, and the other pair of wires TPB+, TPB– is intended for the transmission of a synchronization signal, which

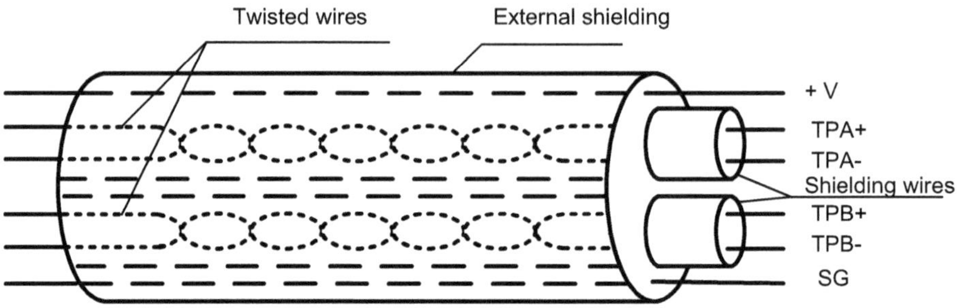

Figure 3.20 Shielded IEEE-1394 cable.

changes its level only with two consecutive identical levels of the data signal. Transmitters of data signals generate *differential voltages* with a range of 200 mV with a common component, given by approximately half of the supply voltage, which is standard 25 V. The use of shielded twisted pairs of conductors on the one hand significantly limits the penetration of the alternating interfering electromagnetic field into the signal paths, and on the other hand, it makes it possible to effectively suppress the influence of the common voltage on the reliability of data transmission.

The individual units of the IEEE-1394 interface are connected by crossed pairs of signal wires so that the TPA (*Twisted Pair A*) data wires of one unit are connected to the TPB data wires of the other unit.

A six-pin connector, known as an *alpha connector*, is used to connect these cables, which also enables the connected units to be powered. A typical supply voltage is 25 V, but it can range from 24 to 30 V with a load of up to 8 W. *Apple* laptops use a supply voltage of 9 V due to the batteries used. Figure 3.21 shows the logo of the IEEE-1394 and the six-pin connector of this interface.

The assignment of the contacts of the six-pin connector to the signals of the IEEE-1394 interface is shown in Table 3.5.

The IEEE-1394 interface variant, introduced to the market in 2000, makes it possible to choose either an asynchronous or an isochronous data transfer method compared to the previous variant.

Asynchronous data transfer is used in those cases where it is necessary to transfer data with maximum transfer speed. According to the urgency of the data transfer method, two data transfer methods can be selected for this data transfer method.

In *non-urgent data transfer* mode (*fair mode*), data is transferred between units with the same priority. In *urgent data transfer* (*urgent mode*), data transfer between units is determined by the frequency of requests for

a) IEEE 1394 interface logo b) six-pin IEEE-1394 interface connector

Figure 3.21 IEEE-1384 interface logo and six-pin connector.

Table 3.5 IEEE-1394 six-pin connector pin assignments.

Pin	Designation	Importance
1	+V	Power supply
2	SG	Signal ground
3	TPB−	Data wires
4	TPB+	
5	TPA−	Data wires
6	TPA+	
Housing	GND	Shielding

their transfer. Although this method of data transmission allows to increase the overall transmission speed of the system, it can also cause collisions, the elimination of which is resolved by a short restart of the interface bus. Asynchronous data transmission can therefore be used in those cases where a guaranteed transmission speed is not required, e.g., when transmitting audio or video signals.

Isochronous data transfer takes place in cycles according to Figure 3.22.

In this method of transfer, first one of the units (*cycle master*) is assigned the highest access priority. This unit generates a start code (*start cycle*) every 125 μs, which synchronizes data transfers between units. If data transmission takes place during this interval, the transmission of the start code is delayed, and each unit is subsequently assigned the so-called isochronous time channels (*isochronous gap*), which enable isochronous data transmission between units. At the end of data transfer between all units, the bus is put into a state of rest (*subaction gap*), which can be used for asynchronous data transfer.

The IEEE-1394b-2002 standard, referred to as *Fire Wire 800*, allows data to be transmitted in full duplex mode (full duplex), in which data is transmitted in both directions at the same transmission rate. This method of transmission makes it possible to achieve transmission speeds of up to 786.432 Mb/s up to a distance of 100 m with a standard shielded cable with twisted pair of category 5e. The latest IEEE-1394-c interface makes it possible to increase the transmission speed up to 3.2 Gb/s. These interfaces use an eight-pin RJ-45 connector and a quadruple shielded twisted-pair cable, which is also used in the *Ethernet* network, for data transmission.

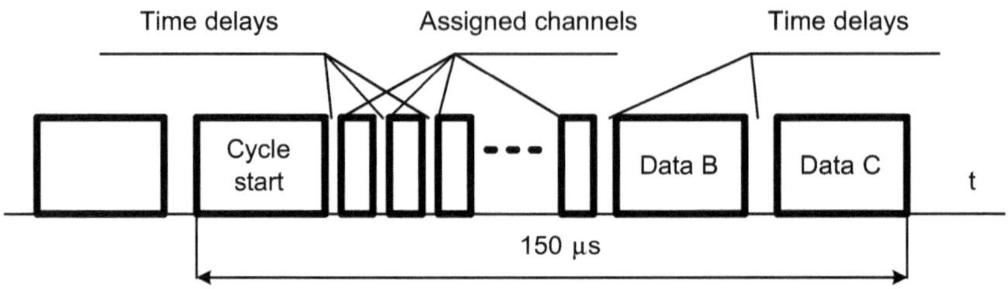

Figure 3.22 Time course of signals during isochronous data transfer via the IEEE-1384 interface.

For the galvanically isolated method of data transmission, glass or plastic fibers with appropriate signal converters can be used.

The software of measuring systems with the IEEE-1394 bus does not require a subsequent restart of the computer after connecting the units, because after connecting them, the system automatically initializes, during which individual functional units are assigned identification numbers and addresses (*plug and play*). Therefore, the header of each transmitted data file always contains the IDN (*identification number*) of the data receiver and transmitter.

3.3.3 Universal serial interface

Universal serial bus (USB) was originally developed to connect external devices such as printers, keyboards, joysticks, cameras, modems, and memory drives to personal computers. The first version of this USB1.1 interface, released on the market in 1996, made it possible to transfer data at a speed of 1.5 Mb/s, i.e., 187.5 kB/s, in the low-speed version. The fast version of this high-speed interface had a transmission speed of up to 12 Mb/s, i.e., 1.5 Mb/s.

In 2001, the USB 2.0 version was introduced with a transfer speed of 480 Mb/s, i.e., 60 Mb/s. Subsequent versions of USB 3.0, USB 3.1, and USB 3.2, released on the market between 2011 and 2017, allowed data transfer rates of 5–10 Gb/s, i.e., 625 to 1.25 Mb/s. So far, the fastest versions of the USB interface, introduced in 2019 and 2020, are the USB 4 and USB 4 2.0 versions, which allow data to be transferred at speeds of up to 40 Gb/s, or 80 Gb/s, i.e. 5 Gb/s, or 10 Gb/s.

For USB 1.1 and USB 2.0 busses, a shielded cable with one pair of shielded twisted wires, supplemented with a +V power supply wire and an SG signal ground wire, is used to connect the units. The drivers of this bus, supplied with a voltage of 5 V, excite with their complementary output voltage impulses with a swing of 200 mV, the corresponding voltage of which

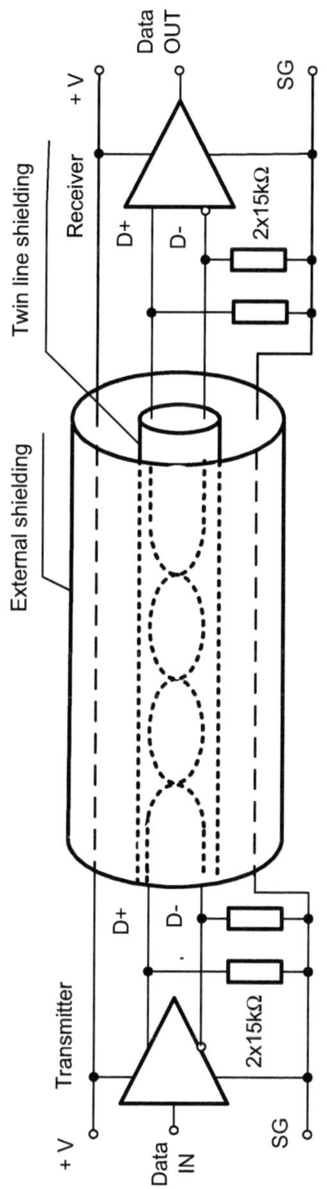

Figure 3.23 USB 1.0 and USB interface cable. 2.0.

is 2.5 V, while the log. 0 corresponds to a voltage of 0.3 V and log. 1 voltage 2.8 V. To avoid reflections of voltage pulses, caused by impedance mismatch of the cable, terminating resistors of 15 kΩ are used at both ends of the data wires, which correspond to the wave impedance of the two twisted wires (Figure 3.23).

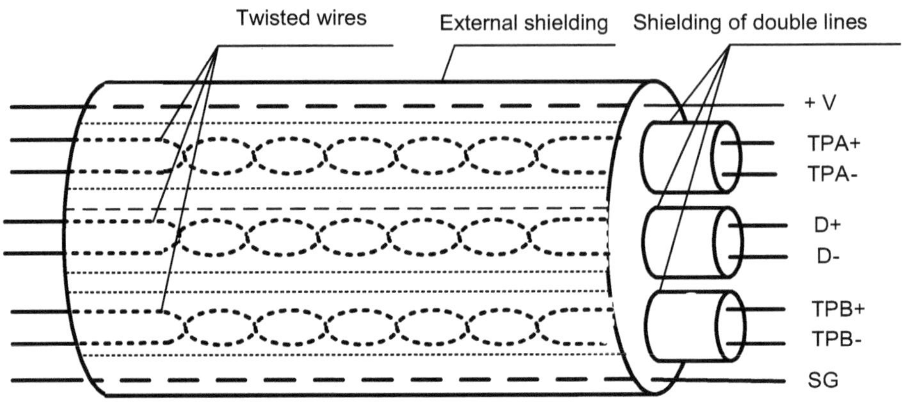

Figure 3.24 USB 3.0 and USB interface cable. 3.1 Powered.

With this connection, transfer speeds of up to 12 Mb/s can be achieved with a cable length of up to 5 m. This version of the interface uses four-pin flat connectors.

The more modern versions of the USB interface, USB 3.0 and USB 3.1 Powered, use a shielded cable with three shielded twisted wires and a nine-pin connector (Figure 3.24).

Unlike the cable for USB 1.0 and USB 2.0 interfaces, the cable for USB 3.0 and USB 3.1 interfaces still contains two pairs of shielded twisted pairs, intended for the transmission of fast complementary signals TPA+, TPA−, TPB+, and TPB−. In addition, each of these double lines is shielded by a separate shield, and the entire cable is shielded by an external shield.

Figure 3.25 shows the logo of the USB interface and the most used types of USB connectors and the assignment of their contacts to individual signals (Table 3.6).

Unlike other developmentally older interfaces, data transfer via the USB bus takes place in the so-called frames, lasting exactly 1 ms. Within each of these *frames*, data packets for several devices can be processed sequentially. At the same time, both slow (*low-speed*) and fast (*Full-speed*) packets can occur together.

Coding and decoding of signals is handled purely by hardware with this interface. NRZI (*non-return to zero inverted*) encoding is used, in which levels of log. 0 signals cause the level of data signals and logs to change. 1 leave their levels unchanged. If, for example, the original data stream contains six consecutive logos. 1, then the data transmitter automatically adds one log at the end of the data message. 0 to force a change in the data signal level. Receiver of this log. 0 will be removed from the data stream again. This

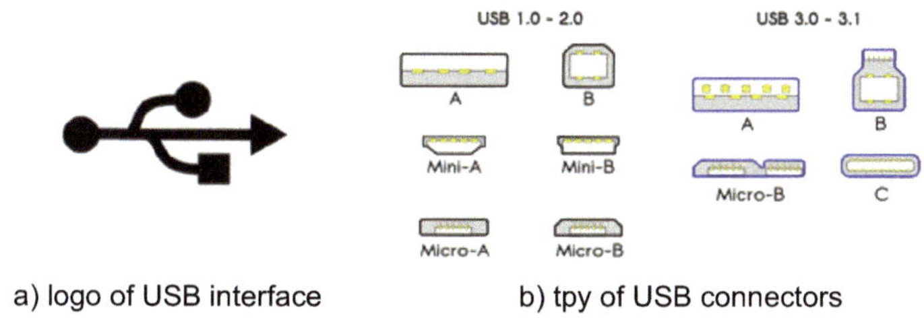

Figure 3.25 Types of USB connectors.

Table 3.6 Assignment of contacts on USB 3.xx connectors.

	USB 1.0 and USB 2.0	USB 3.0 and USB 3.1	Importance
1	+V	+V	Power supply
2	D+	D+	Differential signals
3	D–	D–	USB 1.0–USB 2.0
4	SG	SG	Signal ground
5	–	STD A+	Differential signals
6	–	STD A–	USB 3.0–USB 3.1
7		G	Internal shielding
8	–	STD B+	Differential signals
9	–	STD B–	USB 3.0–USB 3.1
Cover		G	External shielding

so-called *bit-stuffing* makes it possible to significantly increase the transfer speed of the USB bus.

Each data packet has a special LLLLLLLH boot byte for synchronization purposes. As a result of NRZI coding, the data receiver evaluates eight alternating bit states to which it can time synchronize. Synchronization must be maintained during the subsequent data transfer. All these processes take place only in the corresponding hardware components. This data operation is implemented by SIE (*serial interface engine*) units, which include FIFO (*first in, first out*) memories formed by circular data registers. The USB interface makes it possible to create relatively large data networks of up to 127 units in a tree structure (Figure 3.26).

Similar to the IEEE-1394 interface, the USB software allows you to activate the connected units without restarting the computer (plug-in-play). For this purpose, so-called hubs, called HUBs, are used, which, after connecting the unit, send a 10 ms reset signal and at the same time release a supply current of 100 mA for each connected unit. Before the connected unit receives its own USB bus address from the computer, it is possible to communicate with it via the default address 0.

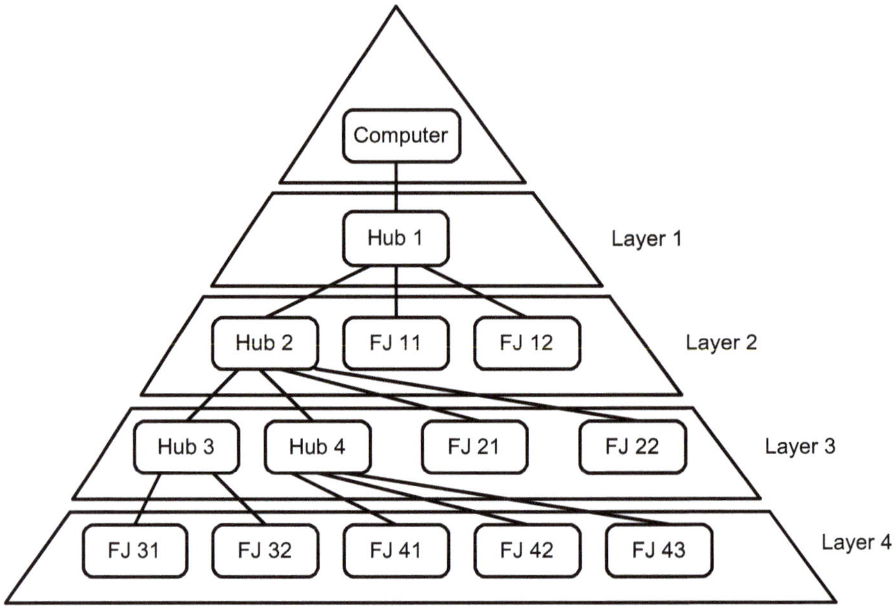

Figure 3.26 Tree structure of units with USB interface.

Figure 3.27 Seven-terminal USB hub without power supply.

In terms of power supply, we distinguish between hubs that take the power supply of the connected units directly from the computer's USB interface, which is usually limited to a maximum of 500 mA, and hubs with their own power supply, to which it is possible to connect, for example, up to seven USB ports with a total consumption of, e.g., 3.5 A. Figure 3.27 shows a seven-terminal hub with an external power supply.

Each hub always has one connection, the so-called *upstream port* UP, through which it is connected either to a computer or to an output, the so-called downstream port DP of the hub connected in front of it. Each hub has its own 1 IN interrupt endpoint (the address of the 1 interrupt endpoint that points from the hub to the master). By connecting the unit to the hub, the address of the port to which the unit was newly connected is reset, and its new (non-zero) address is subsequently generated by the superior unit.

References

[1] Ebelin, J. *Measurement systém application and design*. Mc. Graw Hill Publishing Company 1990. ISBN: 0-07-017338-9.
[2] Garrett, P. H. *Advanced Instrumentation and Computer I/O Design*: Real-Time System Computer Interface Engineering, John Viley, 2000. ISBN: 978-0-780-36013-6.
[3] Anderson, D., Dzatko, D. *Universal Serial Bus System Architecture*, Addison-Wesley, 2001. ISBN: 978-0-20130-975-1.
[4] Boyes, W., (ed.), *Instrumentation Reference Book*, 3rd ed., Boston, MA: Butterworth-Heinemann, 2003. ISBN 978-0750671231.
[5] Park, J., Mackey, S. *Practical Data Acquisition for Instrumentation and Control Systems,* Amsterdam, the Netherlands: Elsevier, 2003. ISBN: 978-0750-65796-9.
[6] Nawrocki, W. *Measurement systems and sensors*. Second edition, Artec House 2016. ISBN 1-58053-945-9.
[7] Dawoud, P. Serial communication protocols. River Publishing 2022. ISBN 9781-0-0333-949-6.
[8] EIA Standard RS-232-C. Electronic Industries Association, Washington 1969.
[9] *IEEE Standard for parallel inteface.* IEEE Std. 1284.
[10] *IEEE Standard for a High-Performance Serial Bus*. IEEE Std. 1394-2008.

4

Measurement Systems Based on IEEE-488

This serial–parallel bus was developed by *Hewlett-Packard* as HP-IB Bus (Hewlett-Packard interface bus). It was standardized in 1975 by IEEE Std. 488. During the following years, this bus was widely used by leading manufacturers of measuring and testing equipment. In 1987, this standard was revised under the designation IEEE-488.1-1978, in which the basic mechanical and electrical parameters of the bus were defined. In 1992, IEEE-488.2 introduced additional command syntax and format conventions to be independent of the circuit design of the bus control circuitry, including the format of data structures and error logs. In parallel with this activity, the IEC (*International Electrotechnical Commission*) introduced the IEC 60 625-1 and IEC 60 625-2 standards, which were later replaced by the IEC 60 488 standard. At that time, the *National Instruments* company introduced the HS-488 standard, which it allowed, in contrast to the transfer rate of 1 MB/s of the IEEE-488 bus, to increase its transfer rate up to 8 MB/s. In 2004, the IEEE-488.2 and IEC 60 488 standards were merged into a single IEEE/IEC 60 488-1 standard.

4.1 IEEE-488 Bus

The IEEE-488 interface bus is an 8-bit *bidirectional serial–parallel bus*, enabling the connection of up to 15 functional units at a distance of up to 20 m, while the maximum distance between individual functional units can be up to 2 m. The bus consists of a 24-wire shield, eight of which are data cables (DIO1–DIO8), three are for data transmission control (DAV, NRFD, and NDAC), five wires are dedicated to single-wire message transmission (ATN, IFC, REN, SRQ, and EOI), and eight wires are the signal ground of the data wires. A typical transmission speed of this bus reaches up to 1 MB/s in the basic IEEE-488 variant; in practice, it reaches a speed of 250–500 kB/s.

The connection cables of this bus are terminated with double-sided 24-pin double-row connectors from the Amhenol company, enabling their

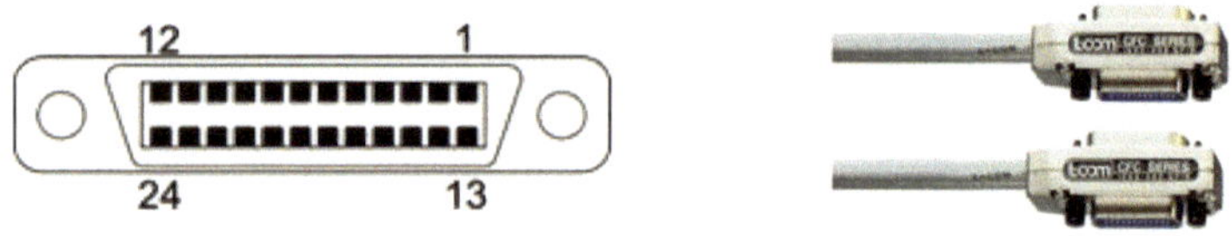

a) distribution of contacts on the connector b) double-sided double-row connectors

Figure 4.1 IEEE 488 bus connector and connection method.

Table 4.1 Pin assignment of IEEE-488 interface signals.

Pin	Signal	Importance
1	DIO1	Data 1 input/output
2	DIO2	Data 2 input/output
3	DIO3	Data 3 input/output
4	DIO4	Data 4 input/output
5	EOI	End of identify
6	DAV	Data valid
7	NRFD	Not ready for data
8	NDAC	Not data accepted
9	IFC	Interface clear
10	SRQ	Service request
11	ATN	Attention
12	S	Shielding
13	DIO5	Data 5 input/output
14	DIO6	Data 6 input/output
15	DIO7	Data 7 input/output
16	DIO8	Data 8 input/output
17	REN	Remote enable
18–24	GND	Ground

mutual connection directly to their housing. This way of connecting buses allows you to create only linear data structures.

Figure 4.1 shows the pin layout on the IEEE-488 bus connector and double-sided double-row connectors.

Table 4.1 shows pin assignment of IEEE-488 interface signals.

4.2 Communication in the IEEE-488.1 Interface

The standard arrangement of a measurement system with an IEEE-488.1 interface is made up of one control unit, called the *system controller*, which is usually a personal computer equipped with a plug-in board for this interface. This control unit sends a one-wire IFC (*interface clear*) message, which sets the interface of all connected functional units to the initial defined state,

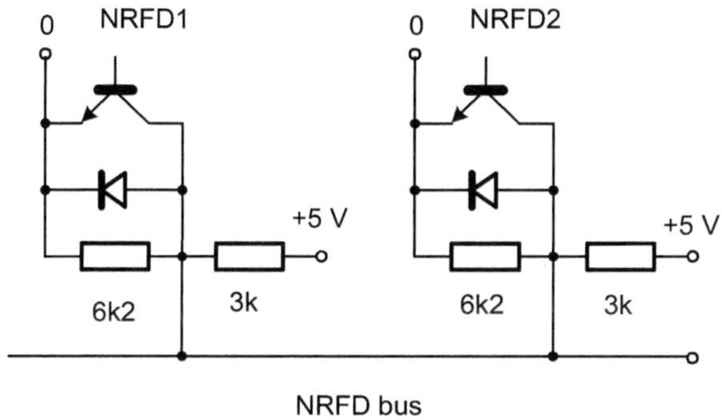

Figure 4.2 Exciting the bus with an open-collector TTL circuit.

and a REN (*remote enable*) message, which switches the functional units to remote control. The control unit also enables the ATN (*attention*), EOI (*end of identity*), and other control messages, specifying, for example, the addresses of the units, to control the communication between the units of the measuring system.

The measurement system with the IEEE-488.1 bus allows the existence of several control units, but only one of them has the function of a system controller. The latter can hand over rights to another controller for the duration of certain functions, but after they have been performed, the original controller will take back control of the bus. The IFC signal is used to interrupt the control activity of the sub-controller.

Data transfer between individual functional units and the controller is carried out using data wires DIO1–DIO8, while the transmitter unit (*talker*) and the receiver unit (*listeners*) must always be defined. The asynchronous transmission of 8-bit data messages is controlled by the signals DAV (*data valid*), confirming the existence of data on the bus, NRFD (*not ready for data*), indicating that the data is not ready for transmission, and NDAC (*not data accepted*), confirming that the data has not been received.

The data transmission system uses negative logic, i.e., when standard TTL logic circuits are used, powered by a voltage of +5 V, it corresponds to log. 1 (*true*) voltage level $L < 0.8$ V and log. 0 (*false*) voltage level $H > 2.0$ V. The advantage of using negative logic is the fact that the quiescent level of the signal corresponds to the voltage on the open collector TTL circuit, which in this case is +5 V. Figure 4.2 shows the method of driving the IEEE-488 bus by TTL 74Lxx circuits with an OC (*open collector*).

An active signal (log. 1) leads to the opening of the transistor of one of the functional units and thus to the voltage drop to zero. However, if one of

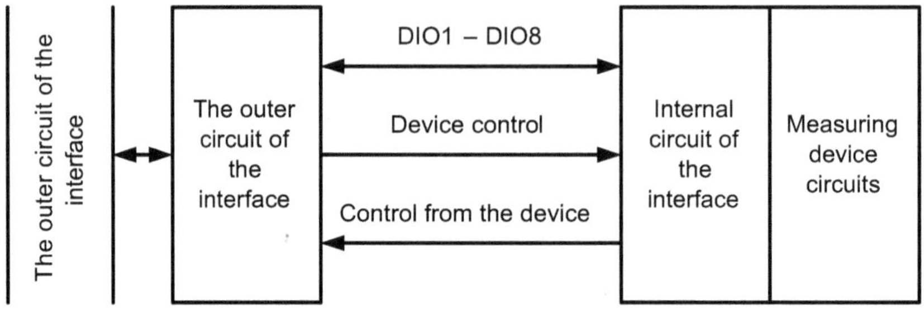

Figure 4.3 Interaction of the functional unit with the bus.

the transistors is closed, then when the other transistors are open, there is a positive voltage of +5 V on the bus.

The single-wire signals ATN, REN, SRQ, and EOI have the following meanings.

The ATN (*attention*) signal determines the meaning of the bytes transmitted on wires DIO1–DIO8.

The IFC (*interface clear*) signal puts all functional signals in the idle state. This signal is generated only by the control unit.

The SRQ (*service request*) signal is an interrupt signal sent by the functional unit when the control unit receives this signal. The controller then asks all units if other units have requested an interrupt.

The EOI (*end of identity*) signal specifies the last byte of a serial transfer. Some manufacturers derive the end of transmission from the CR and LF characters instead of this signal.

The cooperation of the functional units with the control unit is in mutual communication. At the low level (L) of the ATN signal, the controller sends messages to the function unit via the DIO data wires. Multi-wire messages are sent at log. to the H level of the ATN signal. The functional unit also sends a response via the DIO wires, either again in the form of a string of characters in the ISO7 code or a block of binary data at the L level of the ATN signal, or requests the unit for service with a single-wire SRQ message.

The interaction of the functional unit (measuring device) with the bus is shown in Figure 4.3.

In addition to these single-wire messages, multi-wire messages (*commands*) are used in communication between units, which are divided into universal, addressed, and secondary.

Universal messages determine the corresponding activity in all connected units without the need to address them. Table 4.2 provides an overview of universal multi-wire messages.

Table 4.2 IEEE-488.1 universal interface messages.

Message	Code	Name
LLO	11H	Local Lockout
PPU	15H	Parallel Poll Unconfigure
DCL	14H	Device Clear
SPE	18H	Serial Poll Enable
SPD	19H	Serial Poll Disable
UNT	5FH	Untalk

Table 4.3 IEEE-488.1 interface attached messages.

Message	Code	Name
GTL	01H	Go to Local
SDC	04H	Selected Device Clear
PPC	05H	Parallel Poll Configure
GET	08H	Group Executive Trigger
TCT	09H	Talk Control
MLA	–	My Listen Address
MTA	–	My Talk Address

Table 4.4 IEEE-488.1 interface secondary messages.

Message	Code	Name
PPE	60H to 60FH	Parallel Poll Enable
PPD	70H	Parallel Poll Disable

Addressed messages generate activity only for those devices that are addressed to receive them. Table 4.3 provides an overview of addressed messages.

Secondary messages specify secondary device addresses. Table 4.4 shows the secondary messages of the IEEE-488.1 interface.

Since the message file of the IEEE-488.1 standard does not contain a signal that would determine the direction of data transfer between two units, this deficiency is solved by assigning two different addresses to each unit, i.e., by assigning a receiving address MLA (*My Listen Address*) and a transmitting address MTA (*My Talk Address*). These addresses differ only in the two most significant bits (seventh and sixth), while the other bits of both addresses are identical. Therefore, if a unit receives, for example, an MLA message, it becomes a message receiver (listener), and if it receives an MTA message, it becomes a message transmitter (*speaker*).

In the MLA message, the 7 bit is the log level. 0. If, for example, the device is set to address 8, which corresponds to the binary word 01000, then

Table 4.5 Interface functions of the IEEE-488.1 interface.

Function	Name
SH	Source Handshake
AH	Acceptor Handshake
T (TE)	Talker (Extended Talker)
L (LE)	Listener (Extended Listener)
RL	Remote/Local
DC	Device Clear
DT	Device Trigger
SR	Service Request
PP	Parallel Port
C	Controller

the MLA8 message corresponds to the binary word 0101 000 (40) and the MTA8 message corresponds to the binary word 1001 000 (72). It follows that message 31 (0011 111) cannot be used to address units and is reserved for UNL 0(011 111) and UNT messages (0101 1111).

The UNTL (*unlisten*) message is intended for addressing all receiving units and the UNT (untalk) message for addressing all transmitting units.

The interface functions of the IEEE-488.1 interface are listed in Table 4.5.

4.3 Communication in the IEEE-488.2 Interface

Unlike the IEEE-488.1 interface, the data formats in this interface are divided into receive and transmit. Both kinds of formats can receive any data in any format and transmit it in a well-defined format. For these purposes, a 7-bit ASCII code is used for instrument messages, and an 8-bit binary code is used for expressing numbers.

Numeric formats are preferably expressed in ASCII decimal and NR1 numeric format. The optional formats are NR2 and NR3. Optional formats also include *hexadecimal numeric* data such as #H01F2, octal numeric data such as #Q7035, and *binary numeric* data such as #B011101. Only capital letters can be used to broadcast messages.

4.3.1 Numeric and character formats of commands in the IEEE-488.2 interface

The NR1 *number format*, representing integer-type numbers, consists of decimal digits including a sign, e.g., 324, +234, –432.

The NR2 *number format* uses a decimal number system including a sign with one decimal point placed anywhere between the numbers, e.g., 3.24, +2.34, –43.2.

The NR3 *number format* consists of decimal digits including a sign that forms the mantissa and a decimal point placed between any two digits of the mantissa. This format is supplemented with the letter E, or e, which forms the exponent of the number, e.g. 3.24E–3, or 3.24e–3.

The *character format* of the message consists of uppercase letters and numbers including the _ character (code 95) and must always start with a capital letter, e.g., R3_F7.

The *message string* format begins and ends with quotation marks ("). This data format allows the use of lowercase and uppercase letters, including spaces, periods, and commas. This format also allows the insertion of additional chain messages.

The *block format* transmits a block of data using 8-bit DAB (data byte). At the beginning of each data block, there is a so-called *block header*, consisting of the character # and subsequent digits in the ASCII code, which indicates the number of subsequent characters in the ASCII code. These subsequent characters, expressed in NR1 format, indicate the number of data bytes that the transmitted data block contains, e.g., #12 <DAB>, or #3001 <DAB>. This format allows a data block of undefined length to be transmitted using 8-bit data bytes. In this case, the block header consists of # and zero in ASCII code, and the data block is terminated by the LF character also in ASCII code together with the EOI signal, e.g., #0<DAB><DA^EOI.

The *syntax of the messages* in the IEEE-488.2 standard is determined either by the separator of individual commands at two different levels or by a colon, e.g., INP:IMP120, indicating the setting of the input impedance of the data transmitter to 120 Ω. It is also possible to use a separator of commands of the same level in a multiple command using a semicolon, for example, "INP:IMP120;INP:FILT 1". In this case, any character in the range 0–9 or 11–32, referred to as white spaces (*white space*), is used, e.g., TRIG:SOUR EXT, indicating that the device will be externally triggered.

A comma is often used to separate commands, e.g., CONF:VOLT:DC 100.1E-3, indicating the setting of the voltmeter to measure DC voltage on the 100 V range with a resolution of 1 mV. The end of the query is a question mark, e.g., "MEAS:VOLT:AC?," indicating the measurement of alternating voltage and the saving of the measurement result. The terminating character of the string during reception is the LF character, or LF EOI characters. The terminating character of the string when transmitting is the LF character with the EOI character.

4.3.2 State models of the IEEE-488.2 interface

The IEEE-488.1 standard contained only two interface functions SR and PP to apply the service request of functional units. Therefore, each of the units of this interface was equipped with a status register which, when requested via the SR function in the seventh bit at log.1 level, requested service from the system controller and at log.0 level did not request service from the controller. The remaining bits of this command could have been used for other purposes.

The IEEE-488.2 standard extends the status register to at least two duplicate status registers, which are the standard status word register SBR (*standard byte register*) and the standard event register (SER). The status word auxiliary register EBR (*enable byte register*) and the event register EER (*enable event register*) are assigned to these registers, enabling activation, or masking individual bits of these status registers. This allows the ability to specify which bits of the status registers the functional unit will respond to by sending an SRQ service request message.

Figure 4.4 shows the mutual connection of the status registers SBR and SER and the masking registers EBR and EER and the connection of the event register SER to the status word register SBR.

Three bits are defined in the SBR according to IEEE-488.2. The seventh bit has the meaning of the start bit MSS (*master summary bit*). If this bit is set to log. 1, then there was at least one reason for the service request, namely that either a MAB (message available bit) message is ready to be sent in the output queue (*output queue*), or one of the bits in the SES (*standard event status*) register has been set of the ESB registry on the log. 1. The other bits in the SRE register are not defined. Commands for setting and reading registers are listed in Table 4.6.

The SBR also uses the function of parallel reporting. In this case, this register is extended by another 8 bits, of which the first 8 bits correspond to the contents of the original SBR. The next 8 bits define the instrument conditions. This entire 16-bit register is connected via product elements to the PPER (*parallel poll enable register*) register, which is set by the PRE command. In this way, the generation of an output message can be influenced during program operation, which can be obtained either by parallel reporting according to IEEE-488.1 or by direct reading using the IST? command.

Common commands are messages that are sent by the control unit and are used to control devices and identify their status. They use simple syntax

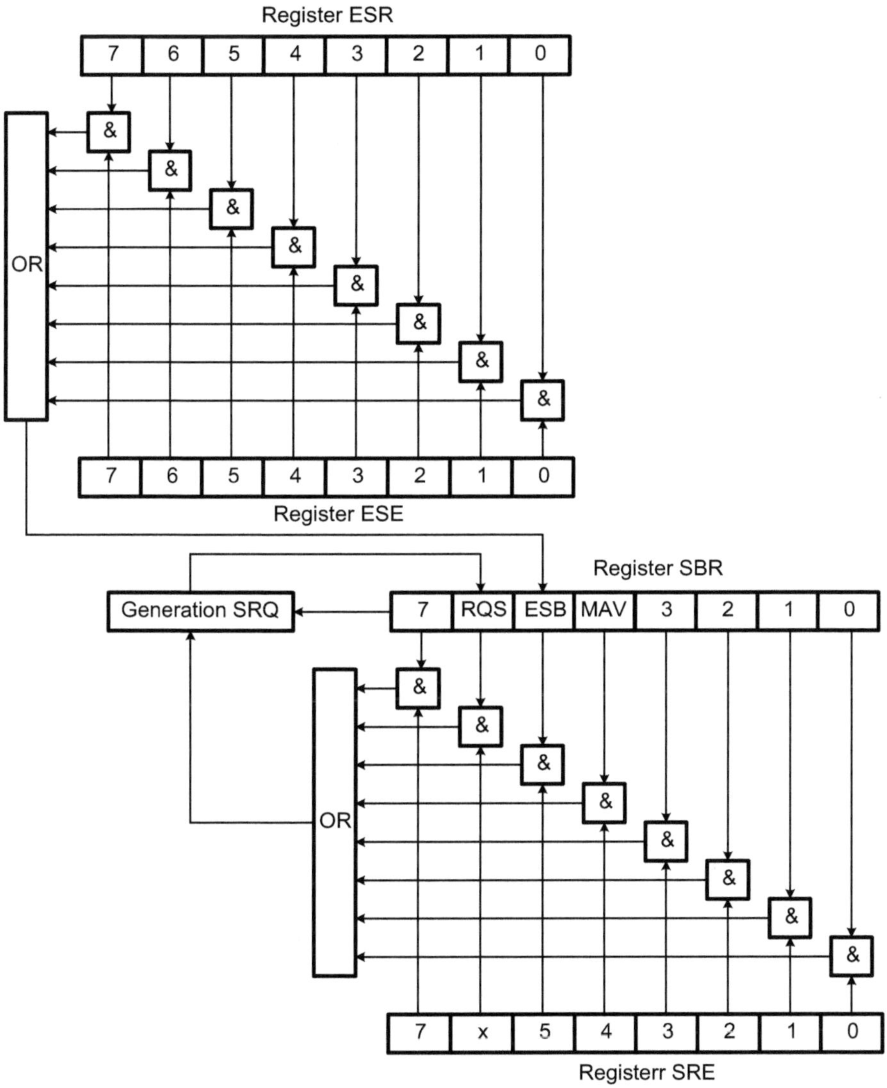

Figure 4.4 Status registers according to the IEEE 488.2 standard.

like *XXX or *XXX?, where XXX is the name of the command. It is made up of three characters from the letters of the uppercase alphabet. The question mark at the end of the command changes the command to a query (*query*), i.e., the device is obliged to respond after decoding it. An overview of mandatory orders is given in Table 4.7.

Table 4.6 Meaning of bits in the SER event status register.

Bit	Function	Importance
0	Operation Complete	The device has finished the operation and is ready to accept new commands
1	Request Control	Controller indication that the device is requesting to take over the bus
2	Query Error	Attempting to read data from an empty output queue or losing data from that queue
3	Device Dependent Error	The device cannot complete the operation due to internal error conditions, e.g., overflow, overload
4	Execution Error	An error occurred while executing the command, e.g., the data is outside the allowed range
5	Command Error	Command syntax or semantic error
6	User Request	Indicates local intervention of the user in controlling the device
7	Power On	Indicates that the device is turned off and on again

Table 4.7 Overview of mandatory orders.

Command	Name	Importance
RST	Reset Command	Setting the device to an idle state
CLS	Clear Status Command	Reset status registers
IDN	Identification Query	Instrument specification inquiry
STB	Status Byte Query	Query the contents of the registry
SRE	Service Request Enable	Query the contents of the registry
ESE	Event Status Enable Query	Registry setting command *ESE*
OPC	Operation Complete	Activation ending of peration report
WAI	Wait to Continue	Waiting for previous operations to complete
TST	Self-Test Query	Command to start the internal test of the instrument

Device message protocols determine the rules not only for their transmission but also for their processing methods. This principle also applies in cases where errors occur during the transmission of messages, e.g., when an incomplete or incomplete message is transmitted.

4.3.3 System control unit IEEE-488.2

This unit must meet the following requirements according to the IEEE-488.2 standard.

Interface functions including circuitry, interconnect cables, and connectors must be compatible with the IEEE-488.1 standard. Drivers with an open collector OCD (*open collector driver*) must be used to generate the SRQ,

NRFD, and NDAC signals. DAV, EOI, ATN, REN, and IFC signals must be generated by tri-state drivers. Data wires DIO1–DIO8 must also be driven by these drivers. The unit must generate an IFC signal at the log level. 1 for longer than 100 ms. Furthermore, it must allow setting the REN signal to log. 0 and log. 1 and identify the logic levels of the SRQ signal. Another requirement for the unit is the determination of individual bits of the status byte, the detection of error conditions during communication between units, and the identification of exceeding the TO (*time out*) time limit during unit communication.

Furthermore, the control unit implements 15 mandatory and 5 optional command sequences (Tables 4.8 and 4.9).

4.4 Communication in the HS-488 Interface

The fundamental disadvantage of both variants of the IEEE-488 interface is its limited transmission speed, which is nominally 1 MB/s, but in practice reaches a maximum of half the value. That is why National Instruments has developed a new data transfer algorithm called HS-488 (*High Speed GPIB*), which theoretically allows the transfer speed to be increased up to 8 MB/s.

The essence of this algorithm is the transition of *asynchronous data transmission* to *synchronous transmission* so that the transmitting unit can notify the receiving unit that it is able to transmit data according to the HS-488 standard and at the same time to indicate that the listener is able to confirm this possibility. If this capability is confirmed, then the transfer of further data already takes place synchronously and the DAV signal takes over the function of the clock signal. If the listener does not confirm readiness for synchronous data transfer, then communication between devices continues asynchronously according to the IEE-488 standard.

Figure 4.5 shows the time diagram of the transmission cycle during asynchronous data transmission.

The implementation of the HS-488 interface is based exclusively on a circuit solution, usually using the TNT 4882 C controller from *National Instruments*, which also allows easy implementation of its software.

The advantages of data transfer according to the HS-488 standard are particularly evident when transferring large data files, e.g., larger than 10 kB, while the acceleration of the transfer of smaller files is practically negligible compared to asynchronous transfer. For example, when using a plug-in board with an IEEE-488.1 interface in a computer with a PCI bus, transfer speeds of up to 0.7 MB/s can practically be achieved. Transfer speeds of up to 7.5 MB/s can be achieved when using a plug-in board with the HS-488 interface plugged into the same bus.

Table 4.8 Mandatory sequences of the IEEE-488.2 standard.

Sequence	**Importance**
SEND	Sending a device message to the selected device
SEND COMMAND	Sending a sequence of messages
SEND SETUP	Addressing selected devices to receive and control unit to transmit device messages
SEND IFC	Sending IFC signal (can only be done by the system control unit)
SEND DATA BYTES	Sending device message sequences of data bytes on the bus
RECEIVE	Receiving an instrument message from the specified instrument
RECEIVE SETUP	Addressing the selected device for transmission and the control unit for reception
RECEIVE RESPONSE MESSAGE	Receipt of a device message from the device addressed for transmission by previous sequences
DEVICE CLEAR	Putting all or selected devices into active state
ENABLE LOCAL CONTROLS	Putting all or selected instruments into local control state (LOCS or LWLS)
ENABLE REMOTE	Putting the entire system or selected devices into remote control state (*Remonte State*)
SET RWLS	Putting selected devices into remote control status with blocking
SEND LLO	Sending a multi-wire LLO (local lock out) message
READ STATUS BYTE	Determining the status byte of the specified device by serial message
TRIGGER	Group launch of selected devices
PASS CONTROL	Transfer of proceedings
PERFORM POLL CONFIGURE	Configuration of selected devices for parallel reporting
PERFORM POLL UNCONFIGURE	Unconfiguring all devices for parallel reporting
RESET	It initializes the system at three levels
BUS RESET	Execution of the ENABLE REMOTE and SEND IFC sequences
MESSAGE EXCHANGE RESET	Sending a DCL message
DEVICE RESET	Sending a general RST command
ALLSPOLL (*Serial Poll All Devices Protocol*)	Determines the status bytes of the devices using the serial reporting method

Table 4.9 Optional sequences of the IEEE-488.2 standard.

Sequence	Importance
FINDRQS (*find device requesting service protocol*)	Transferring control to another control unit
PASSCTL (*pass control*)	Request to take over the proceedings
FINDLSTN (*find listeners protocol*)	Identification of all IEEE-488.1 device addresses
SETADD (*set address protocol*)	Detection of address – configurable devices and their addressing
TESTSYS (*self-test system protocol*)	Running autonomous device tests

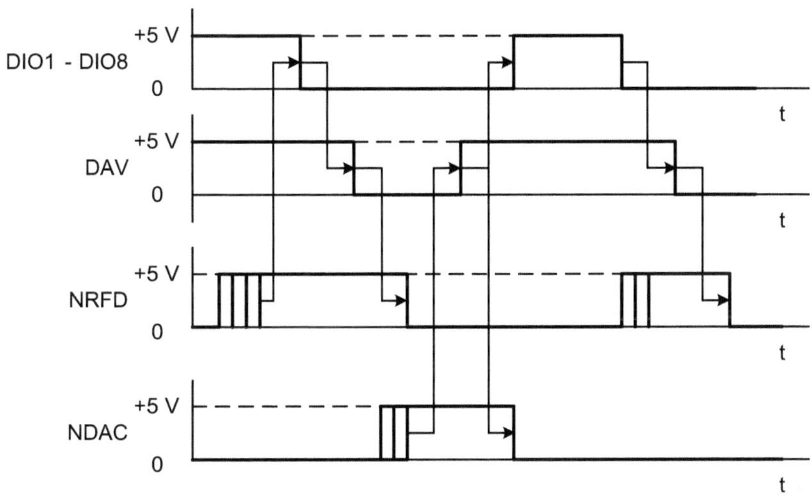

Figure 4.5 Timing diagram of asynchronous transmission cycle.

4.5 IEEE-488 Plug-in Cards

The properties of the IEEE-488.1, IEEE-488.2, and HS-488 interfaces are defined by the respective standards, which is why these interfaces are commonly supplied, for example, in the form of plug-in cards for personal computers.

Plug-in cards for computers are produced either in a version with internal EPROM memory, in which the driver (*Driver*) is loaded, ensuring communication between the user program running in the computer and the bus, or they are produced without memory, in which the driver is stored in the hard memory of the computer, from which it is then transferred to the computer's operating memory.

Typical plug-in GPIB boards for computers that do not have EPROM memory are, for example, GPIB-PCII boards from *National Instruments*, which also enable communication in the IEEE-488.2 standard.

In the version for PC XT/AT computers, these boards reached a data transfer speed of up to 400 kB/s, and in the PC AT version up to 1 Mb/s. These boards had adjustable I/O port base addresses of 2B8H or 3E1H, five interrupt levels ranging from IRQ2 to IRQ7, and three fast DMA (*direct memory access*) channels. These boards came with NI 488.2 software, which was the world standard in the field. It contained *Language Interfaces* for the languages *QuickBasic* and *Microsoft C* and the universal language interface UL1, enabling the programming of boards in any language using standard I/O operations.

Other frequently used plug-in boards were *Keithley's* PC 488 boards, containing a 16 kB EPROM memory, while its own utility occupied approximately 12 kB. In the memory address space, this program was located in the range of segmented addresses C000H–EC00H. For these boards, the base address of the I/O ports can be set in the range of 208H–2B8H, the trigger level in the range of IRQ2–IRQ7, and a maximum of three DMA channels can be used. The maximum transfer speed of these plates reached up to 80 kB/s.

The software of the boards also allows the use of all advanced programming languages supported by *Borland* and *Microsoft*, GPIB-PCII emulation programs, utilities for redirecting serial and parallel interfaces CM1, COM2, LPT1, and LPT2, PC interface to IEEE-488, and auxiliary test and configuration programs. The boards allow the use of more extensive programs for analysis and monitoring of the IEEE-488 bus and support for the Microsoft Windows program in the form of DLL libraries. In modern computers with a USB serial interface, these boards can be replaced with a USB-IEEE-488 converter.

Figure 4.6(a) shows *Keithley's* KPCI-488LPA plug-in card and *Agilent's* Type 82 357B USB-IEEE-488 converter.

4.6 Extension of the IEEE-488 Measurement System

The standard number of connected participants (devices) to the IEEE-488 bus, amounting to 15 devices, can be expanded by inserting expanders, which can be connected in series or parallel to the primary (main) bus

The series connection of bus expanders allows one device connected to the *primary bus* to be connected to an expander, which creates a *secondary bus*, to which either another 15 devices can be connected, or other expanders

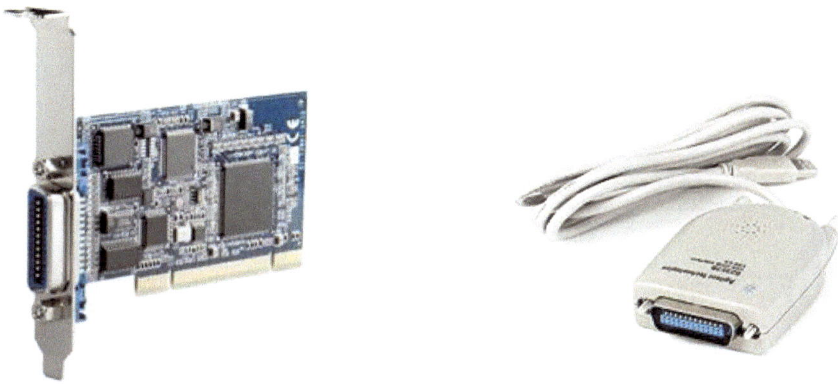

a) 488LPA plug-in card b) USB converter–IEEE-488

Figure 4.6 KPCI plug-in card and USB-IEEE-488 converter.

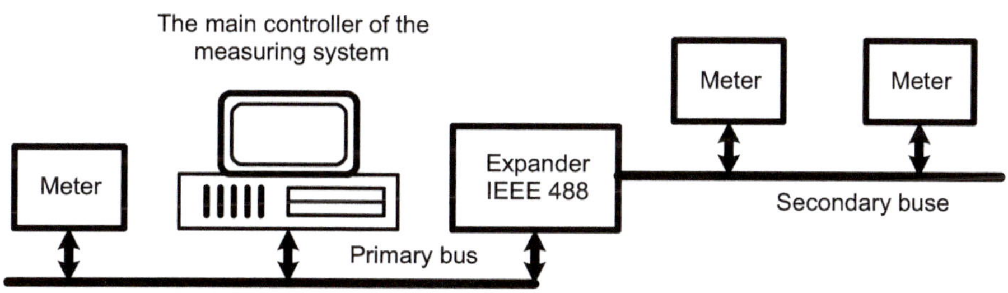

Figure 4.7 Serial connection of IEEE-488 bus expanders.

can be connected to it again, and another 15 devices can be connected to each of them again (Figure 4.7).

The disadvantage of this expansion is the different reaction times between the devices connected to the primary bus compared to the reaction times of the devices connected to the secondary buses, while as the number of expanders connected in series increases, the reaction time of the devices connected to their buses decreases.

The *parallel connection* of the bus expanders makes it possible to connect several expanders to the primary bus of the measuring system in addition to devices, to whose secondary buses up to 15 other devices can be connected (Figure 4.8).

With this method of bus expansion, devices connected to the main bus always have shorter reaction times compared to devices connected to the secondary busses of the expanders.

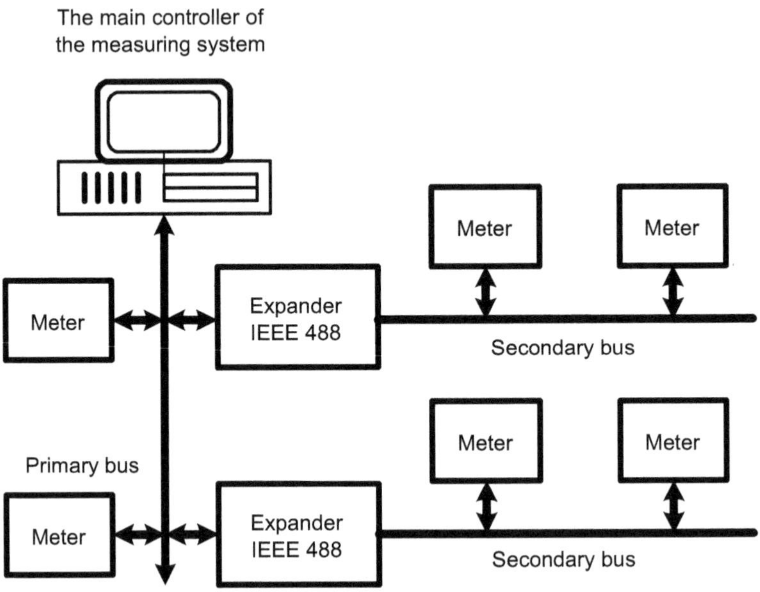

Figure 4.8 Parallel connection of IEEE-488 bus expanders.

Therefore, devices requiring the transfer of a large volume of data, e.g., digital oscilloscopes, are always connected to the primary bus, and devices that do not require the transfer of a large volume of data, e.g., digital voltmeters, are connected to the secondary buses.

Another option for expanding the number of connected devices with the IEEE-488 interface is the use of bus expanders and IEEE-488–USB converters.

A typical IEEE-488 bus expander is, for example, the *National Instruments* GPIB-120A expander, which contains two independent *galvanically separated* IEEE-488 buses, which enables IEEE-488 measurement systems to be expanded to up to 28 participants. Another advantage of this system is the galvanic separation of the two buses from each other, while the insulation strength of this separation is at least 60 V. This expander can also be used to extend the distance between individual devices up to 40 m.

Another option for expanding the number of connected devices with the IEEE-488 bus is the use of IEEE-488 – USB converters, a typical representative of which is, for example, the USB 4388 converter, which uses *Microsoft Windows 2000* and *Windows 10* operating systems in 32- and 64-bit versions for its operation (Figure 4.9).

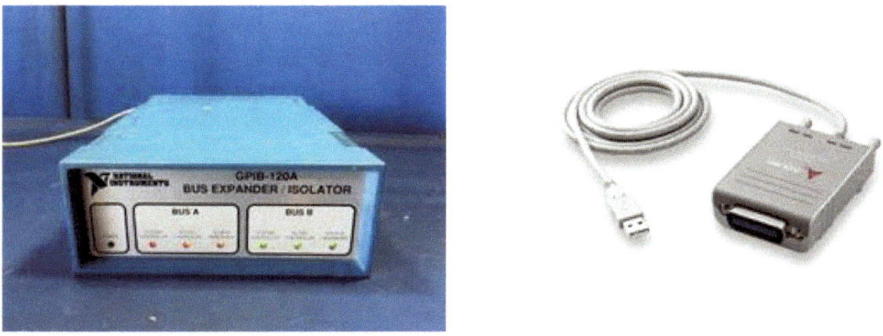

a) IEEE-488 bus expander

b) IEEE-488 - USB converter

Figure 4.9 IEEE-488 isolated bus expander and IEEE-488 –USB converter.

References

[1] Oebelin: *Measurement systém application and design*. Mc.Graw Hill Publishing Company 1990, ISBN 0-07-017338-9.

[2] Garrett, P. H. *Advanced Instrumentation and Computer I/O Design*: Real-Time System Computer Interface Engineering, New York: IEEE Press, 1994. ISBN 978-07803-6013-6

[3] Tooley, M. *PC-Based Instrumentation and Control,* 2nd ed., New York: Elsevier, 1995. ISBN 9780080479903.

[4] IEEE-488. *StandardDigitalInterfaceforProgrammableInstrumentation,* New York. IEEE Press, 1992. ISBN 978-0-7381-3934-0.

[5] Boyes, W., (ed.), *Instrumentation Reference Book*, 3rd ed., Boston, MA: Butterworth-Heinemann, 2003. ISBN 978-00804-78531.

[6] Park, J., Mackey, S. *Practical Data Acquisition for Instrumentation and Control Systems,* Amsterdam, the Netherlands: Elsevier, 2003. ISBN 978-07506-57969.

[7] Nawrocki, W. *Measurement systems and sensors*. Second edition, Artec House 2016. ISBN 1-58053-945-9.

[8] *The Measurement and Automation*, Naional Instruments, Catalogs, 2022.

[9] keithley.com

[10] keysight.com

5

Plug-in Card Measuring Systems

In the early 1980s, the effort to economically use personal computers to collect and process signals in measurement and control technology led to the construction of data acquisition systems based on plug-in cards that are inserted into computer slots.

With the use of the now obsolete ISA buses with a transfer speed of 8 Mb/s, it was possible to achieve actual transfer speeds of several Mb/s. With the more modern *PC Card* and *PC Card Bus* buses still used today, transfer speeds of up to 20 and 132 Mb/s can be achieved, while the actual transfer speeds are always approximately half.

From the point of view of use, we distinguish between single-purpose cards, designed, for example, for digitizing very fast signals, and multifunctional cards designed for multi-channel signal collection and processing, especially signals from typical sensors of physical quantities (temperature, pressure, etc.). The assortment of plug-in cards is further expanded to include types that allow measuring direct and alternating voltages and currents, frequency and time parameters of signals, generating harmonic and modulated signals, performing their amplitude or complex spectral analysis, processing signals of resistive, incremental and coded position sensors, vibration sensors, mechanical stress, pressure, flow and resistance, semiconductor and thermoelectric temperature sensors.

In the following chapters, the properties of typical plug-in cards are presented, the principle of operation of which is similar for the functional units of centralized and distributed measuring systems.

5.1 Plug-in Cards for Fast Signals

The block diagram of the plug-in card (functional module) intended for the digitization of time-varying signals is shown in Figure 5.1.

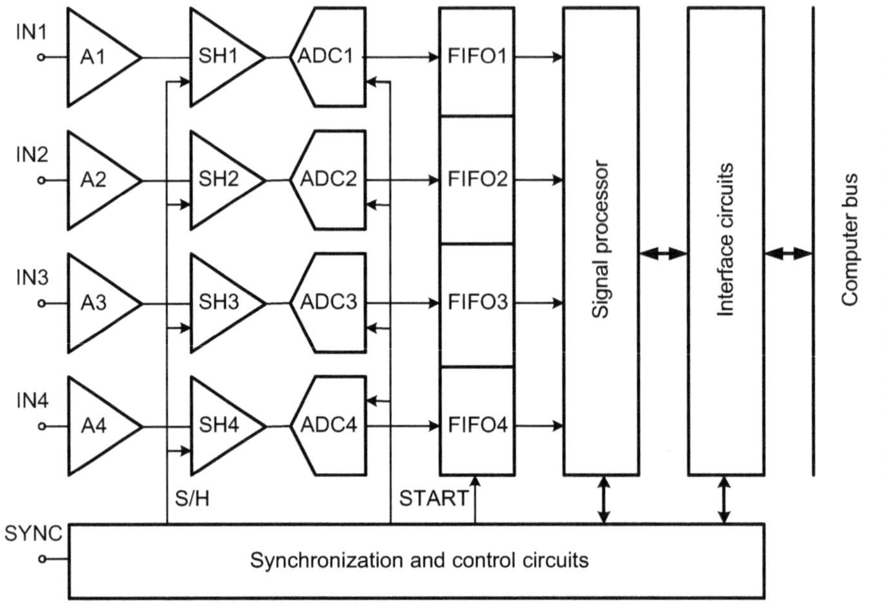

Figure 5.1 Block diagram of a fast four-channel plug-in card.

In these cards, signals are digitized by separate extremely fast S/H sampling circuits and *flash* or *cascaded A/D converters* with 8-bit resolution, enabling signal sampling with sampling rates up to tens of MS/s.

The input voltage range of these cards is typically ±1 to ±5 V. The input impedance of these cards consists of switchable input resistances of 50 Ω or 1 MΩ and input capacitances in the range of 15–25 pF.

Digitized data is written into fast FIFO (*first in, first out*) type buffers with a capacity of several tens of MB. After possible digital pre-processing of the digitized signal by signal processors or gate arrays, the data is transferred to the computer via interface circuits. Synchronization circuits are also part of these boards, enabling the synchronization of the start of sampling circuits and A/D converters in the individual channels of the board. Since the signals are sampled at the same instants of time for these cards, there is no time shift of the sampling instants of the signals in their channels when the signals are sampled.

Figure 5.2 shows a typical PCI plug-in fast measuring card with a frequency range of 100 MHz, containing four 8-bit A/D converters with sampling rates up to 250 MS/s and with an internal data memory of 64 kB.

These cards, sometimes referred to as *oscilloscope cards*, require oscilloscope probes to connect to signal sources with high input impedances.

Figure 5.2 Fast four-channel PCI plug-in card.

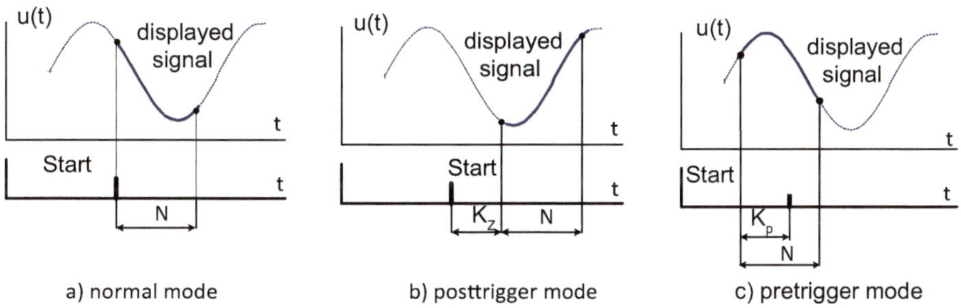

Figure 5.3 Methods of writing data to memory.

5.1.1 Methods of writing data to memory

Due to the method of writing the data of A/D converters into the FIFO memories, in which the data is gradually renewed, three methods of displaying digitized signals are used (Figure 5.3).

In *normal mode*, the time courses of the signals are displayed, starting with the arrival of the trigger pulse. Usually, only a part of the waveforms with a capacity of N samples is always displayed, while the rest of the data is available based on a selection from the data window. If the data memory is large enough, then its entire content can be viewed continuously in *rolling mode*.

In *posttrigger mode*, those parts of the waveforms whose beginning is time-delayed by the N_D of the samples relative to the trigger pulse are displayed.

When writing in advance (*pretrigger mode*), those parts of the waveforms whose beginning precedes the N samples relative to the trigger impulse

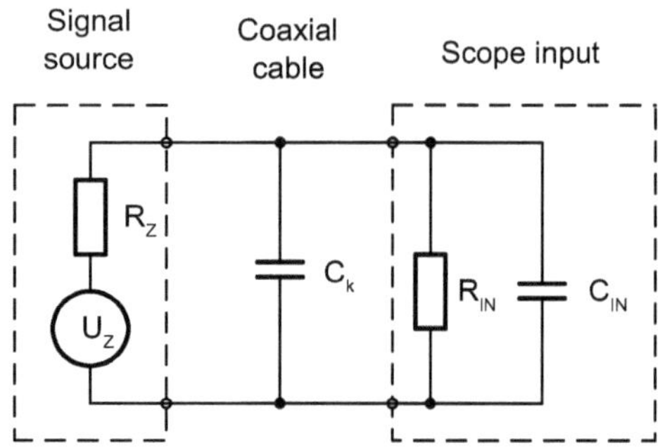

Figure 5.4 Connecting the signal source to the input of the amplifier.

are displayed. This way of displaying waveforms is suitable in those cases where we need to capture signal waveforms before the arrival of the synchronization pulse.

5.1.2 Ways of connecting signals to fast cards

The direct connection of signals to the inputs of these cards with a *coaxial cable* is in principle unsuitable for several reasons. If, for example, the source of the measured signal has an internal resistance R_z and the drain capacitance of the connecting cable is C_k and the input resistance and capacitance of the amplifier are R_{IN} and C_{IN}, then the corresponding time constant of the circuit

$$\tau_m = \frac{R_{IN} \cdot R_z}{R_{IN} + R_z}\left(C_{IN} + C_s\right) \tag{5.1}$$

determines its cutoff frequency $f_m = 1/(2 \cdot \pi \cdot \tau_m)$, at which a drop in the amplitude frequency characteristic of the circuit occurs by 3 dB, i.e., to $1/\sqrt{2}$ of the DC voltage transfer value (Figure 5.4).

If, for example, the input resistance and capacitance of the amplifier $R_{IN} = 1$ MΩ and $C_{IN} = 20$ pF, the leakage capacitance of the cable $C_k = 50$ pF and the internal resistance of the signal source $R_z = 100$ KΩ, then the time constant of the circuit $\tau_m = 7$ μs which corresponds to its cutoff frequency is $f_m = 23$ kHz which reduced when the DC voltage is reduced to 0.9.

At the input resistance of the amplifier $R_{IN} = 50$ Ω, although the time constant of the circuit is reduced to 3.5 ns and its cutoff frequency increases to 45 MHz, the DC voltage transfer drops to a value of $5 \cdot 10^{-4}$. Therefore, voltage probes are used for measurements on sources with high internal resistance, reducing the influence of the internal resistances of the signal sources on the frequency range of the measurement.

The most commonly used voltage probes are passive probes (*voltage passive probe*), which contain the upstream member R_1, C_1, which with the

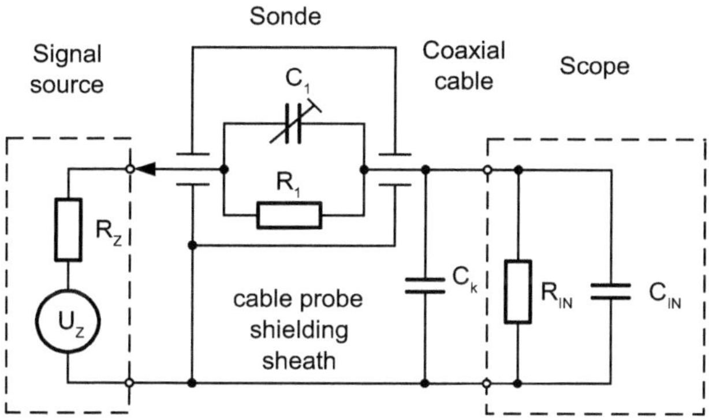

Figure 5.5 Connecting the passive voltage probe to the measured voltage source.

input resistance R_{IN} and the capacity C_{IN} of the amplifier forms a voltage divider with a division ratio

$$p = \frac{R_{IN}}{R_{IN} + R_1}, \tag{5.2}$$

which is usually scaled decimally in ratios of 1:10, 1:100, and 1:1000 (Figure 5.5).

If $R_1 \cdot C_1 = R_{IN} \cdot (C_{IN} + C_k)$, then this voltage divider is frequency independent and the time course of the measured voltage is amplified without distortion.

The frequency compensation of the probes is set by changing the capacity of C_1 when the probe is connected to the rectangular signal generator. If the condition $R_1C_1 = R_{IN} \cdot (C_{IN} + C_k)$ is satisfied, the probe is frequency compensated and the rectangular voltage waveform of the signal is amplified without distortion. When $R_1 \cdot C_1 > R_{IN} \cdot (C_{IN} + C_k)$, the probe is frequency *undercompensated* and the rectangular voltage waveform of the signal is displayed with exponential overshoots. When $R_1 \cdot C_1 < R_{IN} \cdot (C_{IN} + C_k)$, the probe is frequency *overcompensated* and the rectangular voltage waveform of the signal is displayed with exponential docmits (Figure 5.6).

To assess the effect of the input capacitance of the probe on the measured object, its *equivalent input capacitance* is given

$$C_e = \frac{C_1 \cdot (C_{IN} + C_k)}{C_1 + C_{IN} + C_k)} \approx C_1, \tag{5.3}$$

which is approximately equal to the capacity of C_1, since the capacities of C_{IN} and C_k are at least an order of magnitude greater than the capacity of C_1. This capacity, together with the internal resistance of the signal source

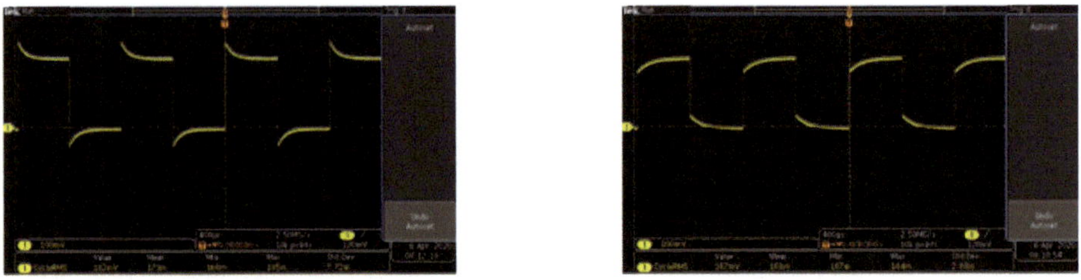

a) undercompensated probe b) overcompensated probe

Figure 5.6 Setting the frequency compensation of the probe.

R_z, determines the time constant of the measuring circuit $\tau_e = Rz \cdot C_e$ and its related limiting frequency $fe = 1/2\pi\tau_e$.

With the same input parameters of the amplifier R_{IN} = 1 MΩ, C_{IN} = 20 pF, drain capacitance of the cable C_k = 50 pF and with a probe division ratio of 1:10, the output resistance in the probe is R_1 = 9 MΩ and capacitance C_1 = 2.2 pF.

The equivalent input resistance of the probe and its capacity is then R_e = 10 MΩ, C_e = 2 pF. With the internal resistance of the source of the measured signal R_z= 100 kΩ, the time constant of the circuit is τ_e = 200 ns and the cut-off frequency is f_e = 796 kHz. With the internal resistance of the source R_z = 1 kΩ, the time constant of the circuit is τ_e = 2 ns, which corresponds to the limiting frequency of the circuit fe = 79.6 MHz.

Figure 5.7 shows the frequency characteristics of the input impedance of the probes for their different equivalent input resistances and capacities.

It follows from the course of the frequency characteristics that as the equivalent input resistances of the probes R_e increase, their equivalent input capacities C_e decrease and their limiting frequencies fe increase. Probes with low input resistances, on the order of hundreds of Ω, reach cutoff frequencies of hundreds of MHz to units of GHz

Knowing the limit frequency of the f_{ms} ·probe, its *rise time* can be determined

$$T_{ns} = \frac{0.35}{f_{ms}}. \tag{5.4}$$

If the amplifier has a cutoff frequency of f_{mz}, which corresponds to the rise time

$$T_{nz} = \frac{0.35}{f_{mz}}, \tag{5.5}$$

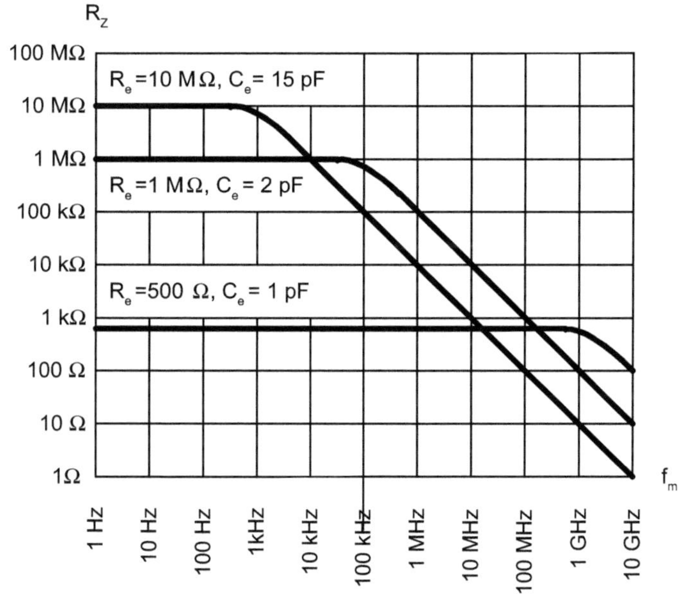

Figure 5.7 Frequency characteristics of probe input impedances.

then the *resulting rise time* of the probe amplifier is approximately given by

$$T_n \approx \sqrt{T_{ns}^2 + T_{nz}^2}, \tag{5.6}$$

It then applies to the corresponding resulting cutoff frequency of the amplifier with the probe

$$f_m \approx \frac{1}{\sqrt{1/f_{ms}^2 + 1/f_z^2}}. \tag{5.7}$$

If, for example, a probe with cutoff frequency f_{ms} = 100 MHz is connected to an amplifier with cutoff frequency f_{mz} = 100 MHz, then the resulting cutoff frequency of this arrangement is 70.7 MHz, which corresponds to a rise time of 4.9 ns.

So that the frequency range of the probes does not significantly affect the frequency range of the cards, we choose their frequency range at least five times the frequency range of the amplifier.

Low-voltage passive probes with switchable division ratios of 1:1 and 1:10 are manufactured up to frequency ranges of several GHz. In addition to these passive probes with input resistances of 10 and 100 MΩ, *active voltage probes* containing fast amplifiers with input resistance up to 1 GΩ and

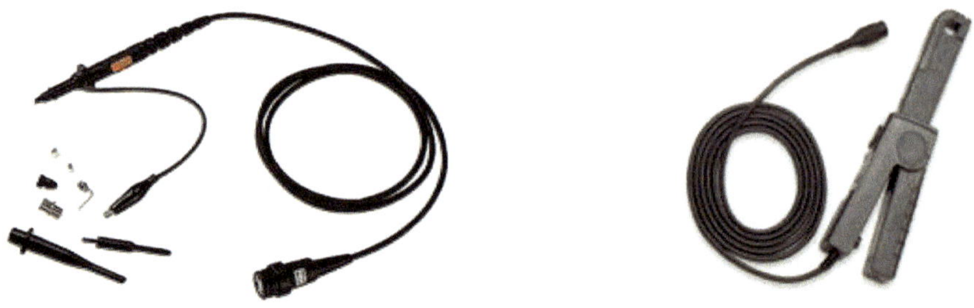

Figure 5.8 Oscilloscope probes.

switchable gain in the ranges 1:10 and 1:100 are used to measure voltage on high-impedance circuits. The power supply of these probes is solved by an external power source, galvanically separated from the signal ground of the probes. To measure small voltages with a high positive component, *differential probes* are used, allowing to measure differential voltages in the range of tens of mV with a positive voltage of the order of V, or tens of V.

To measure direct and alternating currents, *current probes* with a *Hall probe* are used, allowing to measure currents in the range of units of mA to hundreds of mA in the frequency range up to 120 MHz (Figure 5.8).

5.2 Multiplexed Plug-in Cards

These cards, also referred to as DAQ (*data acquisition*) cards, enable the digitization of usually 8 differential or 16 asymmetric voltage signals, which after amplification and filtering are switched by an *analog multiplexer* to an *instrument amplifier* with a binary or decade switchable gain in the range of 1–1024 or $1–10^3$. These cards use *approximation A/D converters* with an internal sampling circuit to digitize the signals, allowing signals to be sampled up to several hundred kS/s (Figure 5.9).

Data from the A/D converters are written into a *data memory* with a capacity of up to several hundred kB, from which they are transferred to a computer via interface circuits. Some cards also contain several D/A converters with voltage or current outputs, which can be used, for example, to drive voltage or current lines in the voltage range ±5 V, ±10 V, ±10 mA, 4 mA to 20 mA. Furthermore, these cards can also contain several programmable I/O circuits, 16-bit timers, and counters with a frequency range of tens of kHz.

Higher-quality cards communicate with the computer through a galvanically isolated interface, using *capacitively coupled isolation* members

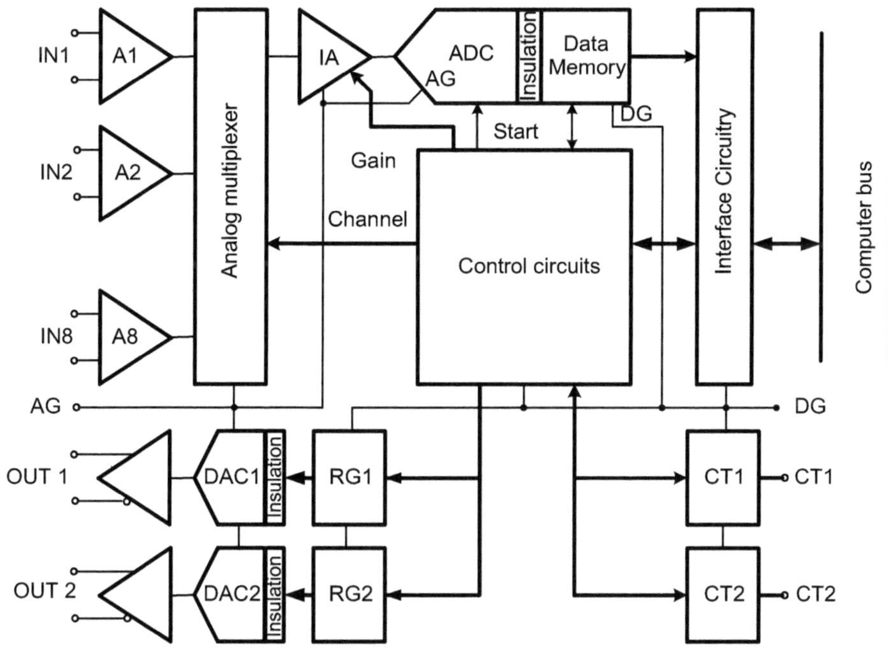

Figure 5.9 Block diagram of a multiplex plug-in card.

for data transfer, allowing data to be transferred at a transfer rate of up to several Mb/s. For these cards, we distinguish between two types of signal countries. The analog ground (AG) refers to the signals of multiplexers, amplifiers, and sampling circuits, including the analog parts of A/N and N/A converters.

Digital ground (DG) refers to the signals of digital circuits, i.e., control circuits, data memories, registers, and counters, including the digital parts of A/D and D/A converters. Isolated *DC–DC converters* are used to galvanically separate both countries.

To galvanically isolate the sources of the measured signals from the analog ground of the cards, the cards can be retrofitted with external isolation amplifiers with transformer or optical coupling. The *insulation strength* of these amplifiers ranges from a few kV to tens of kV, and their frequency range is up to hundreds of kHz.

Figure 5.10 shows a typical 16-channel multiplex measurement plug-in card with a 12-bit A/D converter with a sampling rate of up to 200 kS/s, a program-controlled instrumentation amplifier with a gain in the range of 1–100, two 12-bit D/A converters, and eight digital I/O outputs and four counters and timers.

Figure 5.10 Plug-in multiplex 16-channel card PCI-1602U CR.

5.2.1 Measurement channel switches

Single-pole, two-pole, three-pole, and four-pole switching are used to connect voltage signals to the inputs of multiplexers.

In SMM (*single mode multiplexing*), the signal grounds (SGs) of the signal sources are connected to the analog ground of the board AG, and the outputs of the multiplexers are connected to one of the inputs of the instrument amplifier (Figure 5.11(a)).

This method of switching is suitable for less demanding measurement applications, as interfering signals can be induced into the common signal and analog ground, which directly affect the measurement results.

In DMM (*double mode multiplexing*) two-pole switching, the signal grounds (SGs) of the signal sources are connected to the analog ground of the board AG, but a pair of multiplexers are used to switch the signals, which switch the differential signals of these signal sources to the differential inputs of the instrument amplifier (Figure 5.11(b)).

This method of connection is suitable for more demanding measuring applications, because common mode voltage components u_C of sources of differential voltages u_d are suppressed by the differential instrument amplifier by the common mode rejection ratio (CMMR), which for instrument amplifiers ranges from 90 to 120 dB in frequency range of tens to hundreds of kHz.

A more perfect way of connecting signals to the instrument amplifier allows their *three-mode multiplexing*, in which the signal ground of the signal sources is also switched to the analog ground of the multiplex system (Figure 5.11(c)). The advantage of this connection method is that different common voltages of the switched signals can be suppressed by different

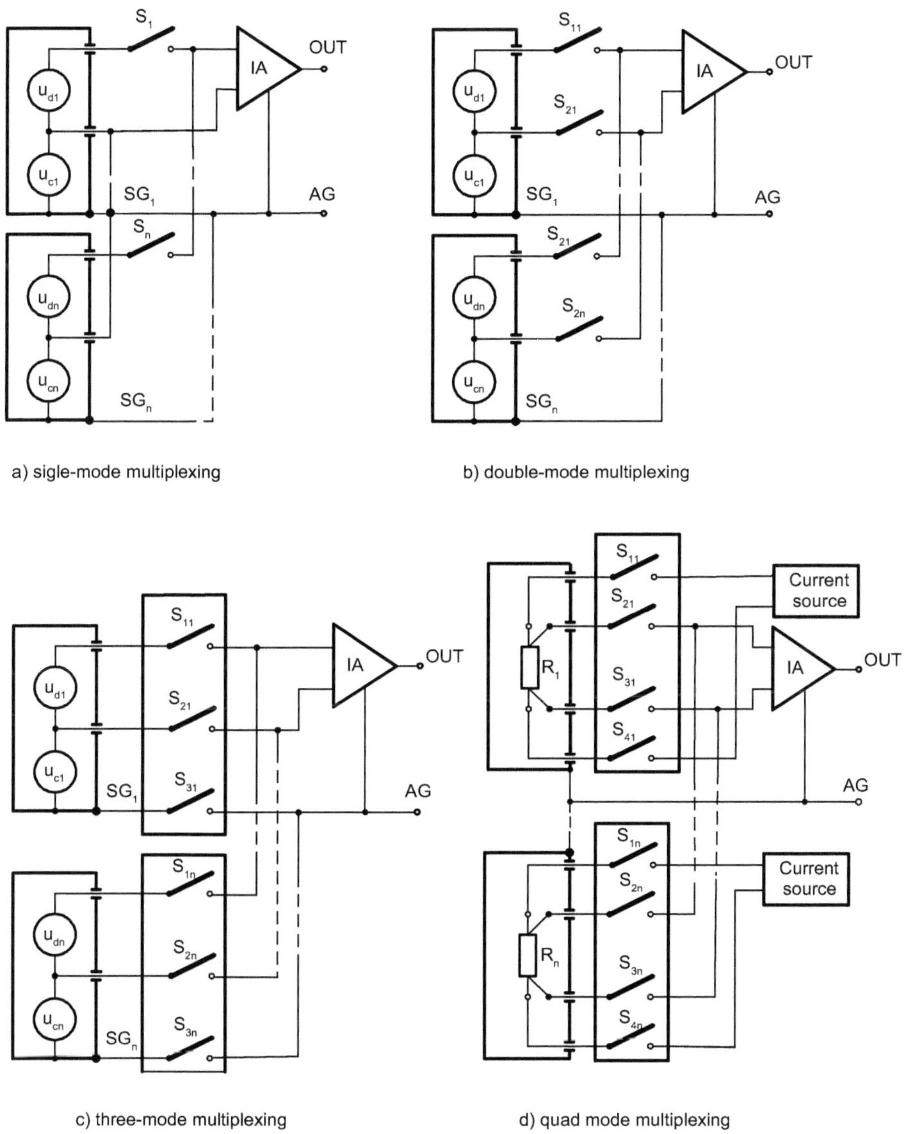

Figure 5.11 Modes of switching signals in a multiplex card.

common voltage suppression factors of the instrument amplifier depending on its differential gain.

Quadrupole switching QMM (*quad mode multiplexing*) signals is used, for example, to measure voltage drops on resistance temperature sensors, which are connected to current sources with their two current terminals and

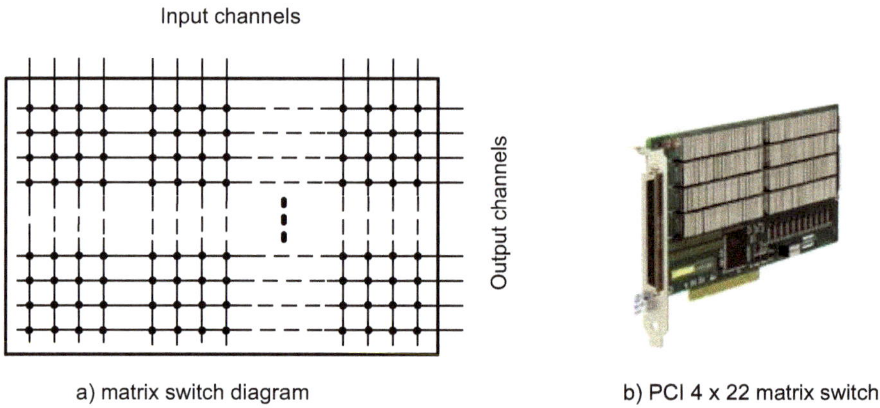

Figure 5.12 Matrix signal switches.

their voltage (sensing) terminals are connected to the instrument amplifier (Figure 5.11(d)). Given the high input resistance of the instrument amplifier, this method of connection enables complete suppression of the influence of the resistances of the supply wires of the current source to the current terminals of the sensors.

Matrix switch MMs (*matrix multiplexers*) are used for multiplexing a larger number of signals, enabling up to 64 signals to be switched in a four-pole connection method. A typical representative of these matrix switches is, for example, the PCI 50-510A plug-in unit from *Pickering* (Figure 5.12).

Multiplex plug-in cards use either contact or *solid-state switches* to switch signals. Semiconductor switches contain CMOS circuits (*complementary metal oxide semiconductor*), whose resistances in the closed state are in the range of tens to hundreds of Ω and whose isolation currents in the open state reach tens to hundreds of pA. The switching times of these semiconductor switches reach hundreds of ns to units of μs.

Contact multiplexers are made up of *reed relays*, whose contacts can be stored in a mercury atmosphere in order to achieve small voltages in the closed state. That is why these multiplexers are used, for example, to switch the voltage of thermoelectric cells.

The resistances of the contact switches in the closed state are several mΩ and in the open state several hundred MΩ. Their switching times reach hundreds of ms and their lifetime is up to several tens of millions of switches.

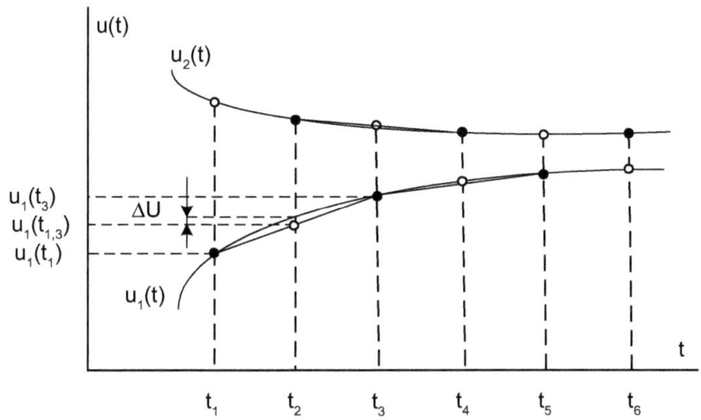

Figure 5.13 Time course of sequentially switched signals.

5.2.2 Elimination of channel switching errors

In multiplex measurement systems, channels are usually switched according to a pre-prepared algorithm, which also includes a possible change in the gain of the differential amplifier. When switching signals at the input of the multiplexer in the so-called *scanning mode*, there is a time delay in the moments of signal sampling in individual channels.

This phenomenon can be eliminated either by connecting a set of sampling circuits in front of the multiplexer and sampling the signals with a common sampling signal, or from adjacent signal samples in individual channels, signal samples that correspond to the sampling times of signals in the other channels can be determined using the linear regression method. Figure 5.13 shows the possibility of determining signal samples in two sequentially switched channels.

With temporally equidistant signal switching, the value of a signal sample can be determined by the mean value of neighboring signal samples

$$u(t_{1,3}) = \frac{u(t_1) + u(t_2)}{2}, \tag{5.8}$$

The error of the samples ΔU determined in this way is determined not only by the uncertainty of the sampling moments but also by the time change of the switched signals.

This method of suppressing the time delay of signal sampling moments in individual channels can be used, for example, when measuring performance parameters.

5.3 Counter and Timer Plug-in Cards

Most multiplex cards contain, in addition to analog channels, several digital channels, which can be used to a limited extent for measurement, or to the generation of impulse signals with variable periods, intervals, and frequencies.

These cards contain 4–8 separate channels, formed by digitally controlled frequency dividers and counters, an internal crystal controlled CLK clock pulse generator, whose temperature and time stability determine the accuracy of determining time and frequency parameters. Control and transfer of data from frequency dividers and counters are ensured by their controllers, which transfer this data to the computer bus through the board interface (Figure 5.14).

Counter and timer cards are used not only for digital measurement of time and frequency parameters of impulse signals but also for evaluating signals from incremental position sensors and measuring the number of revolutions with inductive and optical sensors.

5.3.1 Numerical measurement of time intervals

During the numerical measurement of time intervals, the output signal of the *frequency divider* with the division ratio N is connected to the gating input of the GATE counter, and the pulse signal thus created with the frequency $f_{OUT} = f_0/N$ forms the time base of the period meter.

If the time of the measured time interval is T_p, then after the arrival of the first rising edge of this signal, the counter starts counting pulses with the frequency f_{OUT}, and after the arrival of the first falling edge of the measured pulse signal, the counting of these pulses ends. The state of the N_x counter then determines the duration of the measured time interval

$$T_x = \frac{N \cdot N_x}{f_o} \pm \Delta T_x, \tag{5.9}$$

where $\Delta T_x = N/f_o$ is the quantization error of the time interval measurement, given by the asynchronous arrival of the rising and falling edges of the measured pulse signal with respect to the pulses of the frequency base (Figure 5.15).

The uncertainty of the measurement of time intervals is determined not only by the relative value of the quantization error $\delta T_x = \Delta T_x/T_x = 1/N_x$ but also by the instability of the oscillator frequency δf_0 and by the change in the decision levels of the input circuits of the IN and GATE counter signals

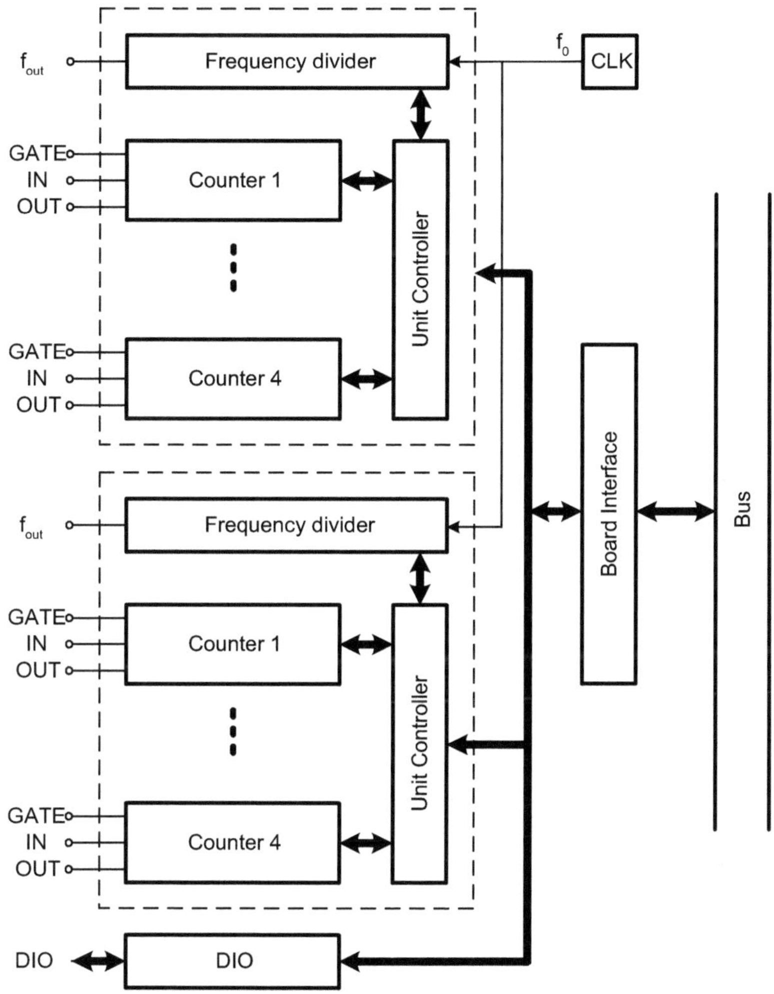

Figure 5.14 Block diagram of counter and timer card.

$$\delta T_x \approx \frac{\sqrt{\delta^2 f_0 + \delta^2 T_x}}{\sqrt{3}}. \tag{5.10}$$

Based on this estimate, it is necessary to select a frequency of the time base that is significantly higher than the reciprocal of the times of the measured intervals to measure the time intervals. For example, when measuring the time of a time interval of 100 ms using a time base frequency of 1 MHz, the time resolution of the measurement is 10^{-5}, corresponding to a quantization error of $\pm 5 \cdot 10^{-6}$.

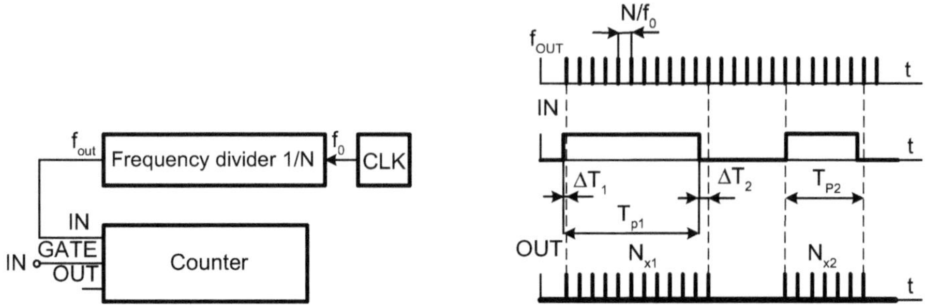

a) connection to measure b) course of signals

Figure 5. 15 Numerical measurement of time intervals.

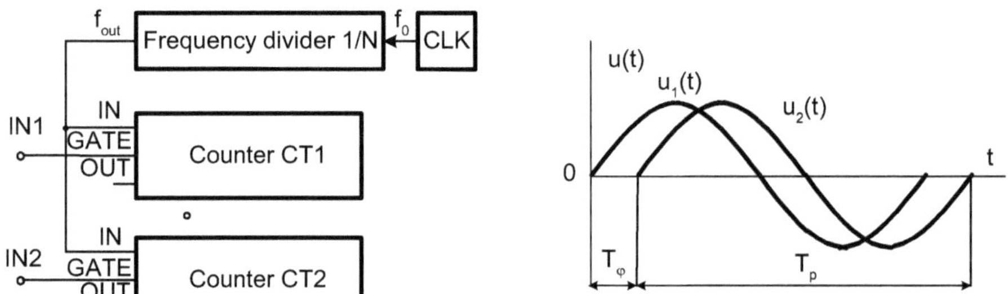

a) connection to measure b) course of signals

Figure 5.16 Numerical measurement of time differences between two signals.

The method of measuring time intervals can also be used to measure the time intervals of impulses between several signals. In this case, two counters are used, whose signal inputs are interconnected and connected to the output of the time base, and to whose gating inputs the measured signals are fed (Figure 5.16).

5.3.2 Numerical measurement of phase shift

A typical application of this method is the measurement of the *phase shift* between two harmonic signals, which is defined by the ratio of the time differences of the moments when these signals pass through the zero level with a positive time change. In this case, the phase shift, defined either in arc rate, i.e., in radians, or in angular degrees, is determined by the ratio of the states of the counters N_{x1} and N_{x2}

$$\varphi = 2\pi \frac{T_\varphi}{T_p}[\text{rad}] = \frac{T_\varphi}{T_p} \cdot 360[\text{grad}] = \frac{N_{x1}}{N_{x2}} \pm 1. \tag{5.11}$$

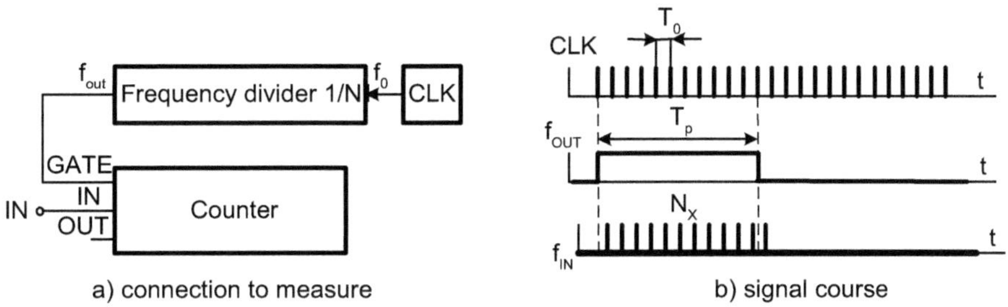

Figure 5.17 Numerical measurement of the frequency.

The uncertainty of the phase shift ratio measurement is approximately determined by the frequency instability δf_0 and the time instabilities δT_φ and δT_p.

$$\delta\varphi = \frac{\sqrt{\delta^2 f_0 + \delta^2 fT_\varphi + \delta^2 T_p}}{\sqrt{3}}. \tag{5.12}$$

5.3.3 Numerical measurement of frequency and frequency shares

In the numerical measurement of the frequency of impulse signals, the period of impulses of the time base $T_p = 1/(N \cdot f_0)$ determines the time during which the counter counts the impulses of the measured signal. The state of the counter then determines the frequency of the measured pulse signal

$$f_x = \frac{N_x}{N \cdot f_o} \pm \Delta f_x, \tag{5.13}$$

where the term $\Delta f_x = 1/(N.f_0)$ determines the quantization error of frequency determination (Figure 5.17).

For example, measuring the AC voltage frequency of the power grid, which is 50 Hz in European countries, would require a measurement time of 10 s with the required resolution of 5.10^{-2} Hz. The measurement result would then correspond to the mean value of the measured frequency during this time. Since frequency fluctuations are also measured in the energy network, which should not be greater than ±0.1 Hz, the reciprocal method is used in these measurement cases, which determines the frequency of the measured signal from the time of its periods, the resolution of which in this case should be less than 40 μs.

The uncertainty of determining the frequency is determined not only by the proportional value of the quantization error $\delta f_x = \Delta f_x/f_x = 1/N_x$ but also by

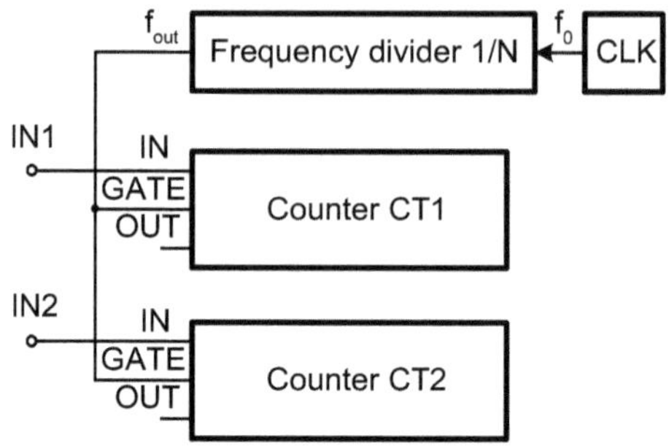

Figure 5.18 Numerical measurement of the frequency ratio.

the instability of the oscillator frequency δf_0 0 and by the change in the decision levels of the input circuits of the IN and GATE counter signals

$$\delta T_x \approx \frac{\sqrt{\delta^2 f_0 + \delta^2 f_x}}{\sqrt{3}}. \tag{5.14}$$

Two counters can also be used to *measure the frequency ratio of signals*, the gates of which are interconnected and connected to the output of the time base, and to whose signal inputs the signals whose frequency ratio is measured are fed. The ratio of frequencies measured in this way is determined by the ratio of states of counters N_{X1}/N_{X2} (Figure 5.18)

$$\frac{f_{x1}}{f_{x2}} = \frac{N_{x1}}{N_{x2}} \pm 1, \tag{5.15}$$

where the term ±1 determines the quantization error of the frequency ratio measurement.

The classic principle of digital frequency measurement determines the frequency of the measured signal by counting the number of periods during the gate opening time, controlled by the time base. After the gate is closed, the counter data is evaluated and displayed, and the measurement is interrupted during this process. The measured signal is actually sampled and information about its frequency is lost during the dead time.

5.3.4 Numerical measurement of frequency changes over time

To measure the *instantaneous frequency*, the reciprocal method are used. In this method, the measurement time is divided into closely connected time intervals chosen to correspond to an integer number of events, or the time between several zero crossings of the signal.

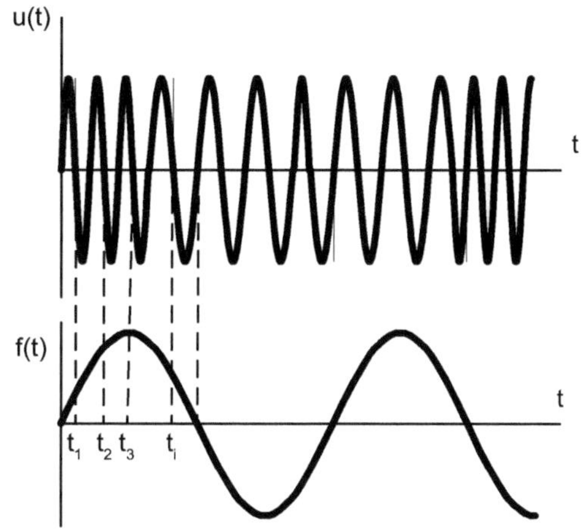

Figure 5.19 Time course of a frequency modulated signal.

The length of the time period that elapses from the start of the measurement to the end of the *i*th sampling interval is expressed by the number of periods T_0 of the reference frequency source, which is stored in the time counter. Similarly, the event counter accumulates the number of events $e(i)$ from the start of the measurement to the *i*th sampling interval. At the end of the *i*th sampling interval, a pair of data $e(i)$ and $t(i)$ are therefore available (Figure 5.19).

For example, when measuring the instantaneous frequency of a frequency-modulated signal with a harmonic signal, a measurement period close to the period of the minimum modulation frequency is chosen. The frequency estimate corresponds to the direction of the function $e(i) = f\,[t(i)]$.

$$f_{od}(i) = \frac{e(i) - e(i-1)}{t(i) - t(i-1)}. \tag{5.16}$$

The frequency estimate is limitedly close to the actual value of the instantaneous frequency, i.e., the derivative of its time course

$$\left(\frac{df}{dt}\right)_{t=t_i} = \sum_{j=1}^{M} \frac{e(i+j) - e(i+j-1)}{t(i+j) - t(i+j-1)}. \tag{5.17}$$

If, for example, the graph of the dependence $e(i) = \mathrm{f}\,[t(i)]$ is a straight line, then at a constant frequency of the signal, the regression line corresponds to this functional dependence (Figure 5.20).

$$f_r = \frac{\sum_{i=1}^{N} e(i)t(i) - \sum_{i=1}^{N} e(i) \sum_{i=1}^{N} t(i)}{N\sum_{i=1}^{N} t^2(i) - \left(\sum_{i}^{N} t(i)\right)^2}. \tag{5.18}$$

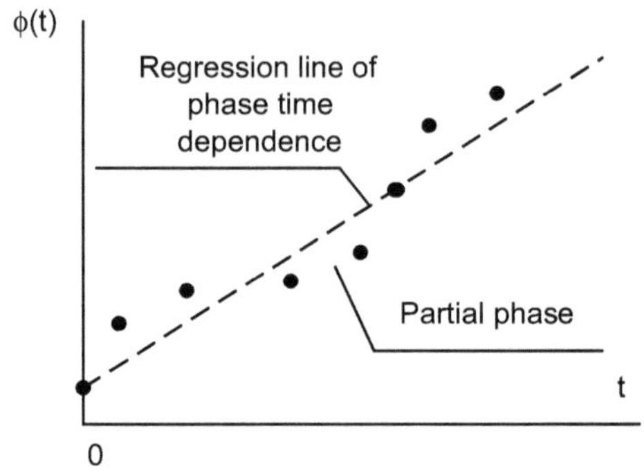

Figure 5.20 Phase progression of a frequency modulated signal.

Since this calculation is time-consuming, a simplified bicentroid method is used to determine the frequency, consisting of arranging the data into three approximately equal groups, with the first and last group having the same number of samples and the middle group being ignored. Furthermore, the mean values of $e(i)$ and $t(i)$ are determined for both groups of data. This creates the so-called center of gravity of each group. The frequency estimate then corresponds to the steepness of the line connecting both of these centroids

$$f_{\text{ctd}} = \frac{\sum_{i=1}^{Q} e(i)_{\text{pos}} - \sum_{i=1}^{Q} e(i)_{\text{prv}}}{\sum_{i=1}^{N} t(i)_{\text{pos}} - \sum_{i=1}^{N} t(i)_{\text{prv}}}. \tag{5.19}$$

The frequency calculation is thus reduced to only data summation and division. At the same time, the quality of the approximation does not deteriorate significantly, as can be seen from the comparison of variances of ordinary and *bicentroid regression*

$$\sigma_r = \frac{q\sqrt{12}}{\sqrt{NT_m}}, \qquad \sigma_{\text{ctd}} = \frac{q\sqrt{13{,}5}}{\sqrt{NT_m}}, \tag{5.20 a, b}$$

where q is the quantization error during time measurement, T_m is the measurement time, and N is the number of data.

5.4 External Plug-in Card Units

These units are intended for external connection of voltage and current signals of resistance and temperature sensors in those cases where the distance of direct connection of these signals to plug-in cards would significantly worsen the immunity of this connection against interference.

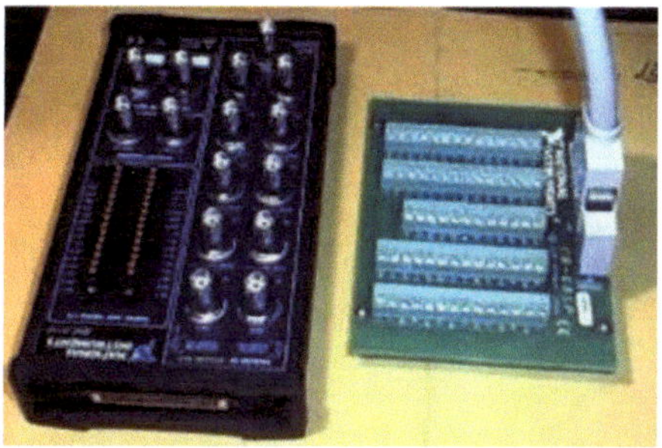

Figure 5.21 Types of DAQ connection units.

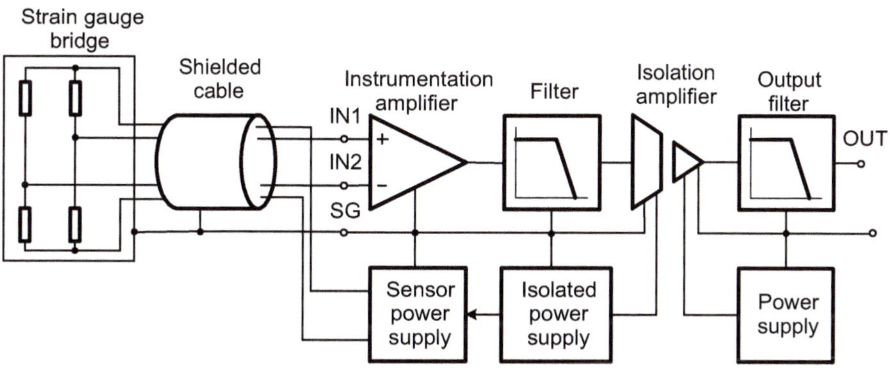

Figure 5.22 Block diagram of the measuring module with strain gauge resistance bridge.

These external *data acquisition units* (*DAQ Board*) contain, in addition to BNC connectors, intended for connection, and/or the generation of voltage and current signals, as well as serial connectors into which measuring modules (*condition modules*) can be inserted, designed to process signals from potentiometric position sensors, tensometric mechanical stress sensors and temperature resistance and thermocouple temperature sensors (Figure 5.21).

The block diagram of the measuring module of tensometric potentiometric and resistance temperature sensors is shown in Figure 5.22.

The input of the module consists of an *instrument amplifier* with protection circuits, ensuring bias protection of its inputs and a lowpass filter with adjustable cutoff frequencies in the range of 4 Hz to 10 kHz. The minimum voltage range of the instrument amplifier of 50 mV corresponds to the voltage range of the differential voltages of the *strain gauge bridge*, which in less demanding applications is powered by direct current and in more demanding applications by alternating current, the use of which removes the influence of direct interference signals, e.g., thermoelectric voltages, arising at different temperatures between individual connection of sensing resistors and connection wires.

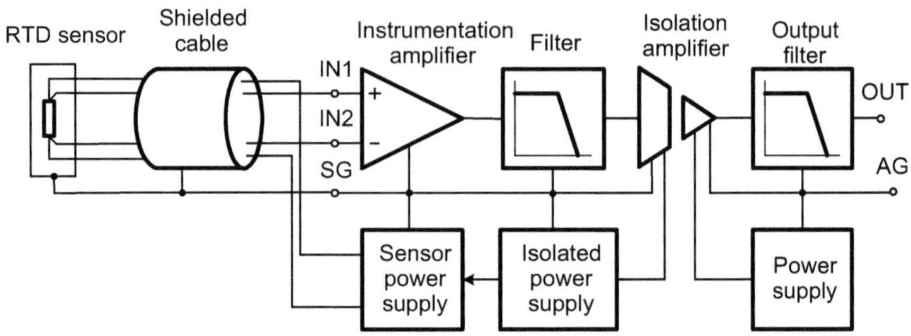

Figure 5.23 Block diagram of a measuring module with a resistance temperature sensor.

The common component of the bridge's differential voltage, which is approximately equal to half of its supply voltage, is suppressed by the CMRR (*common mode rejection ratio*) of the instrumentation amplifier, which ranges from 60 to 90 dB. The galvanic separation of the SG sensors' signal lands from the analog lands of the AG module and multiplex cards ensures the suppression of direct and alternating interference signals between the two lands, which is also determined by the IMRR (*isolation mode rejection ratio*) suppression factor of the isolation amplifier. This rejection factor ranges from 60 to 120 dB, decreasing with increasing frequency.

These modules also allow the connection of potentiometric sensors and resistance temperature sensors. Figure 5.23 shows the circuit diagram of a *resistance temperature sensor*, in which the current terminals of the sensor are again supplied with either direct current or alternating current, and the voltage terminals of the sensor are connected to the inputs of the instrument amplifier.

The block diagram of the measuring module of thermocouple temperature sensors with an isothermal terminal block is shown in Figure 5.24.

In these modules, an isothermal terminal block is used to connect the thermocouples, the temperature of which is measured by a semiconductor (diode) temperature sensor with a temperature coefficient of –2 mV/K, the voltage of which, after amplification, is subtracted in a differential amplifier from the output voltage of the instrument amplifier, which amplifies the voltage of the thermocouples. The shielding sheath of the supply cables of the thermocouples is connected to the signal ground of the module. After galvanic separation, the output voltage of the differential amplifier is then connected by the connection unit to the input of the multiplex plug-in card.

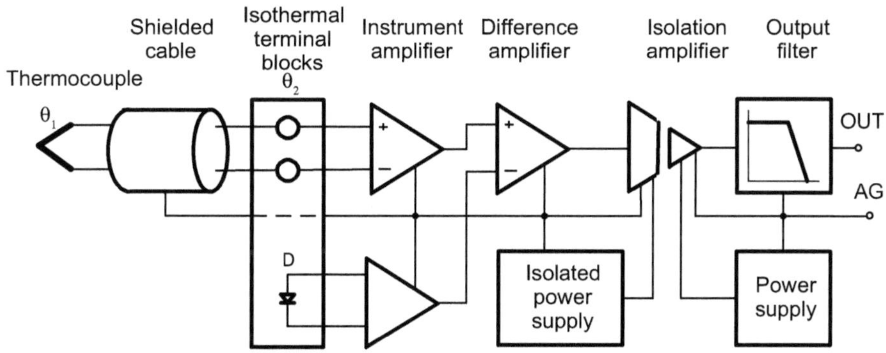

Figure 5.24 Block diagram of the measuring module of thermocouple temperature sensors.

This method of connecting the thermocouple makes it possible, thanks to the high suppression of the *CMRR* of the instrument amplifier, reaching up to 120 dB, to significantly suppress any common voltage of its measuring end to the signal ground. The minimum voltage range of the instrument amplifier of 50 mV makes it possible to measure thermocouple temperatures in their temperature ranges.

In addition to these modules, modules containing isolation amplifiers are used to excite *voltage* and *current lines*, the input and output parts of which are powered by mutually isolated power supplies with different groundings AG and SG (Figure 5.25).

Isolation amplifiers working either on the modulation principle with transformer coupling or optically coupled amplifiers with *delta sigma modulation* are used to isolate signals.

The *three-terminal modulation isolation amplifier* with transformer coupling consists of a synchronous amplitude modulator MOD, modulating the input differential voltage of the amplifier with a modulation frequency, the value of which is several orders of magnitude higher than the limit frequency of the amplifier. This amplitude-modulated signal is transmitted by a transformer, whose output voltage signal is synchronously demodulated by a DMD demodulator (Figure 5.26).

Typical types of these amplifiers are isolation amplifiers AD202, AD204, and AD215 with modulation frequencies of 2 MHz with adjustable differential gain in the range of 1–100 and isolation rejection factors IMRR (*isolation mode rejection ratio*) up to 120 dB. Their insulation strength reaches up to 2.5 kV with an insulation resistance of 1 TΩ and an insulation capacity of 5 pF. Non-linearity of these amplifiers of 0.02% is guaranteed in the frequency range up to 120 kHz.

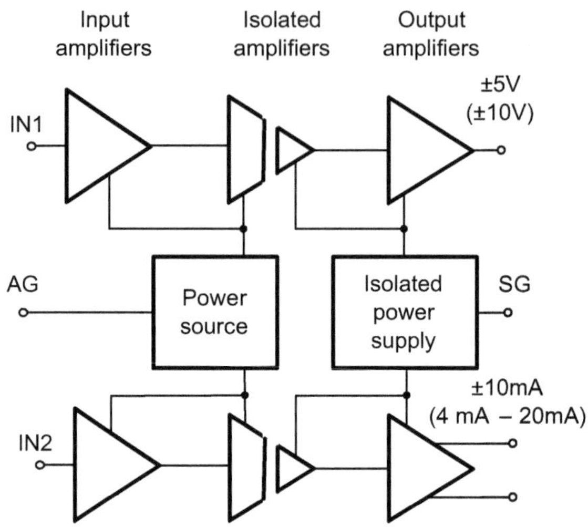

Figure 5.25 Block diagram of the voltage and current line measuring module.

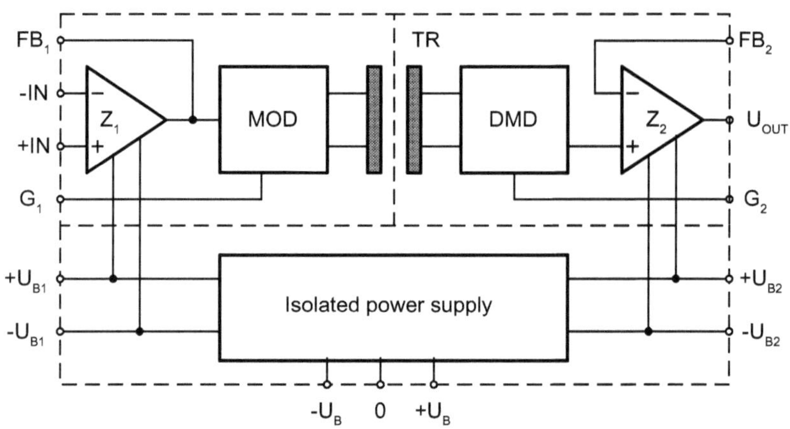

Figure 5.26 Three-terminal isolation amplifier with transformer coupling.

Optically coupled isolation amplifiers use optical coupling elements (optrons) consisting of an LED and a photodiode to galvanically isolate signals. An example of the connection of optically isolated amplifiers with negative optoelectric feedback, linearizing the non-linear conversion characteristic of the optocouplers and eliminating their temperature dependence of the conversion characteristics, is shown in Figure 5.27.

Assuming identical properties of the operons, the transmission of both enhancers is R_2/R_1.

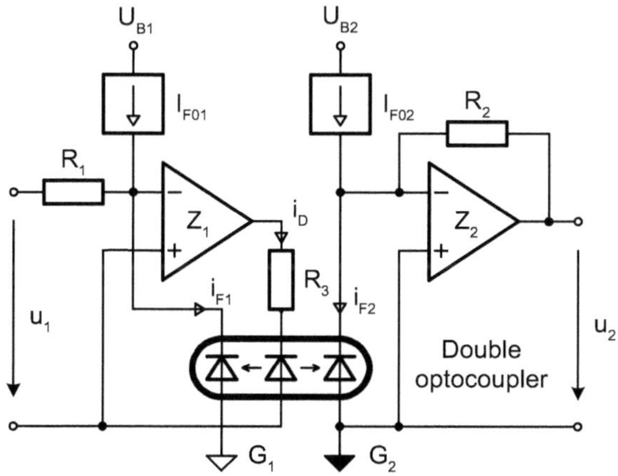

Figure 5.27 Connecting an optically isolated amplifier.

Common isolation amplifiers are, for example, the HPCL780, HPCL782 and HPCL784 amplifiers, having a differential gain of 8, an IMRR rejection factor of up to 70 dB, an isolation voltage of 10 kV, a non-linearity of 0.1%, and a frequency range of up to 200 kHz. Precision isolation amplifiers are, for example, amplifiers of the ISO100, ISO102, and ISO130 series with adjustable differential gain of 1–100, IMRR suppression of up to 140 dB, isolation voltage of 2.5 kV, and nonlinearity of 0.02%. These amplifiers have a frequency range of 60 kHz.

References

[1] Nawrocki, W. *Measurement systems and sensors*. Second edition, Artec House 2016. ISBN 1-58053-945-9.

[2] Boyes, W. *Instrumentation Reference Book*, Boston. Butterworth-Heinemann 2003. ISBN 978-80-08-094188-2.

[3] Park, J., Mackey, S. *Practical Data Acquisition for Instrumentation and Control Systems*, Amsterdam, Netherlands. Elsevier 2003. ISBN 978-8008-047383-0.

[4] Rove, M. *PCI Bus Cards and PCs Must be Compatible*. Test and Measurent World. October 1997.

[5] Winiecki, W. *Virtual Instruments*, Oficyna Wydawnicza of Warsaw University of Technology, 2003.

[6] Mielke, B. *PC-Card Anwender-Lösungen*. Franzis-Verlag GmbH, Feldkirchen. 1997. ISBN 9-783-772-343-131.

[7] Strass, H. *PCMCIA optimal nutzen* [*Using PCMCIA optimally*. Franzis-Verlag GmbH, Poing 1994. ISBN 3-7723-6652-X. 9-783772-366529.
[8] Vedral, J. *Analog signal processing and Digitization in Measurement*. CTU Prague 2020. ISBN 978-80-01-06717-8.
[9] Vedral, J. Digital Measuring Instruments of Electrical Quantities. CTU Prague 2021. ISBN 978-80-01-06860-1.
[10] www.ni.com.

6

Distributed Measurement Systems

In contrast to measuring systems consisting of stand-alone and modular measuring systems, distributed measuring systems are intended for the collection and processing of measured data not only in extensive research, development, and testing laboratories but especially in industrial areas with a high intensity of interference. These systems therefore require the effective elimination of not only DC and AC interfering signals but also error-free data transmission even over long distances to the central acquisition unit CAU (central acquisition unit), which can be not only an industrially resistant computer but also a single-purpose computing system.

6.1 Intelligent Sensors

For data collection and processing, distributed systems use so-called smart sensors, which are located in the immediate vicinity of the place of measurement of physical quantities. These sensors are made up of one or more multi-sensors of physical quantities, which are, for example, sensors of position, vibration, mechanical stress, pressure, flow, and temperature. The signals of these sensors, after being modified and digitized by A/C converters, are distributed along the distribution lines of the measuring system to the central collection unit, where they are written into data memories in the form of data files. The measurement results are often also visualized for the users of the measurement systems.

In the simplest case, *voltage* or *current signals*, modulated signals with continuous and discrete modulation are used to transmit data from these sensors to the central unit. Modern distributed meters use exclusively digitally *coded signals* to transmit data.

In addition to metallic connections, optical cables are also used for the transmission of data from intelligent sensors to significantly suppress compatible interfering signals and also for safety reasons. In this case, the intelligent sensors are powered by a galvanically isolated power supply, powered

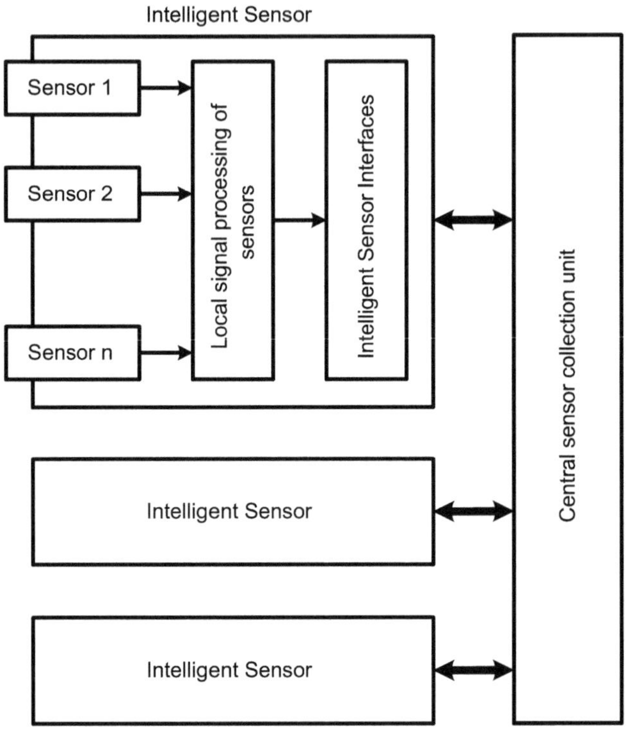

Figure 6.1 Tree structure of the arrangement of intelligent sensors.

either from an accessible power supply network or only by batteries with a long service life.

In more complex distributed measurement systems with a large number of measurement points, various types of their configuration are used, e.g., with a bus, star, ring, or tree structure. Figure 6.1 shows a typical distributed measurement data collection system with a tree structure of intelligent sensors.

6.2 Data Transfer in Distributed Measurement Systems

In addition to voltage and current lines, RS-485 transmission lines and other buses, e.g., ProfiBus, Mess Bus, CAN, InterBus, ProfiNet, DeviceNet, M-Bus, and FlexRay, are used for data transmission in distributed measurement systems.

6.2.1 Voltage and current lines

This historically oldest method of transmitting measured data uses standardized voltage and current levels in the range of a few V and a few mA.

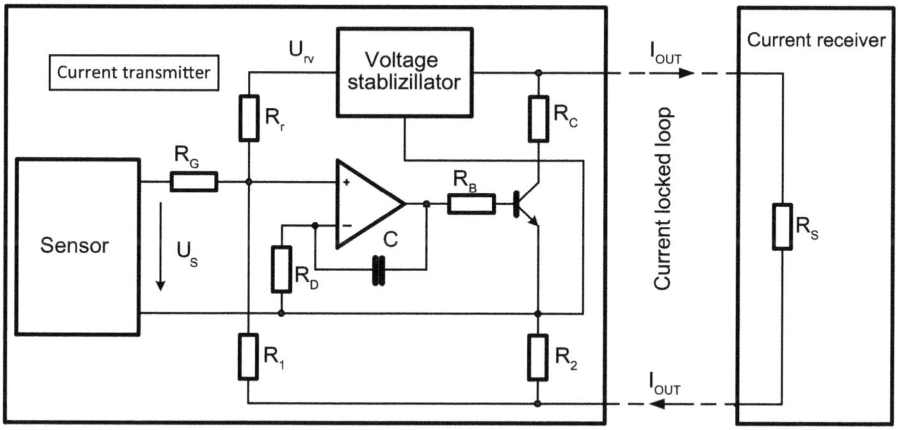

Figure 6.2 Sensor with current transmitter and receiver.

The *industry standard* of this so-called intensive method of data transmission is, for example, a *Current locked Loop*, formed by a current transmitter with a current range of 4–20 mA, while the quiescent current of 4 mA corresponds to the quiescent state of the line. The drop in quiescent current flowing through the line, caused by its interruption, is indicated by the collection unit, at the input of which a resistor with a nominal resistance of 100 Ω is connected.

The *current transmitter* can either be powered by a separate isolated power supply, or it can be powered by this quiescent current, including the other circuits of the smart sensor. Figure 6.2 shows an example of the connection of a sensor with a voltage output that excites a current transmitter whose output current I_{OUT} in the range of 4–20 mA excites a current line terminated by a sensing resistor $R_{\mathrm{S}} = 100\ \Omega$.

The output current of the current amplifier is

$$I_{\mathrm{OUT}} = \frac{U_r}{R_r} + \frac{R_1}{R_2} \cdot \frac{U_s}{R_G}, \tag{6.1}$$

where the first part of the U_r/R_r equation determines the quiescent current of 4 mA and the second part of the equation determines the change in the output current of the amplifier depending on the output voltage of the sensor U_{s}. This current causes a voltage on the sensing resistor R_{SENSE}, which is digitized by the A/D converter.

The advantage of the current line is its principled simplicity and the elimination of the influence of the line resistance on the magnitude of the transmitted current, provided that the supply voltage of the current amplifier is always a few V higher than the voltage required to excite the current line.

Current lines are used not only for intensive information transmission but also as a means of transmitting analog and discretely modulated signals, which are used, for example, by the HART (*highway addressable remote transducer*) transmission protocol. In these cases, the current line must have a sufficiently large frequency range, which requires the use of shielded twisted pair lines with a wave impedance equal to the output resistances of the current source and the sensing resistances of the receiver. With this modification, reflections at the beginning and end of the line do not occur during the transmission of impulse signals.

6.2.2 Data transmission by modulated signals

Of the many types of continuous and discrete modulations, pulse width modulation, delta sigma modulation, and pulse coded modulation are currently used in particular.

PWM (*pulse width modulation*) generates a pulse signal whose duty cycle is proportional to the modulated signal. The PWM modulator, formed by the INT integrator and the K_1 comparator, generates a time-symmetric triangular voltage waveform with a period

$$T = 4\frac{R_1 \cdot R_3}{R_2}. \tag{6.2}$$

The output voltage of the integrator is compared in the comparator K_2 with the input voltage of the modulator U_{IN}. If the output voltage of the integrator decreases below the level of the input voltage of the modulator, then the output voltage of this comparator goes to the log level. 0. On the contrary, if the output voltage of the integrator increases above the level of the input voltage of the modulator, then the output voltage of this comparator goes to the log level. 1 (Figure 6.3).

It applies to the duty cycle of the output voltage of the modulator thus generated

$$\frac{T_1}{T_2} = \frac{u_{IN}}{U_m}, \tag{6.3}$$

where U_m is the amplitude of the generated triangular signal.

The PWM signal generated in this way, the measuring variable of which is the alternation of the pulse signal, is filtered in the collection unit by filters of the LP (*low-pass filter*) type, whose output voltage is directly proportional to the mean value of the transmitted signal. The frequency range of the PWM signal is theoretically limited by half the period T of the triangular signal.

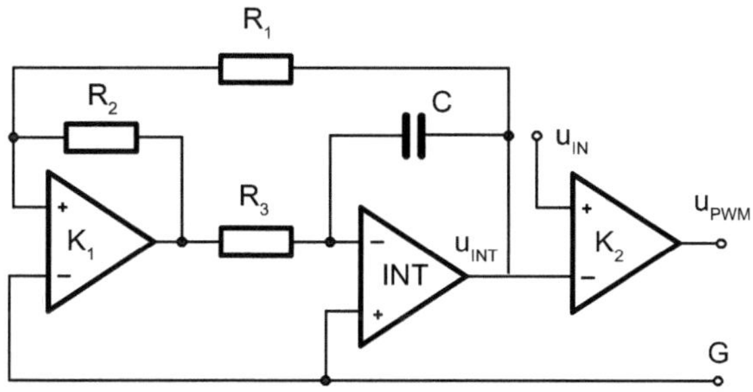

a) circuit diagram of the PWM modulator

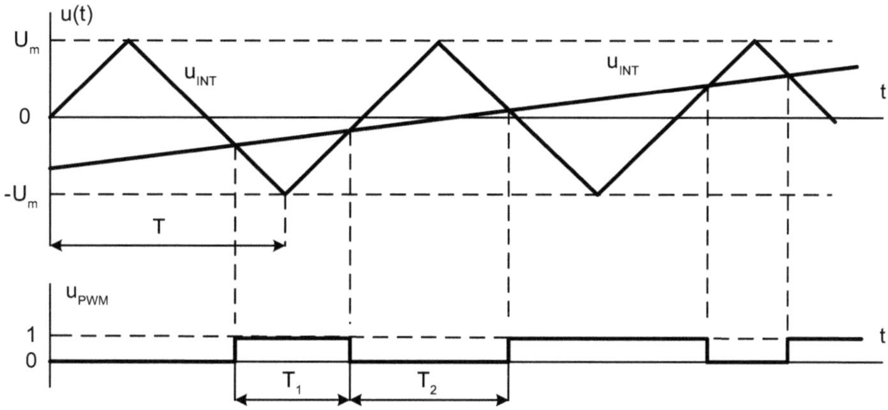

b) time course of PWM modulator signals

Figure 6.3 Pulse width modulation.

In the digital evaluation of the alternating PWM signal, their internal external counters are used in the processor circuits, while the period of the sampling signals T_s must be significantly lower than the measured times T_1 and T_2. Due to the asynchrony of the duration of times T_1 and T_2 and the period of the sampling signal of the counters T_s, a proportional quantization error occurs when determining the shift

$$\Delta\left(T_1 / T_2\right) \approx \sqrt{\delta^2\left(T_1\right) + \delta^2\left(T_2\right)}, \tag{6.4}$$

where $\delta(T_1)$ and $\delta(T_2)$ are the relative errors of determining the times T_1 and T_2, given by the ratio of the period of the sampling signal of the counters T_s to these times.

DSM (delta sigma modulation) is based on the activity of the previous PWM modulation, with the difference that, unlike the PWM modulator, the

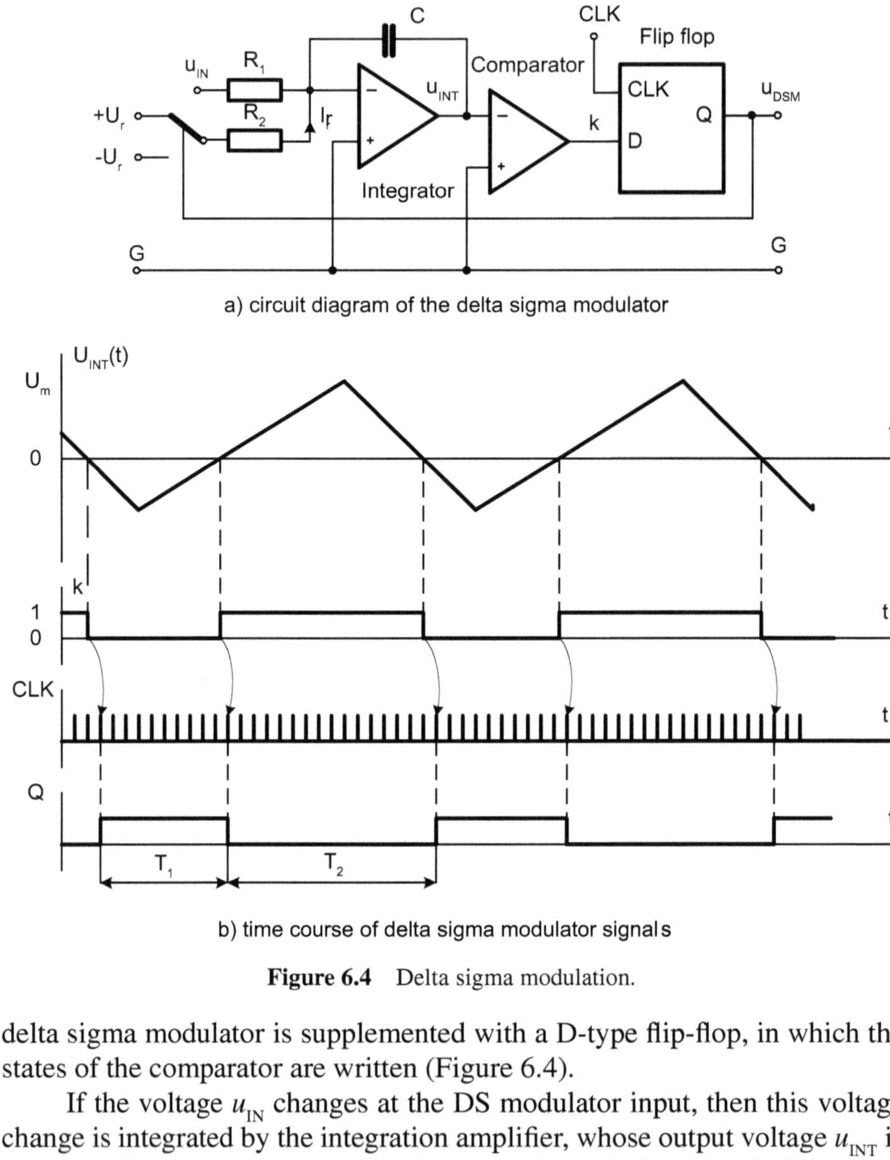

Figure 6.4 Delta sigma modulation.

delta sigma modulator is supplemented with a D-type flip-flop, in which the states of the comparator are written (Figure 6.4).

If the voltage u_{IN} changes at the DS modulator input, then this voltage change is integrated by the integration amplifier, whose output voltage u_{INT} is compared by the comparator with the zero voltage decision level. The output two-valued signal of the comparator is written by the sampling frequency f_{CLK} of the delta sigma modulator into a D-type flip-flop. The turn of the two-valued output signal of the flip-flop is then approximately directly proportional to the mean value of the input voltage

$$U_{IN}(\mathrm{str}) \approx \frac{T_2 - T_1}{T_1 + T_2} U_r, \tag{6.5}$$

where U_r is the reference voltage of the modulator, determining the range of its input voltage.

Thanks to the quantized feedback introduced in the DS modulator, which is given by using the sampled two-value output signal of the flip-flop to switch the reference currents ±Ir to the input of the integrating amplifier, the synchronization of the output signal of the DS modulator with the sample signal CLK is achieved. To evaluate such a modulated signal, similar methods are used as width-modulated signals. The frequency range of DS modulators is again limited by the sampling frequency, which must be up to several orders of magnitude higher than its frequency range for the correct operation of the DS modulator.

Pulse PCM (*pulse code modulation*) is based on the principle of digitizing the modulated signal. In this modulation, the modulated signal is first discretized in time (sampling) by a sampling circuit, controlled by the sampling signal CLK. The sampling circuit usually consists of a switched capacitor *C*, whose voltage corresponds to the input voltage sample at the moment the switch is opened. The thus obtained voltage samples are quantized by a quantizer, usually having equidistant quantization levels. The amplitudes of the impulses obtained in this way are subsequently decoded usually into a binary code. All these processes involve *A/D converters* with corresponding resolution (Figure 6.5).

The frequency range of modulated signals is determined by the *Nyquist sampling theorem*, which limits the frequency range of these signals to less than half the sampling frequency $f_s = 1/T_s$

$$f_m < f_s / 2. \tag{6.6}$$

If this condition is not met, the signal modulated in this way cannot be fully reconstructed and reverberant signals (*aliasing error*) occur, the frequencies of which apply

$$f_a = \text{round}\left(\frac{f_s}{f_m}\right), \tag{6.7}$$

where *round* means the nearest whole number.

The dynamic range of PCM modulation is determined by the SNR (*signal-to-noise ratio*), defined by the ratio of the rms value of the modulated harmonic signal to the rms value of the quantization noise of the *n*-bit A/D converter

$$\text{SNR} = 20 \cdot \log\frac{1/2}{2^{-n} / \sqrt{12}} = 6.02 \cdot n + 1.76[\text{dB}]. \tag{6.8}$$

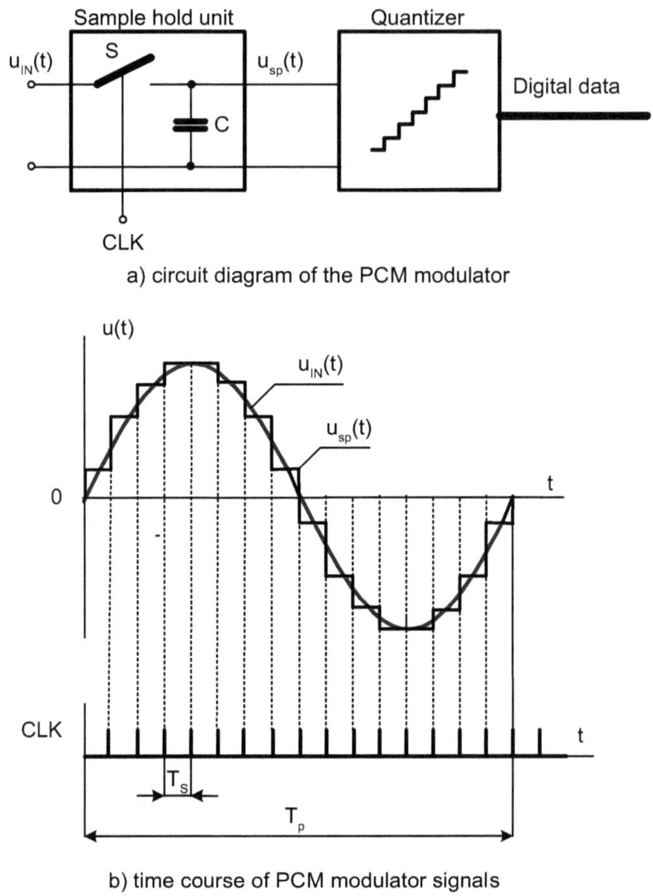

Figure 6.5 Pulse code modulation.

For example, PCM modulator with an 8-bit A/D converter has a theoretical dynamic range of SNR = 49.9 dB. In practice, however, due to uncertainties in sampling times of the modulated signal and non-linearity of the quantizer, smaller signal-to-noise distances are always achieved, e.g., 6.5 dB. As the resolution of the quantizer increases, this difference increases.

6.2.3 Data transmission via RS-422 and RS-485 lines

The RS-485 serial interface is based on the RS-232 interface, the main disadvantage of which is the asymmetric arrangement of the data wires, which does not allow effective suppression of interfering signals induced into this line. To eliminate this disadvantage, the RS-422 and RS-423 interface

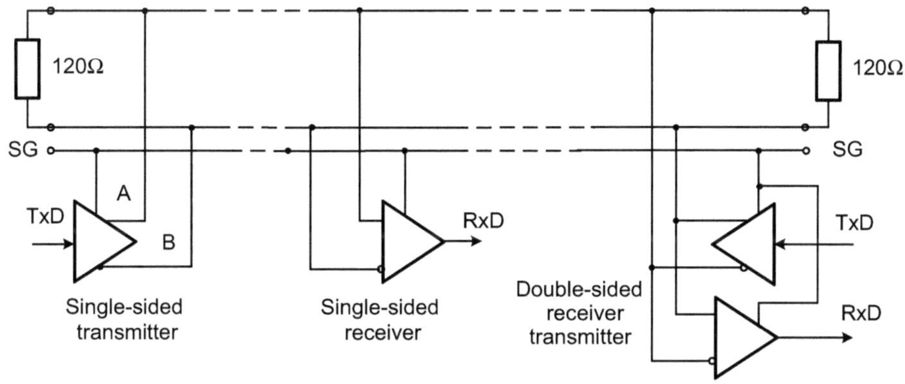

Figure 6.6 Connecting the RS-485 bus.

standards were gradually introduced, of which the RS-422 standard already had a defined ground wire for each signal, and the RS-423 standard had only one shared signal ground. In addition, these standards could no longer use D25 connectors, which had to be replaced by larger D32 connectors. For these reasons, the RS-485 interface was introduced to the market in 1983, which became the standard interface not only for distributed measurement systems but also for other data systems, e.g., LAN systems.

The RS-485 interface specification defines the transfer of data by a *differential voltage* between signal wires that are supplemented by a common signal ground *S*. This arrangement allows for a significant elimination of the induced common voltage that is added to both signal wires. For data signals RxD and TxD, log. 0 corresponds to a voltage range of +3 to +15 V and log. 1 to a voltage range of –3 to –15 V. For control signals RTS, CTS, DTR, and DSR, log. 0 corresponds to a voltage range of –3 to –15 V, and log. 1 to a voltage range of +3 to +15 V. These voltage levels are relative to the signal ground of the SG data wires.

For data transfer, the *simplex* data transfer is used, and for reasons of speeding up communication, the *duplex data transfer* is used, using two-sided exciters/transmitters (Figure 6.6).

Typical double-sided drivers of RS-485 buses are, for example, MAX 488 circuits, which allow up to 32 units to be connected to the RS-485 data network. Connection of up to 128 units in *half-duplex data transmission* mode is possible, for example, with the MAX 487 circuit.

In order to suppress reflections of voltage pulses, damping resistors of 120 Ω are placed at both ends of the line, the value of which should correspond to the wave impedance of the line.

Figure 6.7 Four-pair twisted pair with RJ-45 connector and its connection.

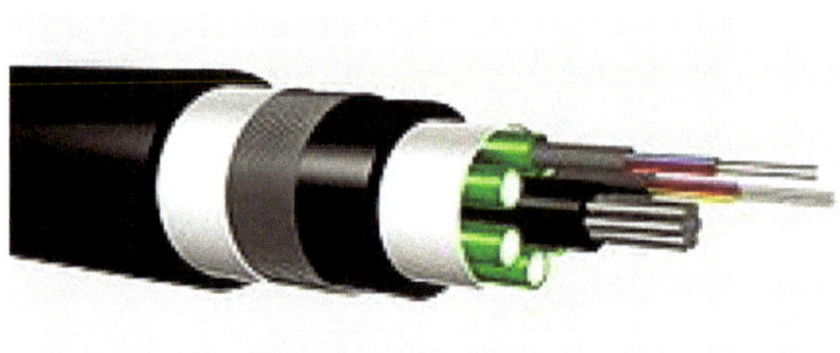

Figure 6.8 Bundle of optical cables.

A shielded twisted pair cable (*twist cable*) is usually used for data transmission, which makes it possible to significantly suppress AC interfering electromagnetic interference. These twisted pairs are produced in *single-*, *double-*, or *four-pair* designs. RJ-45 connectors are used to connect these two lines (Figure 6.7).

The advantage of the RS-485 interface is its simplicity, low price, and the possibility of high suppression of the influence of alternating electromagnetic interference, reaching up to 40 dB in the frequency range from 50 Hz to several tens of kHz for shielded twisted wires. The transmission speed of this interface reaches up to 10 MB/s with a cable length of up to 1.2 km.

6.2.4 Data transmission by optical cables

The transmission of data via optical cables makes it possible to implement data lines with a significantly higher transmission speed over greater distances. For this purpose, cables with glass, or in less demanding applications also with plastic, fibers are used (Figure 6.8).

Glass fibers are produced from quartz glass with a high purity content of silicon dioxide SiO_2. According to the method of propagation of the light beam, we distinguish between single-mode, multi-mode, and gradient optical fibers.

Single-mode optical fibers with a diameter of several μm with constant permittivity and permeability of glass and constant refractive index n between glass and air do not allow partial reflections of the light beam from

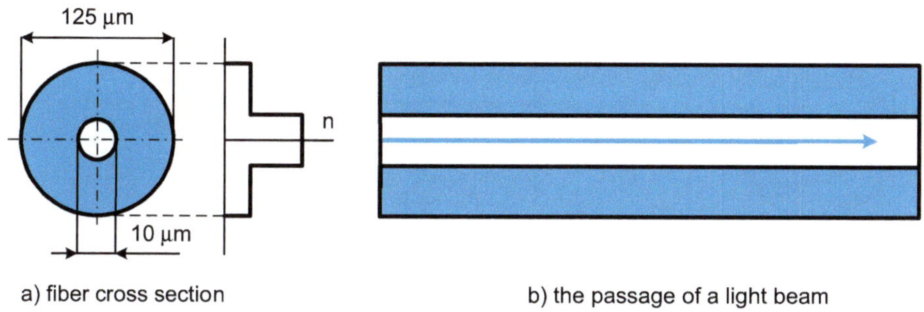

Figure 6.9 Single-mode optical fiber.

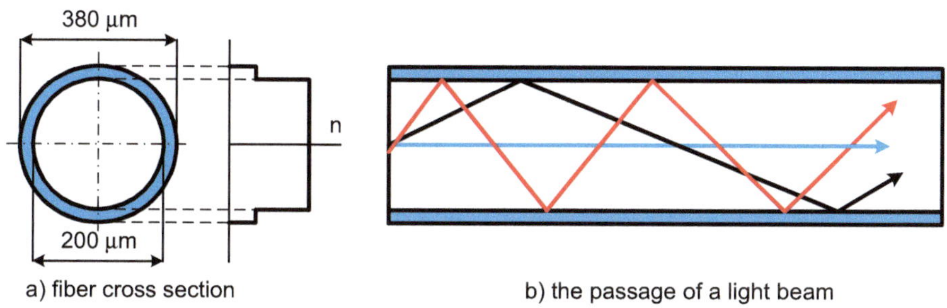

Figure 6.10 Stepindex-mode optical fiber.

their walls. The speed of propagation of the light beam in these fibers is determined by the permittivity ε and permeability μ of the quartz glass

$$v = \frac{1}{\sqrt{\varepsilon \cdot \mu}}, \tag{6.9}$$

which is always smaller than the propagation speed of these rays in a vacuum, which is approximately $3 \cdot 10^8$ m/s. When using glass single-mode optical fibers, the propagation speed of the light beam is achieved in the range of $1.1 \cdot 10^8$–$1.5 \cdot 10^8$ m/s. Cables with these optical fibers are therefore used for high-speed data transmissions (Figure 6.9).

Step-index optical fibers with a diameter of several hundred μm and a step-variable refractive index cause partial reflections of the light beam from their walls, which slow down the propagation speed of the light beams (Figure 6.10).

Gradient optical fibers are formed by a structure of quartz glass with a variable refractive index, which, together with a reduction in the diameter of the fibers in the range of 50–100 μm, make it possible to increase the speed of propagation of the light beam to a value several times higher than that of multimode fibers (Figure 6.11).

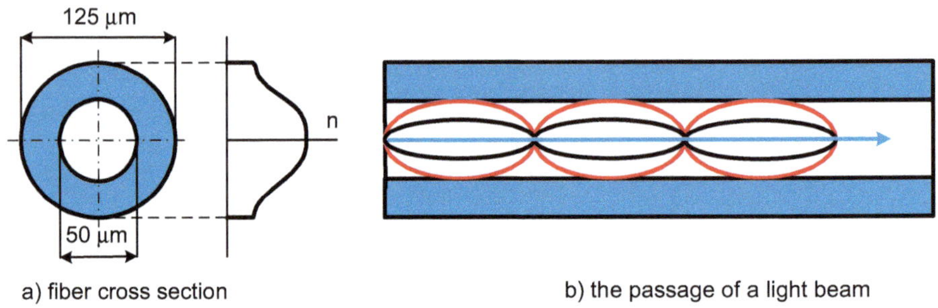

Figure 6.11 Graded optical fiber.

Cables with step-index and gradient optical fibers are therefore used in less demanding data transmissions with transmission speeds in the range of 10 Mbit/s to 10 Gbit/s over distances of up to several hundred meters. The attenuation characteristic of glass optical fibers shows several peaks, between which there are sections with lower attenuation, which are called windows. According to the ITU-T standard, windows are defined in the following wavelengths of light radiation.

- O (Original) in the range of wavelengths 1260–1360 nm
- E (Extended) in the range of wavelengths 1260–1360 nm
- S (Short wavelength) in the range of 1460–1530 nm
- L (Long wavelength) 1565–1625 nm
- U (Ultra) above 1625 nm

To connect and convert electrical and optical signals, special ST connectors are used with corresponding opto-electrical converters, which contain extremely fast light emitting LEDs (*light emitter diode*) and photodiodes (Figure 6.12).

6.3 Communication Layers of Buses of Distributed Measurement Systems

Industrial communication systems with which industrial buses work can be divided according to the type of control and the type of system into three layers:

- *Sensor and actuator layer* is suitable for real-time communication with sensors and actuators. These bus networks form the lowest level of

Figure 6.12 ST connectors and optical line cables.

control. They usually define only the first and second layers of the ISO/OSI model. This arrangement provides a simple and economical technology where data and power are carried over a common transmission medium. Data during communication is transmitted by binary signals. Leading representatives of this layer are, for example, the AS-Interface, Interbus, and HART buses.

- *Devicebus* layer is used for communication at the level of programmable logic controllers. This layer is used to connect I/O devices and control devices in real time. This is a layer of networks with a higher level of control. Defines the first, second, and seventh layers of the ISO/OSI model. These layers use longer transmission frames, allowing the configuration of actuators and sensors. These buses include, for example, the DeviceNet, Interbus, CAN, and other standardized IEC 61158 (IEC 61784) standards.
- *Layer of industrial communication systems* (*FieldBus*) stands highest in the hierarchy of industrial networks. These are mostly multimaster networks with redundancy, using duplicated transmission lines. These buses enable event-driven services and object-oriented data transfers. Typical representatives of these buses are, e.g., ProfiBus, P-Net, and WorldFIP.

This division of layers is based on the seven-layer ISO/OSI reference model, which was adopted in 1984 by the ISO (*International Standard Organization*) under the designation ISO-7498 and by the ITU (*International Telecommunication Union*) in the form of the X.200 recommendation (Figure 6.13).

Each of the seven layers performs a group of well-defined functions required for communication. For its activity, each of the service layers uses

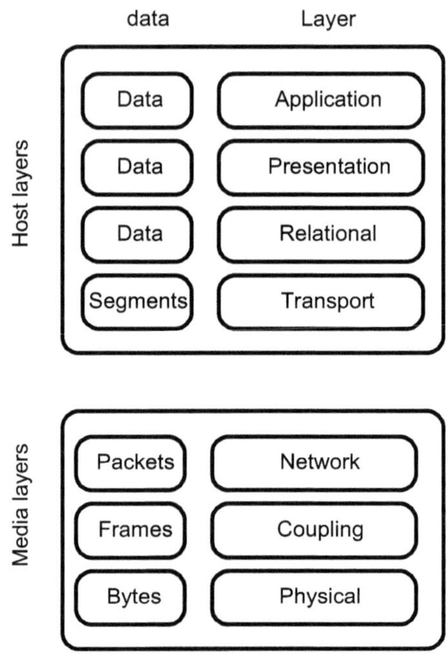

Figure 6.13 ISO/OSI Reference Model.

its neighboring lower layers. It then provides its services to the neighboring upper layer.

According to the reference model, skipping layers is not allowed, but some layer may not be active. Such a layer is called zero, or transparent.

Communication between systems consists of:

- communication between layers of one system, governed by rules that are usually called interfaces;
- communication between the same layers of different systems that follow protocols.

If a request comes from a process in the application layer, then the corresponding subsystem requests the creation of a connection to the presentation layer. Within the application layer, communication with the opposite system is controlled by the application protocol. Subsystems in the presentation layer communicate using a presentation protocol. In this way, one progresses further down to the physical layer, which is intended to connect transmission devices. At the same time, when moving from a higher layer to a lower one, the headers of the individual layers are added to the user (application) data. This is how the original information is gradually encapsulated. At the recipient of the messages, the control information of the individual layers is gradually processed and their functions are performed. The properties of individual layers are as follows.

The *physical layer* specifies physical communication. It activates, maintains, and deactivates physical links (e.g., switched link) between end systems. The physical connection can be two-point, e.g., RS-232 serial line, or multi-point, e.g., *Ethernet*. This layer also defines all the electrical and physical properties of the device. It contains pin layouts and voltage levels, specifies cable properties, and determines how data is transmitted.

The *data link layer* provides the connection between two neighboring systems. It organizes data from the physical layer into logical units known as frames, sorts transmitted frames, takes care of setting link transmission parameters, and reports unrecoverable errors. It formats physical frames and provides them with a physical address and provides synchronization for the physical layer.

A typical example of this layer is *Ethernet*. On IEEE-802-based local area networks and some networks such as FDDI, this layer should be divided into a MAC (medium access control) layer and an IEEE-802.2 LLC (*logical link control*) layer. All data bridges and switches work on this layer. It provides connectivity only between locally attached devices, thus creating a Layer 2 domain for directional and omnicasting.

The *network layer* takes care of network routing and network addressing. It provides a connection between systems that are not directly adjacent to each other. It contains functions that make it possible to bridge the different characteristics of technologies in transmission networks.

This layer provides functions to ensure the transmission of data of varying lengths from source to destination through one or more interconnected networks while maintaining the quality of service required by the transport layer. The network layer provides routing functions and also reports on data delivery problems. All routers operating at this layer send data to other networks, creating a hierarchical address structure. The most well-known protocol working according to this layer is the internet protocol (IP).

The *transport layer* ensures the transfer of data between end nodes. Its purpose is to provide the quality of transmission required by the higher layers. The layer offers connection-oriented and connection-less protocols.

The *connection-oriented* TCP protocol ensures data transfer with the guarantees required by applications where no data packet must be missing. These are transfers of files, emails, WWW pages, etc. The guarantee covers the resolution of transmitted packet losses, preservation of their order, and removal of duplication. The unit of information sent is a data segment at this layer.

The *connectionless* TCP protocol ensures data transfer without guarantees, which is used by applications that would be bothered by network delay caused by waiting for the transfer of all packets, while data loss can be solved in another way, e.g., by reducing their quality and repeating query. This protocol is used for DNS, VoIP, streaming video, internet radio, searching shared files within the DC++ network, online games, etc.

The *session layer* organizes and synchronizes the dialogue between the cooperating session layers of both systems and controls the exchange of data between them, and allows creation and termination of relational connections, synchronization and reconnection, and exception reporting. NetBIOS and RPC networks belong to this layer, for example. It assigns synchronization marks to data packets, which are used in the event that packets are returned, e.g., due to their damage during transmission.

The *presentation layer* transforms data into a form that enables its encryption, conversion, and compression. The format of the data (data structures) may differ on both communicating systems. In addition, data formats are transformed for the purpose of their transmission by lower layers. Other functions of this layer are, for example, the conversion of codes and alphabets, modification of the graphic layout, and adaptation of the byte order. The layer deals only with the structure of the data, but not with its meaning, which is known only to the application layer. Examples of protocols based on this layer are, for example, the SMB (*samba*) protocol.

The *application layer* provides applications with access to communication systems and thus enables their cooperation. This layer includes services and protocols such as HTTP, FTP, DNS, DHCP, POP3, SMTP, SSH, Telnet, and TFTP.

6.4 Bus of Distributed Measurement Systems

Currently, there are a large number of industrial buses that differ in the method of access to the transmission medium, transmission protocol, transmission security, application layer, etc. These buses are used not only in industrial facilities to monitor and control production processes but also, for example, to control equipment in buildings, measuring water and electricity consumption, monitoring the operation of oil and gas pipeline systems, etc. The most common types of industrially oriented buses include *ProfiBus, MessBus, CAN*, *Interbus*, *ProfiNet*, *AS-Interface*, *Device Net*, *M-Bus*, *FlexRay.*

6.4.1 ProfiBus

The industrial bus *ProfiBus* (*process field bus*) is intended for measurement and automation of production lines, for process automation, and for production and technology management. Under this designation, three types of industrial buses are standardized in the European Union by the EN 50 170 standard. They are *ProfiBus FMS* (*fieldbus message specification*) for data communication between PLC systems, *ProfiBus DP* (*decentralized peripherals*) for fast cyclic data transfer with devices in a common industrial environment, and *ProfiBus PA* (*process automation*), used in demanding environments in process automation, e.g., in environments with a risk of fire or explosion.

Data communication in these buses is specified in the IEC 61 158 and IEC 61 784 standards. The ProfiBus standard is based on the ISO/OSI reference model, which uses only the *physical*, *connection*, and *application layers*.

To transmit data over this network, metallic or optical lines with pairs of shielded twisted wires are used with full-duplex data transmission and the number of participants is limited to 128. Data repeaters can be used to branch the network.

When using a metallic line, a network transmission speed of up to 12 MB/s is achievable up to a line length of up to 1.2 km. When using optical cables, the line length can be extended up to 80 km. In both cases, RS-485 interfaces are used for data transfer. Recommendation IEC1158-2 allows the use of current line and synchronous *Manchester coding* for data transmission.

To optimally use the transmission channel, this method uses three types of *messages*, so-called telegrams, which are control, fixed, and variable data frames. Either the logical ring method or the *master–slave method* is used to access the data. In the *token ring method*, the stations connected to the bus gradually transfer credentials (token) to each other in a logical ring. In the *master–slave* approach, the control station always has the bus allocation priority. Both of these methods use a hybrid way of accessing the bus, which is characterized by the structure shown in Figure 6.14.

For this method of accessing the bus, a logical circle is used, in which bus access priorities are gradually passed on to individual network participants MASTER 1, MASTER 2, etc., while these participants control participants of the type SLAVE 1, SLAVE 2, etc.

6.4.2 Mess Bus

This bus, also referred to as *measurement bus*, or *DIN-Bus*, is the result of the activities of manufacturers of measuring systems used in the field of industrial measurement and the PTB (*Physikalisch-Technische Bundesanstalt*) in the BRD. This standard originated in 1986 but was not fully standardized until 1989 as DIN 66 348 Section 2: Interfaces and Control Procedures for

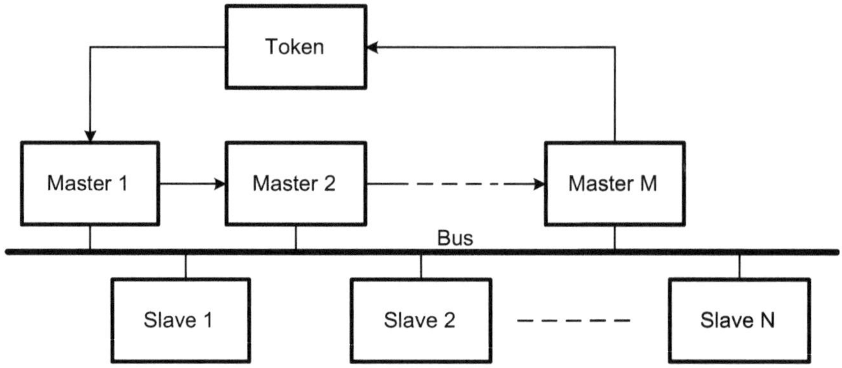

Figure 6.14 Hybrid structure of the PROFIBUS industrial bus.

Figure 6.15 Double-row fifteen-pin connector of the MessBus bus.

Serial Measurement-Data Transfer, Start-Stop Transfer, 4-Wire Bus. In practice, it has been used since 1990 as standard DIN 66348, part 2, which defines the physical and line layers of the ISO-OSI model.

The *physical layer* of the bus corresponds to the RS-485 standard. Up to 32 units can be connected to a four-wire bus with a typical transmission speed of 9.6 kB/s. However, the standard allows the use of transmission rates of up to 1 MB/s over a distance of up to 400 m. The DIN 66 348 standard also defines complete galvanic separation of transmitter and receiver, which significantly limits the influence of interfering signals on the error rate of transmitted data, especially in an industrial environment.

The *Mess Bus system* uses a standard two-row 15-pin CANON connector to connect devices (Figure 6.15; Table 6.1).

The organization of data transfer is *master–slave* type for this bus. Data is transmitted in the classic format for serial asynchronous transmission, which is also used by the RS-232 and RS-485 standards. The data transfer thus begins with a sequence of 7 bits, starting with the start bit with the lowest weight LSB (*least significant bit*), seven data bits, followed by a *parity bit*, and this whole sequence ends with a stop bit.

The disadvantage of this transmission is that the data transmission is more demanding, since this data must be encoded. The advantage is easy readability and control of their transfer. The *parity bit* ensures the protection

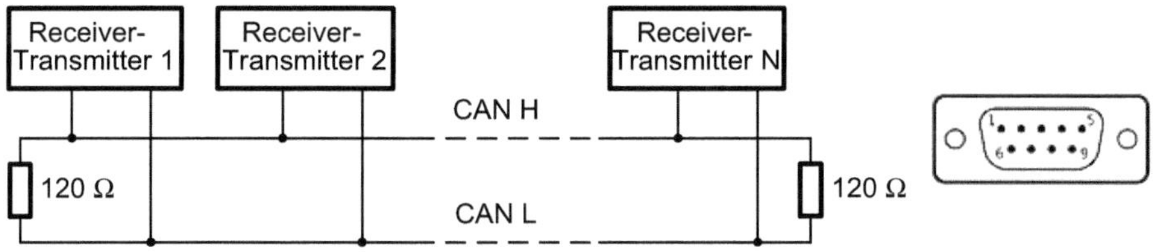

Figure 6.16 Structure of a CAN bus with a nine-pin and two-row connector.

of a block of data during its transmission, or back-check, after their acceptance. The so-called even parity is used here, where it is checked whether the *sum modulo 2* of the data bit values is zero. If this is not the case, the data is retransmitted.

The higher ISO-OSI application layers are defined by parts 3 and 4 of the DIN 66 348 standard, which describe some additional variables, commands, and functions to contribute to the simpler control of measured functions, the transfer of measured parameters, and the representation of data from various applications.

Mess Bus is widely used not only in many industrial operations, but also in other areas, especially in Europe.

6.4.3 CAN bus

The *CAN bus* (*controller area network*) is a data bus, used mainly in the automotive industry. This bus was developed in 1983 by *Robert Bosch*. The first application of this bus was in the BMW 850 Coupe. In this car, 2 km of cables were saved, half the number of connectors. The total weight of the car has been reduced by 50 kg. In 1993, this bus was standardized under the designation CAN 2.0A by the ISO 11 898 standard and subsequently specified by the EN 50 325 standard. Later, the CAN 2.0B specification was developed, which expands the original 11-bit address of terminal devices (*standard format*) with a compatible 29-bit address format (*extended format*).

The physical layer of the CAN bus consists of a shielded *twisted pair line*, which is terminated at its ends with resistors 120 Ω (Figure 6.16).

The standard connector for connecting CAN bus units is a nine-pin double-row connector, which can be replaced by another similar connector (Table 6.2).

A CAN bus network can be configured to work with two different message formats (frames). They are the standard or basic frame format, defined in the CAN 2.0 A recommendation, and the extended format, defined in the CAN 2.0 B recommendation. The only difference between the two formats

Table 6.1 Signal assignment in the MessBus measuring bus connector.

Pin	**1**	**2**	**3**	**4**	**5**	**6**	**7**	**8**
Signal	Shielding S	Broadcast data TxA	Received data RxA	Optional use	–	–	Optional use	Bus lground G
Pin	9	10	11	12	13	14	15	
Signal	Broadcast data TxB	–	Received data RxB	–	–	Optional use	Supply +5 V	

Table 6.2 Assignment of signals in the nine-pin connector of the CAN bus.

Pin	**1**	**2**	**3**	**4**	**5**	**6**	**7**	**8**	**9**
Signal	–	CAN-	Signal ground G	–	–	–	CAN-	–	Supply +5 V

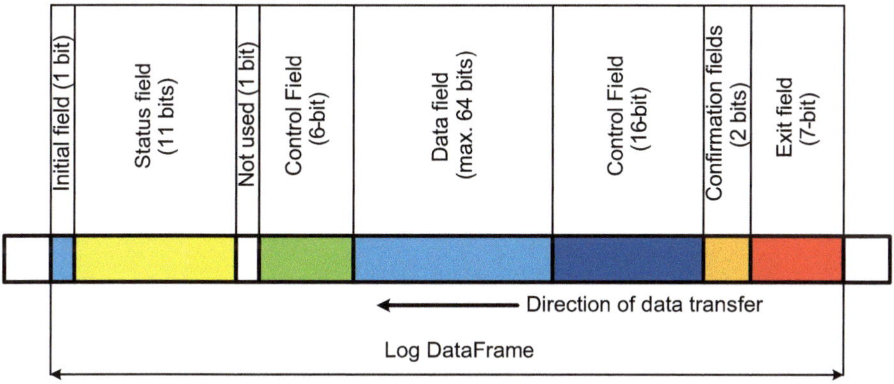

Figure 6.17 Structure of the CAN bus basic frame.

is the length of the identifier, which is 11 bits for the standard format and 29 bits for the extended format.

The CAN bus has four types of *frames*. They are:

- a data frame containing node data for transmission;
- remote frame, requesting transmission of a specific identifier;
- an error frame, transmitted by any node detecting the error.

The *data frame* has two message formats:

- 11-bit basic frame format;
- 29-bit extended frame format.

The CAN standard requires that its implementation accept the basic frame format and be able to accept the extended frame format as well. Figure 6.17 shows the structure of the CAN bus basic frame.

The *message data frame* contains start, status, control, data, control, acknowledge, and exit fields.

- The *start of frame* indicates the beginning of the data protocol.
- The *arbitration field* determines the priority of the data protocol.
- The control field contains the number of information contained in the data field.
- The *data field* transmits information that is important for other control units.

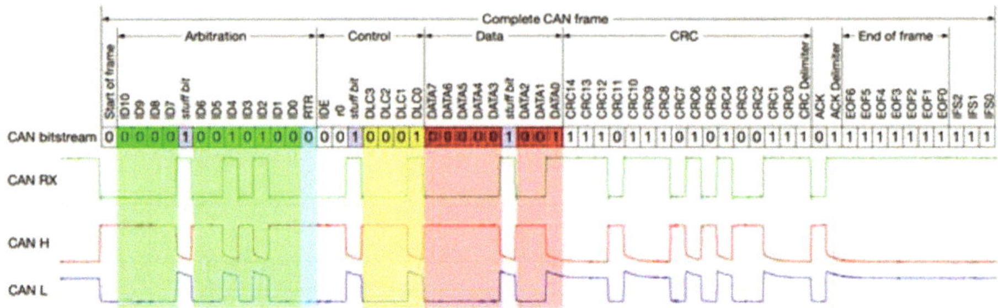

Figure 6.18 Time course of the CAN bus basic frame.

- The *control field* (CRC *field*) detects errors in the transmission, based on the cyclical calculation of the control code before transmission and after the transmission of CRC data (*cyclical redundancy check*).
- The *confirmation field* (ACK *field*) confirms the correct reception of the message by the object that sent it. If there is an error in the message, it will be sent again.
- The *end field* (*end of frame*) confirms the correctness of the message to the transmitting object. If the message is in error, the transmission of further messages will be interrupted and the original message will be restarted.

An example of the time course of the CAN bus basic frame is shown in Figure 6.18.

Access control to individual network nodes is based on *random address allocation* (*random addressing*), which resolves collisions based on priority decision-making. Messages sent over the bus therefore do not contain any information about who they are addressed to and are received by all bus units. Their acceptance is conditioned by the use of the unit indicator during their filtering (*acceptance filtering*).

The CAN bus is designed for a distributed real-time control system with a high degree of transmission security against errors. The CAN protocol is of the *multimaster type*, which means that each node in the network can have the function of *master* and thus control the behavior of other nodes. This type of control enables simpler control and increases the reliability of data transmission. If a fault occurs in one network node, the functionality of the entire network is not limited in any way and only the message with the highest priority is transmitted.

Although the CAN bus was originally intended for measurement, control, and control of functional units, especially of passenger cars, the use of

a) CAN module b) CAN-USB converter

Figure 6.19 CAN modules.

this bus has also spread to other areas, e.g., for control and monitoring of technological equipment and technical equipment of buildings.

Figure 6.19 shows the automotive CAN module and the CAN–USB bus converter.

6.4.4 INTERBUS

This serial bus system is designed to transfer data between measurement and control systems and computers. The *INTERBUS system* was developed in 1978 by *Phoenix Contact* and was subsequently standardized by EN 50 254 and IEC 61158. Currently, more than 600 manufacturers are involved in the implementation of this bus system. Since 2011, the INTERBUS system has been supported by the *ProfiBus* and *ProfiNet International associations*.

The system topology has a tree-like OTRA (*open tree ring area*) ring structure, where each active system participant is connected to the main ring bus by short sub-buses of a lower level (Figure 6.20).

The number of levels in each ring bus is limited to 16. This makes the bus very flexible and easily adaptable to changing topology. Up to 512 devices can be connected to the bus. Each device receives the signal, which it amplifies and sends to the next device on the network.

The *physical layer* of this interface consists of a *twisted shielded double line*, which can also be used to power the system units. When using a metallic connection, a transmission speed of up to 500 kb/s can be achieved over a distance of up to 13 km, while the maximum distance between neighboring units is 400 m. By *optically connecting* the units, this distance can be increased to up to 3.6 km.

The use of *circular busbars* excludes the need to use termination resistors (terminators), which are otherwise necessary for other non-circular busbars.

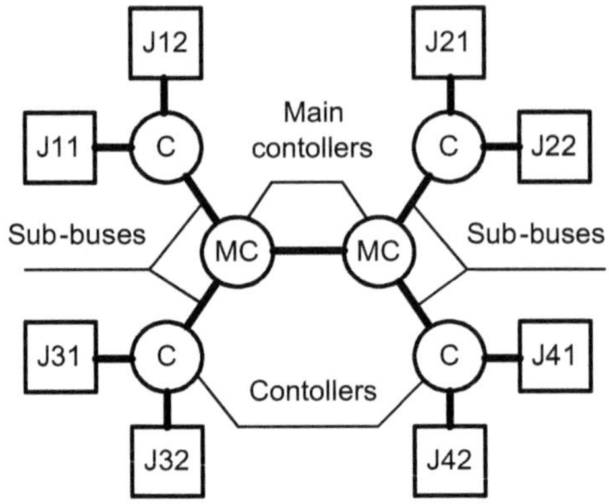

Figure 6.20 Circular bus structure of the INTERBUS system.

Information about the structure of the bus is stored only in the communication processor; so there is no need to physically address individual participants. Complete addressing is done when configuring the communication processor by computer using the CMD configuration tool.

Message communication in the bus is based on the principle of master/slave communication. The main branch of the bus is of the *master type*, which connects all *slave units* and also controls them. The last slave device automatically closes the entire network.

6.4.5 ProfiNet

The *ProfiNet* industrial bus is an open industrial standard, based on the IEEE 802.3 Ethernet 100 Mbit/s standard. In the *European Union*, it is standardized by the IEC 61158 and IEC 61784 standards. This standard is the successor of the *ProfiBus* and *Ethernet bus* and communication systems. The international standard ISO/IEC 11801 and its European equivalent EN 50 173 define the installation of interconnecting cables, the connection method, and bus configuration, including the choice of plugs for special EMC connections, resistance to temperature, dust, and vibrations.

This standard is presented in two modifications:

- *ProfiNet* CBA, intended for modular construction of operations and technological devices, based on components (technology modules) with a defined interface.
- *ProfiNet* IO, intended for the integration of distributed peripherals (inputs and outputs).

With both standards, great emphasis is placed on the availability and reliability of data transmission. This is treated with the MRP (*media redundancy protocol*), enabling the creation of a circular network topology, the reaction time of which is shorter than 50 ms. In this MRPD *smooth data redundancy method*, data is cyclically transmitted in real time in both directions in a ring topology, and the time stamp in the respective data packet in which the error was indicated allows the data receiver to remove redundant duplicates.

To connect the units, this standard recommends the use of either *full-duplex* data transmission with a pair of 100BASW-TX *shielded twisted wires*, or the use of a 100BASE-FX *optical line* according to the IEEE 802.3 standard.

Data addressing is solved by the ARP (*address resolution protocol*), which is extended for this purpose by the detection of duplicate IP addresses. DCP (*discovery and basic configuration protocol*) is mandatory for assigning IP addresses. Optionally, DHCP (*dynamic host configuration protocol*) can also be used for this purpose. No IP addresses are used with the RSI protocol.

The internet protocol can be used in the operating system of the operating device also for other protocols, such as the OPCUA (*unified architecture*) protocol. The LLDP (*link layer discovery protocol*) has been expanded with additional parameters, so that in addition to neighbor detection, the propagation time of signals on connecting lines can also be communicated.

Access to bus application layer services is defined by RTP (*real-time protocol*) for class A and B applications with cycle times ranging from 1 to 10 ms. The IRTP (*isochronous real-time protocol*) for application class C enables applications in the field of drive technology to achieve cycle times shorter than 1 ms. This can also be achieved with the same services via the TSN (*time synchronization net*).

6.4.6 AS-Interface

The AS-Interface (*actuator sensor interface*) standard, also referred to as AS-i, is intended for use in systems with the lowest level of automation. This bus is standardized by the IEC 62 026 standard. This system allows simple sensors and actuators from different manufacturers to be connected to the network using a single serial line. The network structure is open to other higher levels. For example, Siemens supports the development of this standard and also has the widest range of products with this interface.

The *AS-Interface bus* is a *master–multislave* type network, i.e., a single device of the *master* type is connected to this network and the other devices

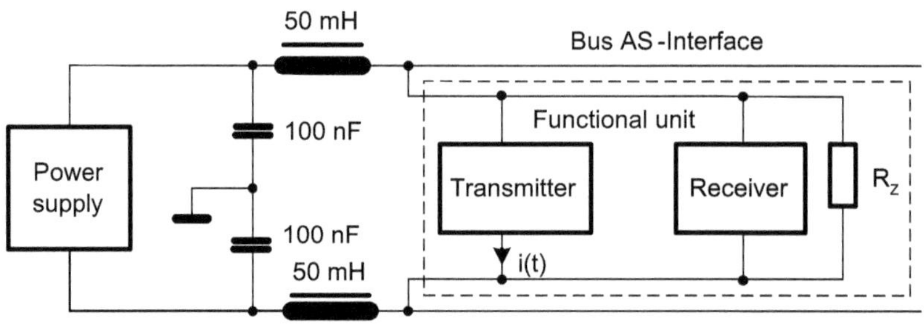

Figure 6.21 Connecting the power supply and functional unit to the AS Interface bus.

are of the *slave* type. In the original version of the AS-Interface 1.0 bus, which was created in the early nineties, it is possible to connect up to 31 slave devices. In 1998, the *AS-Interface* 2.1 specification was proposed, which expanded the number of connected slave devices to 62. Currently, there are devices on the market that support both the older version 1.0 specification and the newer version 2.1.

Another milestone for the development of the bus occurred in 2001, when another version of this bus was created with safety features according to the EN 954-1 standard and SIL3 safety support according to IEC 61 508. In 2004, a new version of the standard version 3.0 was created, which expands the possibilities of the bus for the transmission of 16-bit data both in multiple and in a single cycle at the expense of reducing the possible number of slave devices connected to the bus network.

The AS-Interface bus is fully compatible with *ProfiBus*, *DeviceNet, Interbus*, and the industrial *Ethernet* standard. The AS-Interface bus can be used to connect various components, such as PLC modules, power supplies, sensors, actuators, I/O modules, and safety modules, such as tactile, magnetic, inductive, pyroelectric, and image motion sensors.

An important feature of the bus is the transmission of both communication data and power over a single cable. The connection of the power supply and the method of connecting the functional unit to the bus are shown in Figure 6.21.

The power supply maintains the voltage on the bus at a nominal value of approximately 30 V, while this voltage should not fall below 27 V. A data signal encoded with *Manchester II* code is modulated to this supply voltage, which is transmitted by the unit, addressed to the transmitter function. The coded signal is used to control the current source, which transmits current pulses with an amplitude of 60 mA on the bus with a defined steepness of the rising and falling edges. These current changes induce voltage pulses of

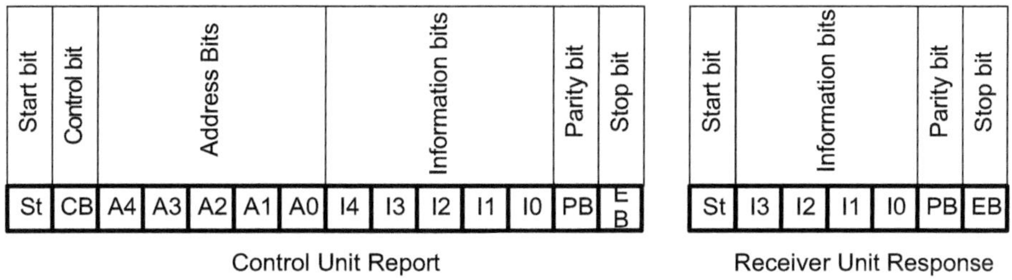

Figure 6.22 Structure of messages sent by the control and receiving unit.

both polarities with a range of approximately ±2 V on the chokes, which are detected by the decoder of the other functional units.

With this bus, transmission speeds of up to 180 kb/s can be achieved with a maximum number of connected units of 31. Data messages have a fixed length (Figure 6.22).

The advantage of the bus is the unlimited choice of its topology, which can be linear, star, ring, tree, or a combination thereof. The maximum cable length can be up to 100 m, and when using repeaters, up to 300 m. Version 1.0 allows you to connect up to 124 binary inputs and outputs, version 2.1 allows you to connect up to 186 binary inputs and outputs, and version 3.0 allows you to connect up to 248 binary inputs and outputs. The maximum transfer time on this bus is 5 ms.

6.4.7 Device Net

DeviceNet is a standard open bus network that is accessible to all manufacturers of automation technology for connection to the system and enables data transfer at the lowest level of sensors and actuators. The Device Net bus was designed by *Allen-Bradley* (*Rockwell Automation*) and is currently used by almost a hundred companies producing sensors and actuators.

The operation of the *Device Net bus* is based on the CAN (*controller area network*) communication protocol, which was originally developed for the *European automotive industry*. The bus enables the transmission of data packets with a length of 1–8 bits with a selectable transmission speed of 125, 250, or 500 kb/s with a data network length of up to 500 m.

6.4.8 M-Bus

The *M-Bus* (*meter bus*) standard is intended for remote reading of consumption meters, such as water meters, gas meters, electricity meters, heating systems, etc. It is also used to connect, e.g., security systems, flexible lighting installation, heating regulation, etc. This bus was developed based on the cooperation of the Technical University in Paderborn with the companies *Texas Instruments Deutschland GmbH* and *Techem GmbH*. The bus is defined by the EN 13 757 standard.

The *physical layer* of this bus consists of a standard shielded two-line telephone line, which can also be replaced by a *wireless connection* according to the *M-Bus Wireless* EN 13757-4 standard. An alternative connection for consumption meters is also a *modem connection*.

The standard configuration of the *M-Bus* system consists of one control station (*repeater*) and a maximum of 250 subscriber stations, while the line length should not exceed 1 km. Larger systems can also use more complex configurations, in which the system is divided into several zones that form system segments, controlled by sub-controllers.

To power subscriber stations along the same line, the M-Bus system uses changes in voltage levels and, in the opposite direction, changes in current consumption for data transmission from the control station to subscriber stations. In the direction from the control station to the subscriber stations, the log corresponds. 1 voltage +36 V at the output of the driver of the control station and log. 0 voltage +24 V. In the direction from the subscriber station to the control stations is log. 1 presented with a current consumption of 1.5 mA and log. 0 consumption in the range of 12–21 mA. Current drawn in log state. 1 can thus be used, for example, to power a galvanically isolated interface as well as the meter itself.

In idle state, there is a log level on the bus. 1, i.e., +36 V, and the consumption from the driver of the control station corresponds to the number of participating stations multiplied by the consumption of each of the participating stations in the log state. 1, i.e., 1.5 mA. With a larger number of subscriber stations, the current requirements for the driver of the control station increase. With a variable number of participating stations during network operation, this station cannot unambiguously detect the value of the consumed current but only its change. Also, the voltage at different points of the bus is variable and depends on the resistance of the line and the number of connected stations. Therefore, subscriber stations do not have to react to the absolute value of the current but only to changes in the voltage on the bus. Due to the large changes in bus voltage and current, this bus system has a high resistance to interference.

The *M-Bus system* uses an asynchronous transmission method for data transmission, which requires that there are no significant time delays between the individual transmitted characters. This method of transmission

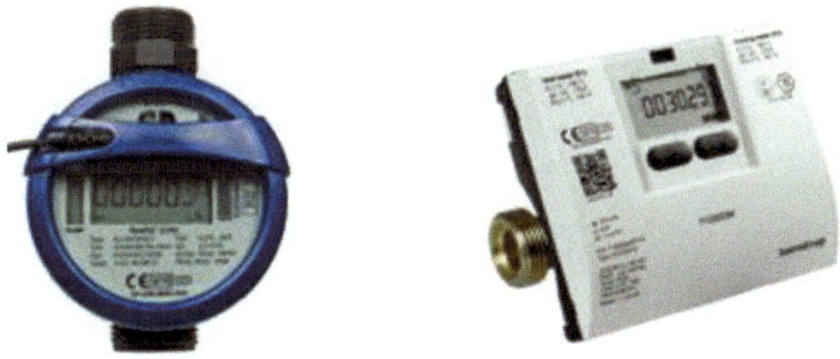

a) water meter b) heat meter

Figure 6.23 Water meter and heat meter with M-Bus data transfer.

therefore requires that the generation of the stop bit is immediately followed by the transmission of the next character. Figure 6.23 shows an ultrasonic water meter with M-Bus wired bus and a heat meter with *M-Bus wireless* transmission.

6.4.9 ARINC

The *ARINC serial interface* (*Aeronautical Radio, Incorporated*) is a serial data protocol for the transmission of data with a high degree of security, used mainly in large aircraft, but also in areas of strategic importance, e.g., military and police. The physical layer of this interface uses a two-wire data bus with Rx and Tx synchronization signals located on separate ports. Data words are 32 bits long and most messages consist of a single data word.

Messages are transmitted at 12.5 or 100 kbit/s to other system elements that monitor bus messages. The bus transmitter continuously transmits either 32-bit data words or a NULL (0 V) state. One wire pair is limited to one transmitter and no more than 20 receivers. The protocol allows for automatic timing at the receiver end, eliminating the need to transmit timing data. The *ARINC 429* interface version is an alternative to the MIL-STD-1553 standard.

Each ARINC 429 protocol word contains five fields:

- Bit 32 is the *parity bit* and is used to verify that the word has not been corrupted or garbled during transmission. Each ARINC 429 channel typically uses "odd" parity – there must be an odd number of "1" bits in a word. This bit is set to 0 or 1 to ensure that the correct number of bits is set to 1 in the word.
- Bits 30–31 are the *sign/status matrix* (SSM). These bits may have different encodings depending on the particular data representation applied

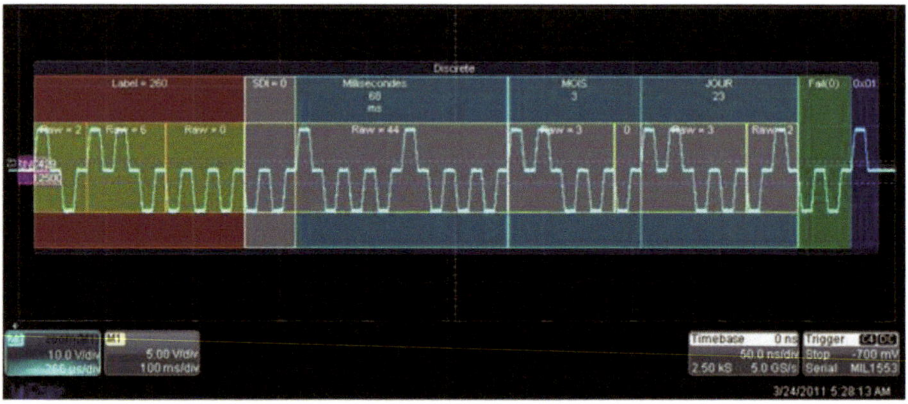

Figure 6.24 Timing of ARINC 429 interface signals.

to the given word. In all cases using SSM, these bits may be encoded to indicate:

- *Normal operation* (NO) indicates that the data in this word is considered valid data.
- *Functional test* (FT) indicates that the data is provided by the test source.
- *Failure warning* (FW) indicates an error that causes the data to be suspect or missing.
- *No calculated data* (NCD) indicates that the data is missing or inaccurate for a reason other than failure. For example, autopilot commands are displayed as NCD when the autopilot is not engaged.

In the case of *binary decade coded* (BCD) the SSM can also indicate the ± sign of the data or some analog sign information. In the case of BSC (*binary signed coding*), bit 29 represents the sign of the number, i.e., the sign indication is delegated to bit 29 in this case.

Bits 11–29 contain the data. Bits 9 and 10 are SDI (*source/destination identifiers*) and can indicate the intended receiver or, more commonly, indicate the transmitting subsystem.

Bits 1–8 contain a *flag word*, expressed in octal (MSB 1 bit numbering), identifying the data type.

Figure 6.24 shows the time course of the ARINC 429 interface signals, where the purple section (Label 200) indicates the label, the blue-green sections indicate the transmitted data, and the dark blue section indicates the parity bit.

The ARINC interface is further supplemented by the following protocols:

- ARINC 615 defines a high-speed data transfer protocol layered on top of the ARINC 429 physical layer.
- ARINC 629 defines a high-speed transmission method TDMA (xxx) extension to ARINC 429, replaced by AFDX.
- ARINC 664.7 defines the use of a deterministic Ethernet network as the avionics data bus in later aircraft such as the *Airbus A380* and *Boeing 787*. This standard defines a point-to-point virtual link implementing the same concept as in *ARINC 429*. Unlike 429, these links do not exist physically but as logical TDMA lines.
- ARINC 708 defines a protocol for transmitting weather radar data. While the 708 transmission is a simplified version of MIL-STD-1553, ARINC 708 component control has been standardized using ARINC 429 labels.
- ARINC 828 defines EFB (*electronic flight bag*) interfaces used in all types of aircraft and includes ARINC 429 interfaces among others.
- MIL-STD-1553, military bus standard for "shared serial bus," often used instead of ARINC 429.

6.4.10 FlexRay

The *FlexRay bus* is a communication bus used mainly in the automotive industry. It is designed to achieve high transfer rates and fault tolerance. The bus is standardized by a set of standards ISO 17458-1 to ISO 17458-5.

The *FlexRay* bus supports both star and line bus topologies. For increased fault tolerance, the bus uses two independent metallic lines with a pair of twisted shielded two-wire lines in the physical layer. The second (redundant) line can be used in more demanding applications to *control data transmission*. The system with this line achieves transmission speeds of up to 10 Mb/s.

Data transfer in this bus is carried out in time cycles, divided into static and dynamic segments. *The time cycle* is divided into two parts: a static segment and a dynamic segment. A *static segment* is reserved in advance for

Table 6.3 Principle of data transmission error correction.

Messages	Comment
000000 11111111 0000	Sent message An error-free transmitted message
00000000 11111111 00	delayed by two cycles
000000 1111**0**111 0000	An error near the middle of the eight-cycle region
00000000 11111111 00	corrected message
000**1**0 11111111 0000	Error near the boundary of the eight-cycle region
000000**1** 11111111 00	corrected message

types of communication that do not require priority. A *dynamic segment* takes control of the bus if the message being transmitted has the highest priority.

The *FlexRay* system consists of a bus and electronic control units (ECU). Each of these units generates a synchronization signal whose deviation from the relative reference signal is less than 0.15%. That is, if the sender unit transmits data signals to the receiver, then for every 300 transmission cycles, the receiver receives 299–301 cycles. The synchronization signal is sent in a *static segment.*

Each bit to be sent is held on the bus for eight clock cycles. The receiver stores the last five samples in its buffer. Thus, transmission errors in one cycle can only affect the results near the bit boundary but will not affect the cycles in the middle of the eight-cycle region (Table 6.3).

Errors in data transmission are thus moved to the following cycles.

The following signals are used in the system for communication:

- TSS transmission start signal, log. 0,
- FSS frame start signal, log. 1,
- BSS0 BSS7 eight-bit word,
- FES frame end signal, log. 0,
- TES end transmission signal log. 1.

If the TSS signal has a log level. 1, the bus is in idle state. If this signal goes to the log level. 1, then with the FSS signal by log. level 1 initiates the transmission of a data frame that contains an eight-bit word. The end of the frame is signaled by the transition of the FES signal to log. 0, followed by the transition of the TES signal to log. 1.

There are data converters for interconversion between *FlexRay*, *CAN*, and *USB* buses. A typical converter for this use is, for example, the *FlexRay CAN-USB* data converter. This converter enables both synchronous and asynchronous data transfer between individual buses. It contains one *FlexRay* channel, two *high-speed channels* and one *low-speed CAN* bus channel and one *USB channel.*

6.4.11 Standard IEEE 1451 and IEEE 1588

IEEE 1451, a collection of standards for the interface of smart sensors and other devices, was created by the TC-9 *Sensor Technology Technical Committee of the IEEE Instrumentation and Measurement Society*. The IEEE 1451 standard defines a set of common communication interfaces to standardize the connectivity of transducers (*sensors and actuators*) to instruments, instrument systems, or control networks that allow transducer accesses over local area networks or the internet. The standard can be used to build sensor networks with devices such as sensors and actuators and can facilitate the interoperability of devices and data in the *Internet of Things* (IoT), Industrial Internet of Things (IIoT), and *cyber–physical systems* for application from smart grids through smart buildings to smart manufacturing to smart healthcare to smart transportation systems to any system that includes sensors and actuators and sensor networks and their applications.

IEEE 1451 standardizes two interfaces, the converter interface, and its network interface. The converter interface is the interface between the converter and the network device, while the converter network interface is the interface between the network device and user applications.

In IEEE 1451 terminology, the conversion device is referred to as an IEEE 1451 TIM (transducer interface module). An IEEE 1451 NCAP (*network capable application processor*) network device is referred to as a network node or gateway device. IEEE 1451 also defines a set of metadata called TEDS (*transducer electronic data sheets*), containing information about transducers. These include manufacturer ID, serial number, measurement ranges, calibration data, location information, and more. TEDS enable self-identification and self-description of sensors into the system or network.

The IEEE 1451 set of standards provides a set of common interfaces that enable sensor manufacturers or users to support different networks, select sensors, and networks for their applications based on merit, and help users address the interchangeability of sensor and network devices. It further allows access to any transmission device on wired or wireless networks or the *Internet* using a common set of commands and responses in standardized message and data formats.

The IEEE 1451.0 standard defines a set of common functions, commands, and sensor electronic data sheets (TEDS) for the IEEE 1451 family of smart sensor standards. This functionality is independent of physical communication media. It includes the basic functions needed to control and manage intelligent transducers, communication protocols, and media-independent transducer electronic datasheet formats.

IEEE 1451.1 defines a set of common network services for smart converters. These common network services provide transmitter discovery, transmitter data access, and TEDS, event notification, and transmitter management services to support the interoperability of networked smart transmitters.

IEEE 1451.2 defines a digital sensor interface for connecting transducers containing sensors and actuators to a network. The previous version of the standard is revised to include serial interfaces such as SPI, I2C, and RS-232/UART. This interface standard works with the IEEE 1451.0 standard for accessing sensors and actuators.

IEEE 1451.4 defines a *mixed interface* (MMI) that allows the transmission of analog sensor signal and digital TEDS on the same wires. The standard specifies both single-wire and multi-wire transmission. In multi-wire data transmission, the TEDS information is first sent in digital mode, and then an analog signal is sent over the same wires (class 1) to instrument systems or networks. In Class 2 operation, TEDS and transducer data are sent simultaneously over parallel wires.

IEEE 1451.5 is a *wireless communication* interface standard for sensors. The standard adopts existing wireless communication protocols such as IEEE 802.11x (Wifi), IEEE 802.15.1 (Bluetooth), IEEE 802.15.4 (ZigBee), and 6LowPAN. 6LowPAN is an IP-driven protocol (compatible with IPv6) that allows direct access of sensors from the *Internet.*

IEEE 1451.7 specifies the communication methods, command set, data formats, and transmitter electronic data sheet (TEDS) for sensors working in conjunction with radio frequency identification (RFID) systems.

IEEE 1451.1.4 defines a method for carrying messages over a network using the *eXtensible messaging and presence protocol* (XMPP) to establish session initiation, secure communication, and distinctive identification between network client and server devices using device meta-identification information. This standard provides significant cost savings and reduces complexity while leveraging current industrial instrumentation and equipment.

IEEE 1588 is a standard known as the *precision time protocol* (PTP). It enables accurate clock synchronization in networked measurement and control systems. It supports system-wide clock synchronization accuracy in the sub-microsecond range. The standard is applicable in many areas such as instrumentation and measurement, industrial automation, power generation, avionics, communications, and more.

References

[1] Dawoud, P. *Serial communication protocols*. River Publishing. 2000. ISBN 978-87-7022-154-2.

[2] Busse, R. *Feldbuss systeme um Vergleich*. Pflaum Verlag Munchen, Munchen 1996. ISBN: 978-37-905-07-225.

[3] Turner, J. *Automotive Sensors*, Momentum Press 2009, ISBN 978-16065-0009-5.

[4] *RS-485 Reference Guide*, archived from the original (PDF) on 2018-05-17.

[5] *Galvanic isolated CAN FD repeater reference design*. CAN Newsletter 1/2019.

[6] *Comparing CAN, CAN FD, and Ethernet. Part 1*. CAN Newsletter 2/2021.

[7] *Comparing CAN, CAN FD, and Ethernet. Part 2*. CAN Newsletter 3/2021.

[8] *Galvanic isolated CAN FD repeater reference design*. CAN Newsletter 1/2019.

[9] http://www.m-bus.com.

[10] Lee, K. IEEE 1451: *A standard in support of smart transducer interface networking*. Proc. IMTC 2000, s. 525–528, 2000.

7

Modular Measuring Systems

These systems are made up of mechanically and electrically standardized *functional modules* that are inserted into the system frame, which, in addition to the system controller module, also contains a power supply including cooling. One of the oldest used modular systems is the CAMAC system, which was widely used in the seventies of the twentieth century for data collection and processing, especially in the field of nuclear technology. Based on this system, modular systems SCXI, VME, and VXI were developed in the following years. Currently, the most powerful modular system is the PXI system. All these modular systems are characterized by precisely defined mechanical and electrical properties and are intended for the collection and processing of large datasets, including the time and amplitude discretization of rapidly time-varying continuous signals.

7.1 CAMAC Measuring System

The CAMAC (*computer automated measurement and control*) measurement system was based on one of the oldest NIM (nuclear instrument modules) modular systems. The CAMAC system has been standardized by the IEC organization since 1975.

The CAMAC system is made up of a 19-inch frame, allowing the placement of up to 24 functional modules. The frame includes a frame controller and a ±6 V, ±12 V, and ±24 V power supply. The functional modules of the system enable amplification, comparison, time, and amplitude discretization of time-varying signals, including their digital processing and data transfer to external devices (Figure 7.1).

Unique board addressing and standard communication commands are used to communicate measuring modules with the controller. The basic version of the frame contains a 24-bit data bus with a maximum transfer rate of up to 1 Mb/s. When using block transfers, transfer speeds of up to 10 Mb/s can be achieved. Frame controllers are used for communication between

Figure 7.1 Modules of the Le Croy CAMAC measuring system.

system units, which also enable the transmission of control signals and data to an external computer via Ethernet 10/100 and USB interfaces. 2.0.

The last innovation of this system took place in 1984 under the name FASTBUS, which instead of transistor-coupled digital circuits TTL (*transistor to transistor logic*) uses emitter-coupled digital circuits ECL (*emitter-coupled logic*) in order to achieve higher transmission speeds. This system was replaced in the following years by the more powerful VME, VXI, and PXI systems.

7.2 VME Measuring System

The VME (*Versa Module Europe bus*) system, which is conceptually based on the SCXI (*signal conditioning extensions for instrumentation*) system, was standardized by IEC in 1987 with the ANSI/IEEE1014-1987 standard. The structural solution of the system uses mechanical elements and connectors of the *Eurocard system* (DIN 41612). The VME system consists of a frame (*mainframe*), into the slots of which up to 21 modules can be inserted (Figure 7.2).

Figure 7.2 VME modular measuring system.

The VME standard defines two plug-in sizes. Single modules have one three-row JP1 connector with 3 × 32 contacts, and the double module has two of these connectors. Signals are defined on this connector, enabling full inclusion of the module in the measuring system.

The assignment of the signals on the JP2 connector is only partially defined, and is mainly used to expand the data and dress bus from 16 to 32 bits. The extreme ends of the contacts of this connector are determined by the user for his own selected functions.

7.2.1 VME system bus

Because the VME system is *multiprocessor*, that means that several processor modules can work in one frame and can share the resources of other modules through a common bus. From a functional point of view, the VME bus can be divided into a data transfer bus (DTB), an arbitration bus DTB (*DTB arbitration bus*), a priority interrupt bus (PIB), and a service bus (*unity bus*) (Figure 7.3).

Buses for data transmission are made up of address, data, and control lines.

The signals belong to the address lines of the bus, which determine the addressing of the modules:

Signals D00–D31 (data bus) belong to the bus data lines.

- A01–A31 (*Address Bus*) of the address bus.
- AM0–AM5 (*Address Modifier*) address modifiers.
- DSO (*Data Strobe 0*) data valid DS0.

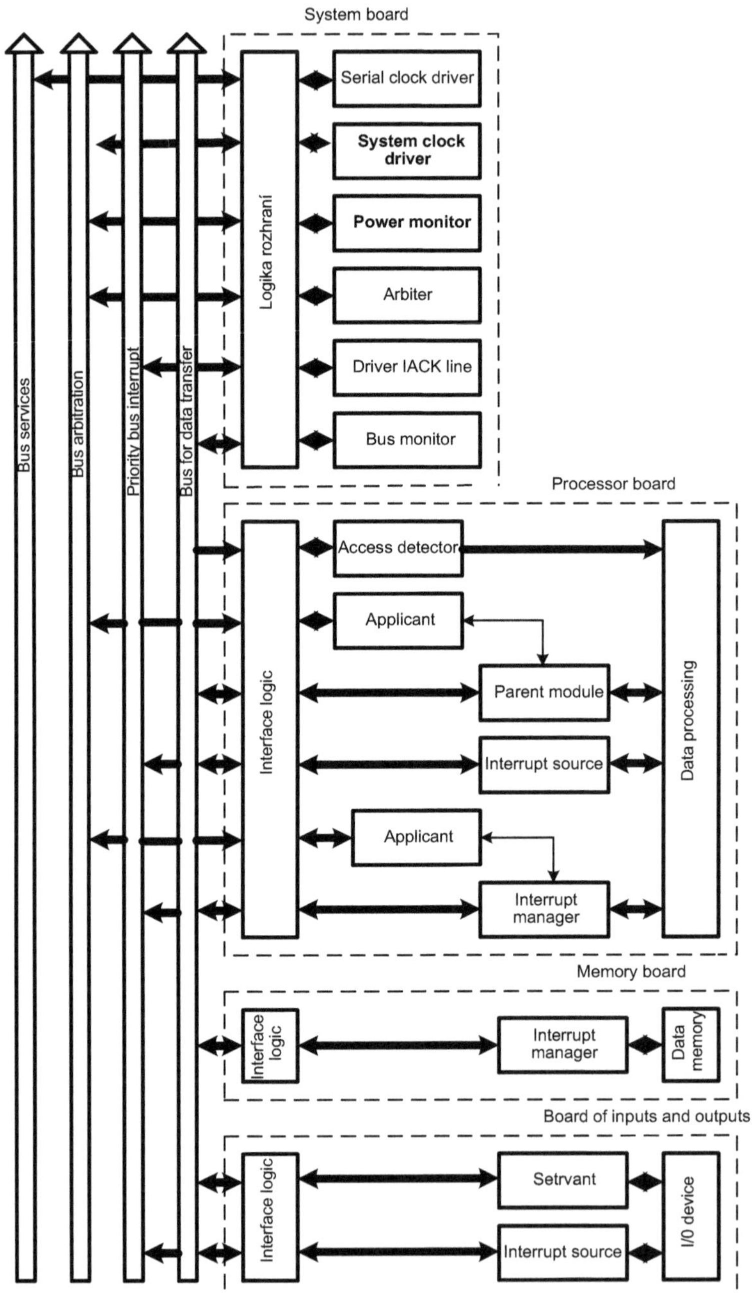

Figure 7.3 Block structure of the VME system with buses and functional modules.

- DS1 (*Data Strobe 2*) data valid 1.
- LWORD (*Long Word*).

Signals D00–D31 (Data Bus) belong to the bus data lines.

The control lines that control the transfer of data include the signals:

- AS* (*Address Strobe*) address valid.
- DSO* (*Data Strobe 0*) data valid 0.
- DS1* (*Data Strobe 1*) gift valid 1.
- BEER* (*Buss Error*) bus error.
- DTACK* (*Data Acknowledge*) confirmation of data transfer.
- WRITTE* (*Writte*) write data.

Note: signals ending with * are active at log level. 0.

The signals DS0*, DS1*, and LWORD*, together with the level of the A0 signal, determine which of the 4 bytes of the 32-bit bus will be transmitted in a given cycle. Thus, one-, two-, three-, and four-byte data files can be transmitted.

Address modifiers define the address space, i.e., the number of decoded address bits for a given DTB cycle and specify the access rights of this cycle. With the VME system, it is possible to create so-called short, designated signals A01–A15, standard addresses, designated signals A01–A23, and extended addresses, designated signals A01–A31. These three address spaces are parallel, that is, the same addresses in different address spaces address physically different locations.

Address modifiers allow defining *non-privileged* and *privileged* (*supervisory*) access to data. The AS* line (*address valid*) indicates the validity of the levels on the address lines, on the address modifier lines, and on the LWORD* line. Lines DS0* and DS1* announce the validity of the data takeover during the *write cycle*, and announce their location on the data lines during the *read cycle*. The BEER* signal is an alternative to the DTACK* signal. Both signals confirm successful data transfer. The WRITTE* signal controls the direction of data flow between modules.

When transferring data, we define a *superior* (*master*), a *subordinate* (*slave*), a bus timer (*bus timer*), and an access detector (*location monitor*). In a VME system, data transfer control is not fixedly assigned to a single module, but can be delegated to other modules as well. The bus monitor detects with an active BEER* signal that the data transfer has not been terminated with a DTACK* signal, e.g., when attempting to write or read data from an address where physical memory is not present.

7.2.2 Data transfer cycles

The basic data transfer cycles are the read and write cycles. The *write cycle* is given by the transfer of data from the master to the slave, the *read cycle* by the transfer in the opposite direction.

During the read cycle, the address, its address modifiers, and the WORD* signal are first placed on the appropriate lines and the AS* signal is made active. First, the *supervisor* decodes the address and, if the address is valid, then activates the transmission system. *Subsequently*, the slave decodes the address, and if this address is valid, then activates the transmission system. The *supervisor* then further specifies the log reading cycle. 1 of the WRITE*a signal, which generates a combination of DS0* and DS1* signals according to the required transfer cycle. *Subsequently*, the slave places the required data on the data bus and activates the DTACK* signal, and the master takes over the data (Figure 7.4).

If it is the last cycle of data transfer, the jersey lines and lines of their modifiers are released with the LWORD* signal. *Subsequently*, both DS0* and SS1* lines and the AS* signal are entered into the log. 1. Finally, the *slave* releases the data bus with the DTACK* signal, and the *master* releases both lines DS0* and DS1* to the idle state, corresponding to the log level. 1.

The *read–modify–write cycle* is conditioned by the use of a multiprocessor VME system, where access to common system resources is coordinated through so-called semaphores. During this type of cycle, handover of data transfer control to another *master* is avoided, and thus consistency of access to the semaphore is guaranteed. This cycle differs from the previous one in that the AS* signal is permanently active from the moment the address is placed when reading the semaphore until the moment the cycle ends.

The *address cycle* is the only cycle in which there is no data transfer. During this cycle, an address is placed on the address bus and the AS* signal is set to the active log level. 0. Then the cycle is finished and another cycle can follow. The use of this addressing cycle makes it possible to increase the transfer speed of processor cards, as it allows generating the following addresses in advance, before the given cycle is implemented.

The *block transfer cycle* (*block transfer*), already used in CAMAC systems, makes it possible to significantly increase the system's transfer speed. The principle of block transfer is that only the address of the first memory location is transmitted and the addresses of other locations must be generated

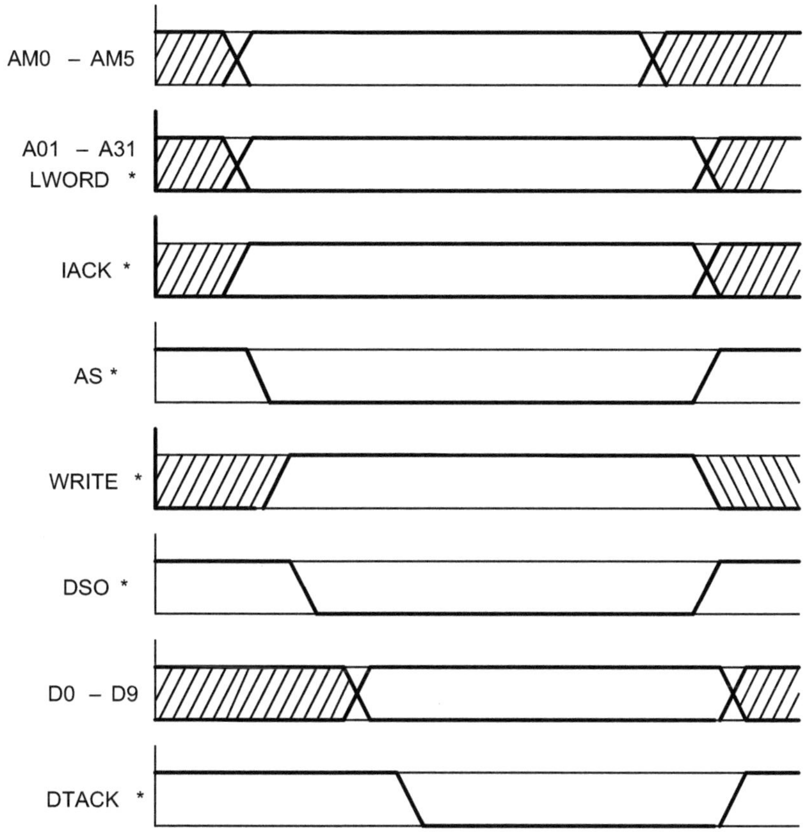

Figure 7.4 Time diagram of the read cycle in the VME system.

by the *slave* itself based on impulses on the DS* lines, which control this transfer with the confirmation signal DTACK*. The usual length of the block transfer is 256 8-bit bytes.

Individual types of cycles can be implemented with *short addresses* A16, *standard addresses* A24, and *long addresses* A32, while it is valid that if the system supports A32 addressing, it must also support A24 and A16 addressing, and if the system supports A24 addressing, it must also support A16 addressing. All cycle types can transfer data between 1 and 4 bytes long. Devices capable of transmitting only single-byte (8-bit) data use only odd addresses and are designated as D8. Devices capable of transmitting double-byte (16-bit) data are referred to as D16. Devices capable of transmitting 4-byte (32-bit) data are referred to as D32, while devices capable of transmitting more bytes of data are capable of transmitting less bytes of data.

7.2.3 Communication in the VME system

Since the VME bus is intended for data transfer of multiprocessor systems, it is necessary to prevent collisions during data transfer between these individual systems. This activity is entrusted to individual *supervisors* through the assignment bus (arbitration).

The *signal lines* of the assignment (*arbitration*) bus are:

- BE0*–BR3* (*Bus Request* 0 to 3).
- BBSY* (*Bus Busy*).
- BCLR* (*Bus Clear*) release of the bus.
- BG0*–BG3* (*Bus Grant 0 to 3*) bus allocation.

Signals BR0*–BR3* determine the request to take over the data transfer. The outputs of their drivers are of the open collector (OC) type, which makes it possible to request the simultaneous acceptance of data transmission by several devices on one data line. The priority of allocation of individual data lines is determined by the type of decision-making. Signals BG0*–BG3* are intended to transmit data transfer control. Although these lines are not bus lines with OC-type drivers, they create a chain (*Daysi-Chain*) between the individual slots of the VME system. Inputs and outputs of individual lines BG0*–BG3* are then designated as BGxIN*–BGxOUT*, where x is 1–3. Line BBSY* at log level. 0 informs the device that is the parent during data transmission that the takeover request has a higher priority.

Figure 7.5 shows the chain structure between VME bus slots.

The arbitration bus functional modules are defined by their *arbiter* and *requester*. In each VME system, there is always only one arbiter in the module, which is inserted into slot 1. Its task is to control the allocation of data buses either to individual superiors or to *interrupt managers* (*arbiter integrant*).

The *requestor function* module is implemented for those devices that require occasional data transfer. These are such devices where the parent function module or the interrupt *manager function* module is implemented.

The VME standard allows for three interrupt operation algorithms, i.e., *priority algorithm*, *cyclic priority algorithm*, and *single-level algorithm*. In addition to these algorithms, the VME system also allows their combination and/or another type of decision algorithm.

The PRI (*priority algorithm*) allocates the data bus on the basis of an unchanging priority from signals BG0* with the lowest priority to BG3* with the highest priority. This algorithm puts the BCLR* signal into log. 0 if a request with a higher priority than the priority of the data block currently

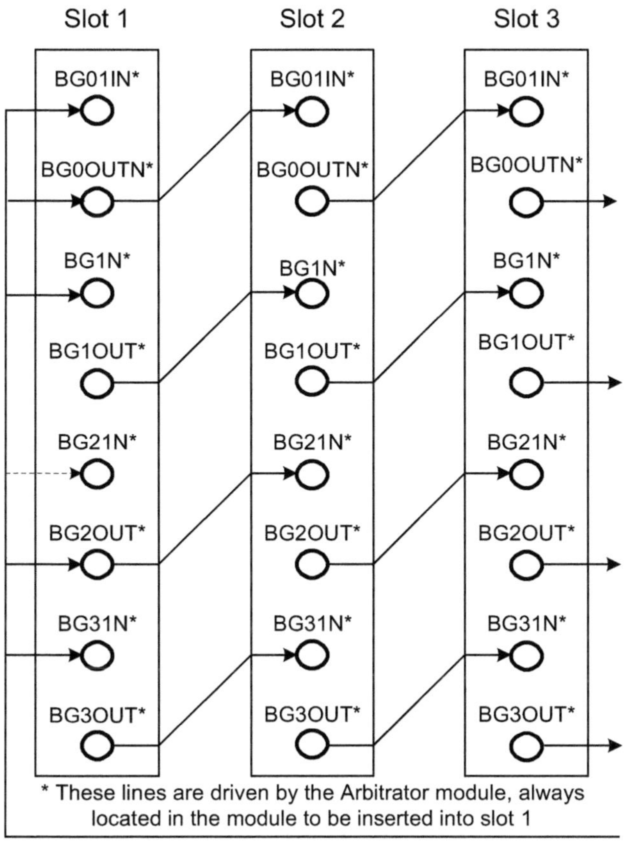

Figure 7.5 Chain structure between VME bus slots.

being transmitted is detected. This algorithm does not limit the time in which a device must give up data transmission with the current priority.

The arbitrator with *cyclic priority* RRS (*round robin select*) shifts the priorities of individual data lines after each data arbitration. If, for example, in the *n*th arbitration the line BR3* has the highest priority and BR0* the lowest priority, then in the arbitration $N + 1$, the line BR2* will have the highest priority and the line BR3* the lowest priority. The SGL (*single level*) algorithm affects only the BR3* line and therefore does not perform any arbitration.

The *requester function* module also ensures that data line management is taken over and participates in its own arbitration through chained BG*x** lines. The VME standard defines three basic ways of behavior of the applicant RWD (*release when done*). The requester releases the BBSY* line in

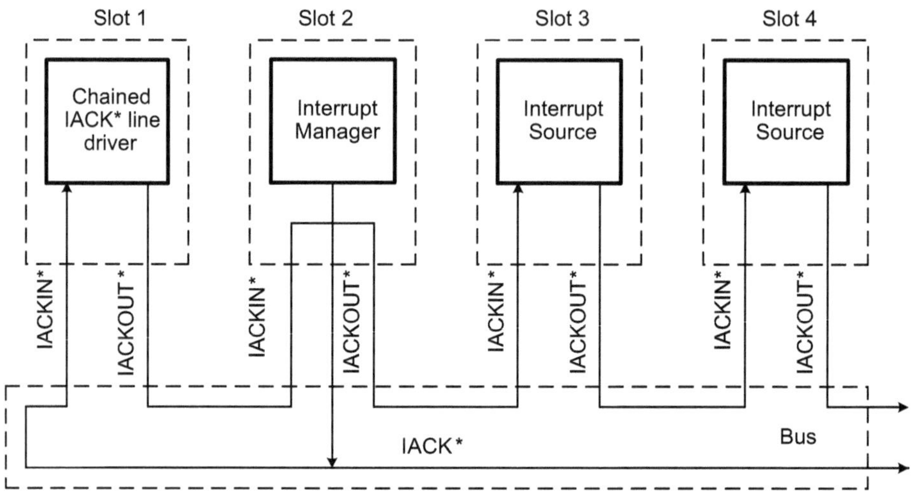

Figure 7.6 Structure of the interrupt system.

the event that the device has completed all necessary data transfers at any level. ROR Requester (*release on request*) releases the BBSY* line when it detects a data transfer request at any level. In this way, the number of arbitration cycles can be reduced. The FAIR requester will not request a bus allocation again if it detects an unacceptable bus allocation request on its BRx* line.

7.2.4 Priority interrupt bus

For the VME bus, either a *centralized* or a *distributed interrupt handling system* can be chosen. The definition of interrupts is defined similarly to the way buses are allocated, by definitions of signal lines, function modules, and their interactions. In addition to the priority bus, the data bus and the arbitration bus also participate in this process.

The signal lines of the *interrupt priority* bus are the signals:

- IEQ1*–IRQ7* (*Interrupt Request 1*) interrupt request.
- IACK* (*Interrupt Acknowledge*) confirmation of request interruption.
- IACKIN*/IACKOUT* (*Interrupt Acknowledge Daisy Chain*) chained interrupt acknowledge line.

The structure of the IACKIN*/IACKOUT* line is identical to that of the BGx* arbitration lines, and these lines have a similar role to that of determining the highest priority interrupt (Figure 7.6).

Lines IRQ1*–IRQ7* are interrupt request lines. The drivers of these lines are again of the open collector type. Therefore, multiple interrupt requests can be active on each of them at the same time. The IACK* line is used to signal the interrupt acknowledge cycle. This cycle also uses data bus lines.

The interrupt priority bus functional modules are the interrupt manager, interrupt source, and interrupt driver modules.

The *Interrupt Handler* defines IRQ1–IRQ7 for each of the lines. This solution allows the implementation of a completely distributed interrupt system in which each of the interrupt requests can be served by a different device. The interrupt manager handles multiple IRQ*x** lines, selecting interrupt requests with the highest priority each time.

According to the moment at which the interrupt source cancels its interrupt request, these sources are divided into two types. The RORA (*release on register access*) interrupt source cancels an interrupt request based on the interrupt handler's access to the specified register. A ROAK (*release on acknowledge*) interrupt source will clear an interrupt request based on the activation of the appropriate interrupt cycle. The line's IACK*/IACKOUT* interrupt driver, which is another module of the interrupt cycle, generates its falling edge at the start of it at any time and any interrupt manager initializes the interrupt acknowledge cycle. This module must always be initialized in the device intended for slot 1.

7.2.5 Service bus

The service bus in the VME interface includes signal lines providing system-oriented functions, a clock signal, and a serial bus.

The *service bus signal lines* are:

- SYSCLK (*System Clock*) 16 MHz system clock signal with a ratio of 1:2 for general use.
- SERCLK (*Serial Clock*) serial clock signal.
- SERDATA (*Serial Data*) serial data, which creates another communication signal with the SERLOCK line.
- ACFAIL (*AC Fail*) power failure, active in case of power failure.
- SYSRESET (*System Reset*) system reset, reset to default state.
- SYSFAIL (*System Failure*) system error, active in the event of a negative self-test result of any of the system elements.

In the event of a power failure, the VME system reacts with a mutual time link of ACFAIL, SYSRESET, and SYSFAIL signals.

The SYSCLK system clock driver is physically located in the slot 1 module. The driver generates a time-symmetric clock signal with a frequency of 16 MHz. This signal is also provided by the clock signal for the serial channel on the SERDAT* line. The specification of this signal is given by the IEEE P1132 standard.

The ACFAI power monitor monitors the supply voltage failure on the mains feed upstream of the source so that the time constant of its filters does not affect its operation. This monitor must also always be placed in the module intended for slot no. 1.

7.2.6 VME 64 measuring system

The VME64 measurement system was created based on the requirement to increase the transmission speed of the original VME system, specified by the IEEE 1014-1987 standard, which is 40 Mb/s for 32-bit block data transmission. The VME64 system allows you to increase the transfer rate up to 80 Mb/s and extends some transfer protocols for automated identification and configuration of this system.

Unlike the VME system, the VME64 system is characterized by changing the originally three-row connectors P1 and P2 to five-row connectors and inserting an additional P0/J0 connector between the P1/J1 and P2/J2 connectors. In addition, the VME64 system has a higher power supply and additional levels of supply voltages. It is also characterized by different slot addressing, increased resistance to electromagnetic interference, and a new definition of the content of the configuration cycle.

The *configuration cycle* of the VME64 system, called 2eVME, differs from the cycle used by the VME system, in that in the VME system the falling edge of the DTACK* signal responds to the falling edge of the DS* signal. For the VME64 system, the falling edge of the DS* signal informs the *master* about the presence of data on the bus. The *slave* confirms their acceptance with the rising edge of the DS* signal, notifying the master of the location of additional data on the data bus. The acceptance of which is confirmed by the *subordinate* by the rising edge of the DTACK* signal (Figure 7.7).

In this way, the transmission speed of the system can theoretically be doubled, and when using the multiple data transmission method MBLT (*multi byte letter transfer*), a transmission speed of up to 160 Mb/s can be achieved.

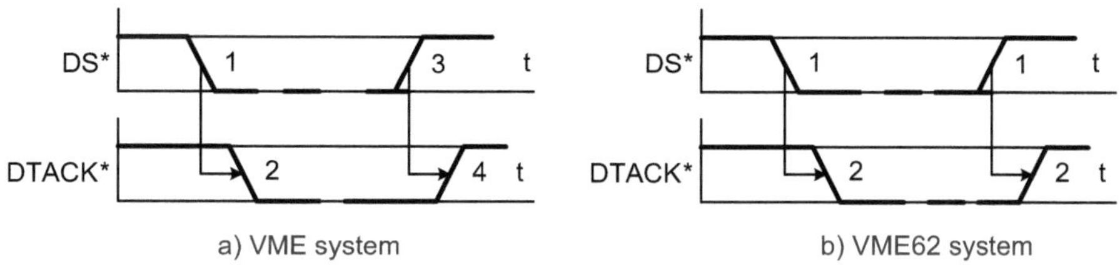

Figure 7.7 Time course of signals during data transmission.

In addition, 2eVME cycles enable *multiphase addressing*, in which not only data but also addresses, their simple and extended modifiers, and the expected sizes of data blocks are transferred during the individual data transfer phases.

In addition, this transfer cycle allows both *superiors* and *subordinates* to terminate their transfer prematurely.

The VME64 system allows, unlike the VME system, to define the use of the so-called *undefined signal lines* of the J2/P2 connectors, protocols for exchanging modules on the fly, new types of controllers, and other extension protocols, e.g., *Autobahn*, *VSB Bus*, and the 2eSST protocol.

The *Autobahn protocol* system was introduced by PEP *Modular Computer* based on the use of fast emitter-coupled ECL (*emitter-coupled logic*) circuits. This system uses NRZ (*non-return to zero*) coding for the serial data transfer method, with log. 1 corresponding to positive voltages and log. 0 negative voltages, which are derived from the supply voltages of the VME system. With the *Autobahn I* version, transfer speeds of up to 200 Mb/s can be achieved, and with the *Autobahn II* version up to 400 Mb/s, which corresponds to transfer speeds of 1.6 and 3.2 Gb/s.

The *VS Bus protocol* uses undefined 32-bit bus lines to create an I/O subsystem independent of the underlying VME bus. In this way, the basic bus is relieved of less demanding operations and is reserved only for operations requiring high computing power.

The protocol *2eSST* (*source synchronous transfer*) works with *synchronous data transfe*r, in which individual data transfers are not confirmed. By this simplification, at the expense of the reliability of data transmission, data transmission speeds of up to 320 Mb/s are achieved.

The *VME64 standard* also defines the procedure for inserting and removing individual functional modules during operation. The range of frames and functional modules of the VME and VME64 measuring systems currently includes several hundred units of various functions.

7.2.7 Functional modules of the VME measuring system

VME mainframe system frames manufactured in the range of 6–20 slots, with a 20-slot frame having a width of 12", which corresponds to standard instrument frames. These frames contain an air-swept power supply that sits below the VME bus. The first few positions of the frame are intended to accommodate the VME system controller, which is usually based on several MC 680xx processors from Motorola or PowerPC. Data memories with capacities of up to several hundred MB are intended for data recording, which are designed as *two-gate memories* with access by *local processors* and a *communication interface*.

Functional modules equipped with A/D converters with a resolution in the range of 12–16 bits enabling the digitization of sensor signals with sampling frequencies in the range of units of S/s to hundreds of kS/s. Another group of functional modules are separate amplifiers, multiplexers, A/D converters, and multiplex DAQ (*data acquisition system*) data collection systems.

The range of VME modules also includes modules enabling communication with external units via RS-232 and RS-485 interfaces and industrial buses *ProfiBus*, *BitBus*, *CAN*, *InterBus*, and *Ethernet*. More complex modules, containing *local processors*, allow working without direct cooperation with the system controller, which gives the possibility of using them for more demanding data and control operations.

7.3 VXI Measurement System

The VXI system, referred to as the extended VME system (*VMEbus Extension for Instrumentation*), is conceptually and structurally based on the VME system. In 1987, based on the cooperation of *Hewlett Packard*, *Colorado Data Systems*, *Racal Data*, *Tektronix*, and *Wawetek*, the VXI standard was launched and the *VXIBus consortium* was subsequently founded. The VXI IEEE Std. 1155 system standard was further gradually supplemented by versions VXI-1 to VXI-11.

7.3.1 Spcification of VXI system

The VXI system, like the VME system, is an open modular system. It consists of a basic unit (*mainframe*), which, in addition to the power supply, also contains a maximum of 13 (*slots*), while, similarly to the VME system, one or several leftmost positions are intended for the location of the system controller (Figure 7.8).

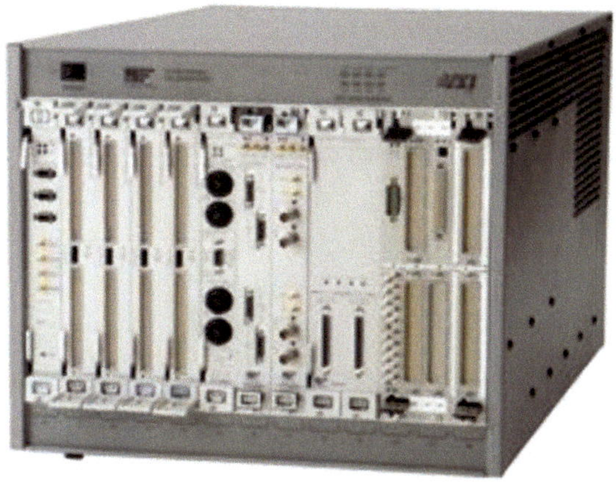

Figure 7.8 VXI measurement system.

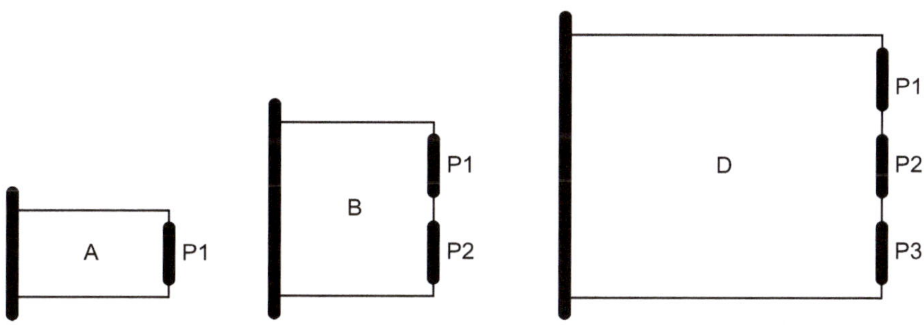

Figure 7.9 VXI system plug-in formats.

The VXI standard defines four types of plug-ins. Type A with dimensions of 100 × 160 mm has one double row connector P1, type B with dimensions of 233 × 166 mm has two connectors P1 and P2, type C with dimensions of 233 × 340 mm also has two connectors P1 and P2, and type D with 366 × 340 mm has three connectors P1, P2, and P3. Unlike the VME system, which has a slot spacing of 20.3 mm, the VXI system has a slot spacing of 30.5 mm, which allows more effective electromagnetic shielding to be used (Figure 7.9).

7.3.2 VXI system bus

In contrast to the VME standard, the VME standard defines the signals on the outer columns of the three-row connector P2 on size B, C modules and

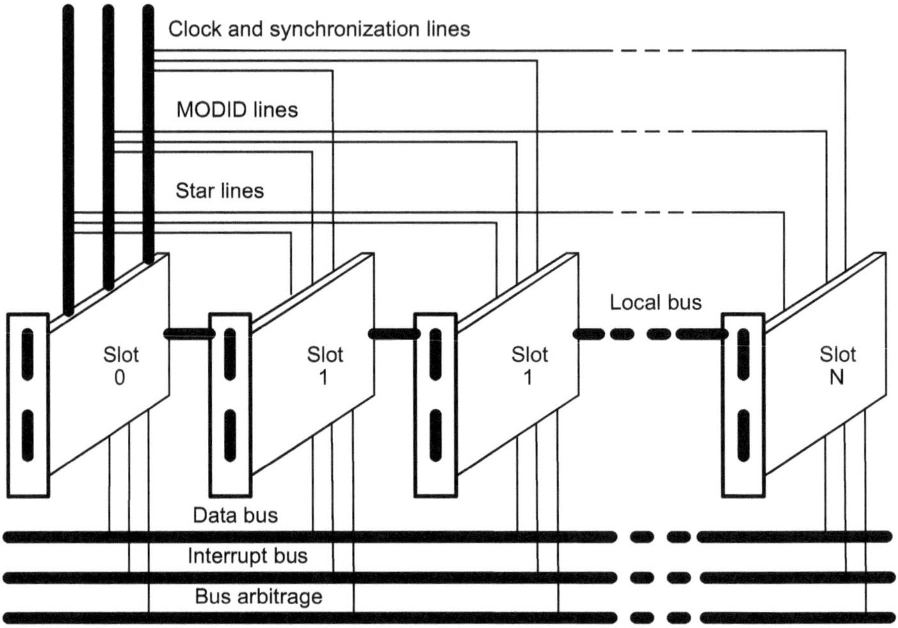

Figure 7.10 Structure of VXI system bus connections.

connector P3 on the size D module. Based on this extension, the VXI system supports additional instrumentation applications, creates additional data transfer paths between modules, and supports automated configuration. The connection structure of the VXI bus slots is shown in Figure 7.10.

At the end of connector P2, the *clock signal* CLK10 is output, the frequency of which is 10 MHz. CLK100 and SYNC100 signals with frequencies of 100 MHz are output on connector P3. These differential signals are generated by ECL drivers, with each position of the system being driven by separate drivers in order to suppress mutual influence. The source of all these clock signals is located in the module at position 0. These clock signals are used to synchronize all modules in the VXI system.

On the P2 connector there are eight TTL synchronization signals TRIG0–RIG7 and six ECL signals RIG0–RIG5. The four remaining TRIG6–RIG9 ECL synchronization signals are located on the P3 connector.

The most important difference between the TTLTRIG* signals and the ECLTRIG* signals is in the achievable transfer rate of the sync pulses. For example, pulse transmission by log. TTL levels are 12.5 MHz, and ECL levels are 62.5 MHz.

For TTLTRIGx signals, the VXI standard defines five protocols:

- *TTLTRG* synchronous trigger protocol* generating trigger signals on one TTLTRG line (bus conductors), which does not require confirmation from the receiver with a maximum frequency of 12.5 MHz.
- *TTLTRG asynchronous trigger protocol* generating two TTLTRG* signals, one of which is a synchronization signal and the other is a confirmation signal with a maximum of 12.5 MHz.
- TTLTRG* *clock transmission*, designed to distribute a clock signal with a maximum frequency of 12.5 MHz.
- TTLTRG* *data transmission* defining several TTLTRG signals used for parallel data transmission, while one of the lines is used as a clock signal, on which they descend, or leading or on both edges, data is transmitted with a maximum frequency of 12.5 MHz.
- TTLTRG* *start/stop protocol* defining synchronous start-up with the possibility of stopping the operation of the modules, while the TTLTRG* signal is used as a qualifier for determining the leading edge of the CLK10 signal.

Similarly, six protocols are defined for ECLTRG signals with a maximum repetition rate of 62.5 MHz. The extended start/stop protocol (ECLTRG) extends the use of the TTLTRG* start/stop protocol to devices that use more precise synchronization of modules using the CLK100 and SYNC100 signals. These protocols are optional and the VXI system manufacturer may use other synchronization protocols.

7.3.3 VXI system local bus

This bus is made up of lines connecting two adjacent system slots.

The local bus creates a fast parallel path for communication between adjacent modules. 12 and 24 local bus lines are defined on the P2 connector and 24 on the P3 connector. Since the signal voltage levels of individual devices (modules) can vary significantly, the VXI standard defines five device classes: TTL, ECL, ANALOG LOW (±5 V), ANALOG MED (±16 V), and ANALOG HIGH (±42 V). In order to prevent incorrect connection of modules to this bus, the modules are equipped with non-interchangeable keys on their front panel.

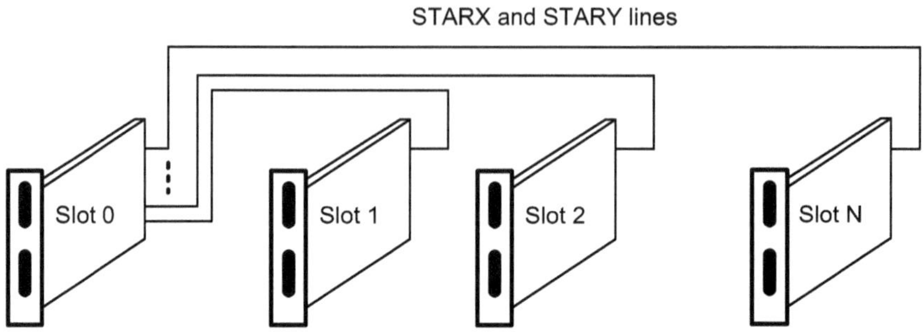

Figure 7.11 VXI system star structure.

7.3.4 SUMUBUS bus

This bus consists of a single line, creating an analog addition node. Both ends of this bus are terminated with 50 Ω resistors to suppress the reflection of fast impulse signals. Individual devices (modules) excite this line with *current pulses* and at the same time monitor its power supply using isolating amplifiers with high input impedance. One of the typical applications is, for example, the *generation of shaped oscillations*, created by combining the signals of two or more generators. This solution is often simpler and therefore cheaper than the generation of these shape signals by special functional *digitally controlled generators.*

7.3.5 STARX and STARY lines

These lines, defined on connector P3, form a *star communication structure*, in the center of which is the device located at position 0 (Figure 7.11).

STAR lines are bidirectional differential lines using ECL logic circuits. A *cross switch* is located in the device on slot 0, allowing any two system modules to be connected together. The STARX and STARY lines thus create two independent systems. Their implementation is not mandatory.

7.3.6 VXI system module identification lines

These lines, also called MODID (*module identification*), create an individual connection between slot 0 and the other slots of the VXI measuring system. These lines are determined by the resource manager, located on slot 0, which devices are inserted into the individual slots of the system, or which devices are non-functional (Figure 7.12).

In addition to the technical extensions mentioned above, the VXI standard also enables system-oriented extensions. These extensions make it

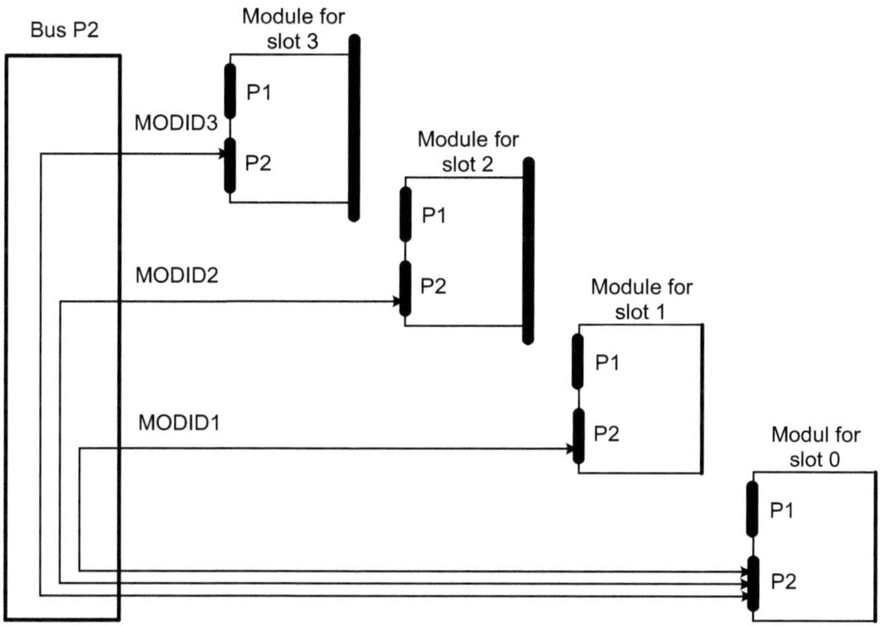

Figure 7.12 VXI system module identification lines.

possible to create hierarchical structures, create system configurations, and standardize communication protocols between individual system elements.

The hierarchical structure of the VXI system defines a *commander* and a *servant*. A device with a declared *master function* can initiate communications with a device with a *servant function*, while the *master* can only communicate with such a *servant* that is subject to it. This structure can also be multi-level. The given device is then a *servant* with respect to the *master device* standing in the hierarchy above it, and conversely a *master* with respect to the devices standing in the hierarchy below it.

At the top of the entire multi-level hierarchy is always the *resource management device*. It always has a logical address of 0 and identifies all VXI devices located in its frame, manages the system *self-test*, configures the A24 and A32 address maps, configures the *master/slave* hierarchy, commands IRQ lines, and initializes normal system operation.

All devices are uniquely identified by their logical address in the VXI system. It is always 8-bit, i.e., the system has 256 logical addresses. Address 0 is always reserved for the *resource manager*, addresses 1–254 are reserved for application purposes, and address 255 is reserved for the *dynamic system configuration algorithm*. In connection with this addressing, it should be noted that a logical address does not identify a physical module, but a

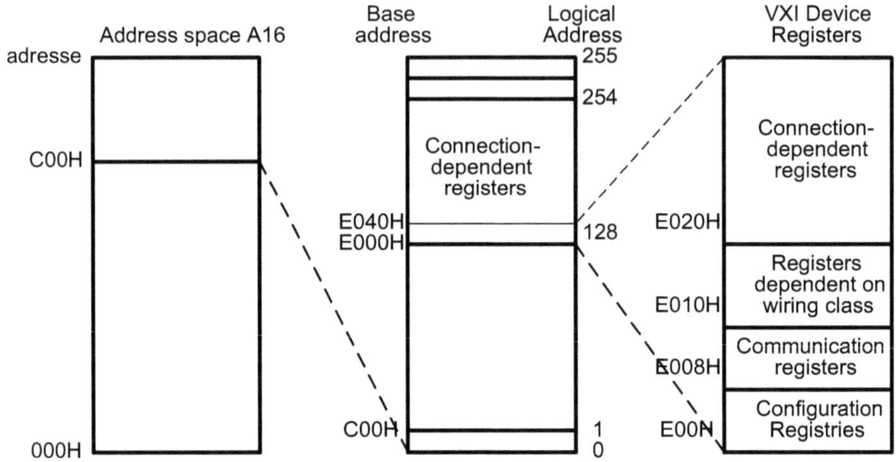

Figure 7.13 Logical and base addressing in the VXI system.

logical device that can be physically implemented on two or more modules. Conversely, multiple devices with different physical addresses can be located on one module. The logical address thus uniquely determines the *base address* of the device according to the logical equation

$$BA = LA \,\&\, 64 + 49152, \tag{7.1}$$

where & stands for logical product and LA is the logical address of the device.

The *base address* of the device determines the beginning of the block of address memory in its A16 address space of 64 B (bytes). In the address memory, each VXI device has implemented sets of registers, enabling standard system configuration, or communication between individual devices (Figure 7.13).

7.3.7 VXI system interface classes

The VXI standard defines two types of devices that can be included in a VXI system. According to the level of supported protocols and purpose, devices are divided into four classes:

- register-based devices,
- message-based devices,
- memory devices,
- extended facilities.

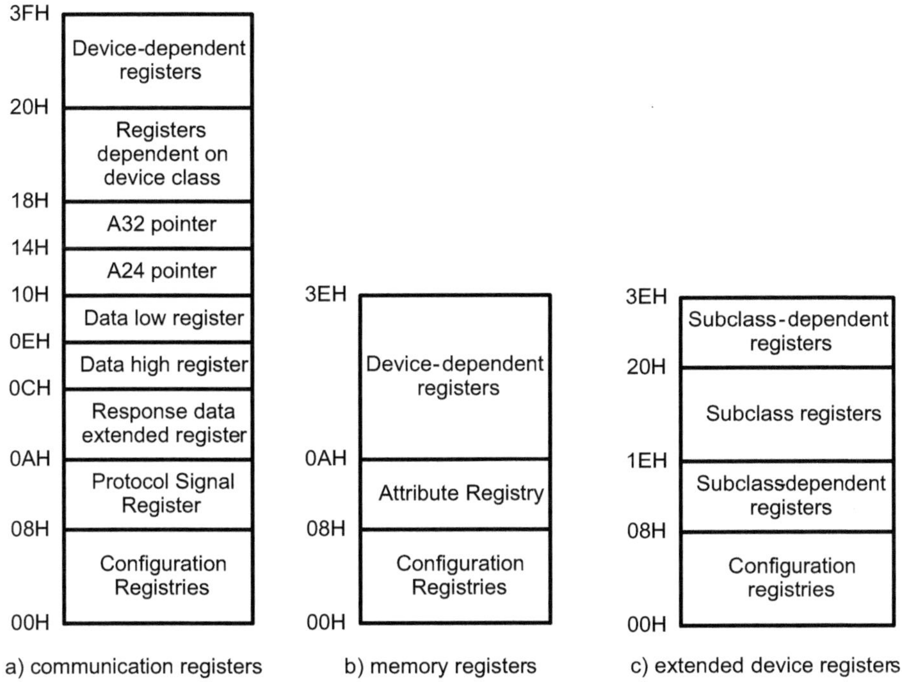

Figure 7.14 VXI bus registers.

Register-based devices have only basic protocols implemented, supporting automatic system configuration. These registers are 16-bit and their addresses are given relative to the base address (Figure 7.14).

Configuration registers have different functions when reading and writing them, which are expressed in their naming, expressed by a slash /, with the part before the slash describing the function of the register when reading it, and the part after the slash when writing data.

On the jersey 00H of the configuration register is the *identification register* of the logical address. This defines the class of the device and the address space that the device is able to address, including the identification number of the device manufacturer. The device type register is located at address 02H. This register contains a request to reserve memory space in address space A24, or i A32 and the device type code defined by its manufacturer.

There is a *status register* at address 04H. The content of this register is information about the result of the device self-test, the status of the relevant MODID line, and whether access to the area in the A24 address space is allowed, or A32. By writing to this register, the device can be initialized, the SYSFAIL* line can be disabled, and access can be enabled or disabled in the address spaces A24 and A32. The so-called offset register is located at address 06H, which contains information about the base address of the registers in the address space A24 or A32. The base address is assigned by the *resource manager* already during system configuration.

Message-based devices contain, in addition to the standard set of configuration registers, also a set of communication registers in which addresses starting with address 08H are placed relative to the base address. This allows devices to communicate at the level of higher protocols, such as the *word serial protocol.*

The *protocol signal/register* is always located at address 08H, by reading which information can be obtained about the protocols supported by the devices. This information includes, for example, information on whether the device supports shared memory protocols (fast handshake mode or whether the device has an *interrupt source* or *master* or *slave* function implemented).

The *response/data extended register* is located at address 0AH, containing information about the state of the device's communication registers, e.g., readiness to receive a command to transfer data and an error during their transmission.

The *data high* and *data low* registers are located at addresses 0CH and 0EH. These registers contain the first and second least significant data or command words.

The 20-bit registers A24 *Pointer Register* and A32 *Pointer Register* containing shared protocols are located at addresses 10H and 14H.

Memory Devices allow the remaining elements of the system to store data information. In addition to configuration registers, data devices are equipped with an *Attribute Register*, located at address 08H.

The configuration registers store, for example, the type of memory (ROM, RAM), its access time, the ability of 32-bit transfers, the implementation of block transfers, and whether access to it is unprivileged or supervised.

Extended devices are intended to define new devices to meet the VXI standard. In addition to the set of configuration registers, these devices have a *Sublass Register*, which defines the appropriate subclass of the device at address 1EH.

There are two kinds of subclasses: the first class is defined by the VXI standard and the second is defined by VXI device manufacturers, and the first class is assigned a logo. 1 and second class log. 0.

7.3.8 Communication between VXI devices

The simplest communication protocol is a protocol using registers. In addition to the configuration registers, these 64-byte registers also contain another set of registers, and their implantation depends on the specific device, which is then controlled by writing and reading into them. This communication is quite complex, as we usually do not have a standard communication protocol available.

Word serial protocol consists of a set of commands and its own protocol for transmitting these commands. Most commands are intended to configure the device within the system; a smaller part is used to control device functions, while the implementation of some of these commands is not mandatory. The ability to use commands is limited by the state in which the system is located. Some commands can only be used during system configuration, and others only during normal operation.

Communication registers also include the *data low register* and the *response register*. The last of these registers contains information about the state of communication. *Word Serial protocol* commands are written to the *data low register*. Before writing the command, the *master* must check whether the *servant* is ready to accept the command by reading the appropriate bit of this register with the *write ready* command. If the command is accepted, the command is written to the *data low register* and it is waited for the *servant* to set another bit in the *response* register with the *read ready* command.

In order to simplify the communication between the units, the VXI standard enables a fast collaborative method of communication (*handshake mode*), in which data transfers are synchronized not by the *response bits* of the register but directly by the VME signals DTACK* and BEER*. In this case, the *servant* will hold the VME transfer for 20 ms to satisfy the write and read conditions. If the conditions for these functions are not met at the end of this time, the transfer cycle is terminated not by the DTACK* signal but by the BEER* signal.

In addition to the *word serial protocol*, the VXI standard defines another *long word serial protocol* with 32-bit transmissions, using *data high register* and *extended long word serial protocol*.

It is also possible to use byte transfer protocol for data transfer, which uses *word serial protocol* commands for this purpose. They are *byte request* for reading data and *byte available* for writing it. Data transfers by this protocol are synchronized by the appropriate bits of the *response register*. Data transferred in this way are often referred to as device-oriented commands, or measured values.

7.3.9 Communication between VXI and IEE-488 systems

The VXI standard allows the use of existing measuring devices and systems with an IEEE-488 interface to connect them to the VXI system. Therefore, the VXI standard additionally defines a special *IEEE-488 – VXI* and the design rules for its application. These two products allow an external IEEE-488 controller to convert IEEE-488 protocols to control protocols of VXI devices. When addressing devices with an IEEE-488 interface, the entire VXI Bus has a single primary IEEE-488 address, and their secondary addressing is used to distinguish individual devices. In addition, *VXIbus instruments protocol* commands can be used to add devices with an IEEE-488 interface, including *Byte Available*, *Byte Request*, *Clear*, *Trigger*, *Set Lock*, *Clear Lock*, and *Real STB*.

Other standards extending the VXI interface are their specifications.

The VXI-2 and VX-3 specifications allow the word serial command to obtain information about the drive's serial number and type. The VXI-4 specification allows the names of *byte available* to be simplified to *BAVaviable*, *control event to CECent*, *logical address* to *LADDress*, and *word serial protocol* to *WSPProtocol*.

The VXI-5 specification defines high-level ASCII commands for the VXI system. They are the addressing command (*select command*), *configuration command*, commands to access registers (*register access commands*), and commands to support the word serial protocol (*word serial protocol support commands*) to the control system. These commands are sent to the *resource manager*. They can be used to obtain system configuration data, read and write to any A16 VXIbus registers, and communicate via the *word serial protocol*.

The VXI-6 specification defines the *extender of the basic unit* of the device, enabling the interconnection of several basic units of the system with each other. The actual connecting means of communication is not precisely defined in this specification, but, e.g., the MXI bus can be used for this purpose, or IEEE 1394 (*Fire-Wire*). From the point of view of the VXI system, this extender is defined as an extended device of a specific subclass, which means that all extenders are compatible with each other from the point of view of the VXI interface. This specification defines methods of transferring data between units, extending the interrupt system, triggering lines, support for the transmission of some service bus signals, the configuration method, and the *resource manager* procedure for configuring a system with multiple base units.

The VXI-7 specification defines the basic system requirements for working with *shared memory*, types of permissible cycles, structuring of data, the method of their transfer, and types of data (numerical, alphabetical, and completely general).

The VXI-9 specification defines protocols for communication and data exchange between devices via shared data memory. It also contains the definition and description of services for the creation, use, and release of shared communication channels that implementations of shared memory protocols must offer to users. It also defines general requirements for devices supporting memory sharing protocols and the use of A32 *Pointer* and A24 *Pointer registers*. Procedures for creating, using, and canceling a shared communication channel are also listed, including exception handling and support for command queues over shared memory.

7.3.10 Functional modules of the VXI measuring system

Similar to the VME measurement system, a large number of frames and plug-in functional modules of the VXI system are produced by leading instrumentation manufacturers, including, for example, precision digital multimeters, measurement channel switches, fast A/D converters, waveform generators, digital oscilloscopes, and spectrum and vector analyzers .

7.4 Modular Measuring System PXI

The *PXI system* is currently the most advanced and least used modular data collection and processing system. It uses a compact PCI bus (*Compact PCI*), which enables integrated timing and synchronization of plug-in modules. The open architecture of this open system allows it to be used in the control of measuring systems by programs with *Windows 7*, *Windows XP*, *Windows 2000*, and *Linux* operating systems. Using the PXI Express protocol, a data transfer rate of up to 6 Gb/s can be achieved in a two-way transfer mode. The expanded version of *PCI Express 3.x* allows you to increase the data transfer speed up to 24 Gb/s. When using the Compact PCI card, transfer speeds of up to 208 Mb/s can be achieved over a distance of up to 200 m using metallic or optical cables.

7.4.1 PXI measurement system specification

The main difference between the *PXI system* buses, or *PCI Express and Compact PCI* computer buses are adding timing and synchronization functions.

The *PXI system* has eight shared trigger buses, a 10 MHz star reference clock signal CLK100 that can be used to synchronize and time the function modules. The rise and fall times of this signal are less than 1 ns. The system also offers a 100 MHz differential clock signal, designated as PXIe_CLK100, whose rise and fall bursts are shorter than 100 ps. In addition to these clock signals, a differential synchronization signal PXIe_SYNC100 is also available, which synchronizes the clock signals CLK100 and PXIe_CLK100 every 10 hours.

The *boot bus* of the PXI system consists of eight separate lines. These lines enable the transmission of trigger, synchronization and timing signals from one module to another, whereby one of the modules can act as the main transmission module (master) that determines timing and synchronization to other modules (slave). The trigger bus has limited bandwidth and is not recommended for clock frequencies higher than 20 MHz. The star structure of these lines, whose center is located on slot 2 of the system frame, allows the distribution of trigger signals with a time uncertainty of 1 ps.

The *PXI system local bus* consists of a 13-line daisy-chained bus from one slot to adjacent system frame slots. This bus enables the transmission of digital and analog signals between individual modules and also supports voltage signals up to 42 V.

The *system controller* (*controller*) is always located on the leftmost slot (slot 1) of the frame. The controller can be either an embedded computer or an external laptop. Built-in controllers contain processors, data and operational memory, hard drives, and peripherals (*USB, Ethernet*), which ensure that the PXI system can also be integrated into a larger measuring system. Built-in controllers most often use the Windows operating system and other operating systems, e.g., *Linux*.

The *PXIe measurement system* allows the use of a wide range of PXI and PXIe modules. These standards specify two dimensionally different modules. The 3U module uses a 100 × 160 mm (3.94 × 6.3 in) card. The 6U card measures 233.35 × 160 mm (9.19 × 6.3 in). The smaller 3U module is the most commonly used configuration. More complex modules, such as controllers, usually requiring multiple cards stacked on top of each other, occupy more than one slot. Both types of cards have mechanical keys preventing modules from being inserted into inappropriate slots.

Figure 7.15 shows the frame of the PXI measurement system with the controller and typical functional modules.

7.4.2 Functional modules of the PXI measurement system

The functional modules of the *PXI measuring system*, like the functional modules of the VME and VXI systems, are produced by many global

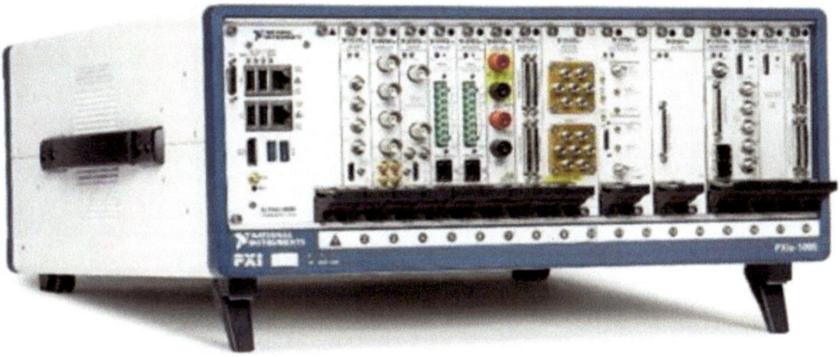

Figure 7.15 *National Instruments* PXI measuring system.

manufacturers of instrumentation, which include, for example, the companies *National Instruments*, *Tektronix*, *Kesyght*, *Pickering Tes*, *Apex*, and others.

The offer of these companies includes frame controllers, digital multimeters, signal amplifiers of typical sensors of physical quantities (temperature, pressure, mechanical stress, etc.), analog signal switches, counters, digital multimeters, oscilloscopes, digitizers of continuous signals, signal generators, spectrum, and vector analyzers including microwave coaxial switches and data transfer modules.

The *frames* of the *PXI measuring system* contain 4–18 slots with connectors P1, P2, and P3 and enable the power supply of functional modules in the range of 32–84 W and with additional air cooling of up to 2.4 kW (Figure 7.16).

The *frame controllers* of *PXI systems* enable the independent operation of functional modules without the use of an external computer. In addition to integrated four- to eight-core processors with sampling rates of up to 24 Gb/s, they contain operational and data memories, including hard SSD disks with a capacity of up to 512 GB. The frame controllers are equipped with video signal drivers and USB and Ethernet serial interfaces. Typical controllers of PXI system frames are, for example, controllers of the PXIe-884x, PXIe-886x, and PXIe-888x series by *National Instruments* (Figure 7.17).

To transfer data between the PXI system controllers and the computer, so-called *MXI lines* are used, which, in addition to metallic wires, also use optical cables for data transfer. When connecting a computer using a PCIe(x1) *plug-in card* with a *single-channel metallic* data connection to a PXI controller, transfer speeds of up to 110 Mb/s can be achieved. When using an ExpressCard, the transfer speed can be increased up to 214 Mb/s.

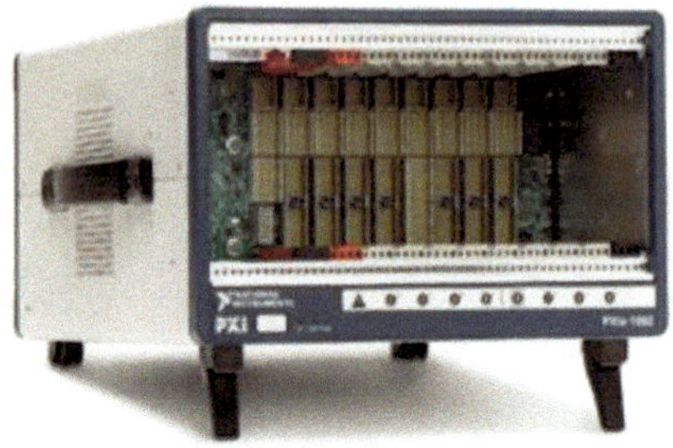

Figure 7.16 PXIe-1092 measuring system frame with power supply.

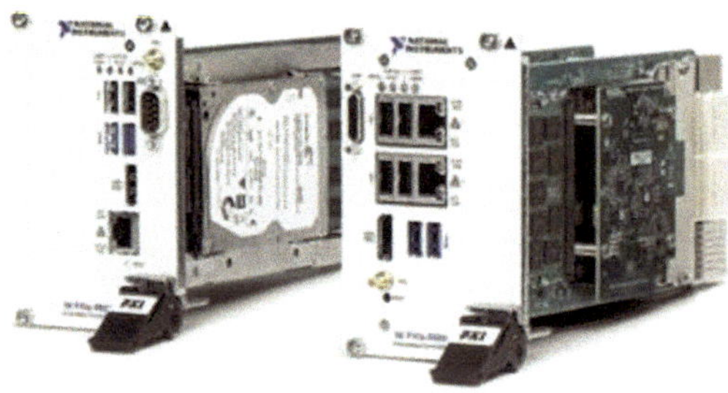

Figure 7.17 PXIe-886x and PXIe888x Series PXI System Controllers.

Transfer speeds of up to 516 Mb/s can be achieved using a 16-channel metallic PCIe(x16) link, and transfer speeds of up to 838 Mb/s can be achieved using a *duplex Fiber Optic PCIex4 link.*

The advantage of optical data links is not only the increase in their transmission speeds but also the galvanic separation of the measuring system from the computer and a significant reduction of the influence of electromagnetic interference on transmission reliability. Typical modules for the metallic connection of PXI measuring systems are, for example, modules from the company *National Instruments* PXI-8370 and for *optical connection* PXI-8375.

Digital multimeter modules are manufactured with a resolution of 6.5–7.5 decades and allow the measurement of DC voltages and currents in the range of up to 1 kV and 3 A and resistances in the range of 50 μΩ to 5 GΩ. The refresh rate of the measured data for these multimeters is 12–0 data per 1 minute. Typical modules of these devices are, for example, multimeters PXI-4070 and PXI-4082 from National Instruments (Figure 7.18).

The *analog input modules* contain 8–32 differential channels and A/C converters with a resolution of 8–24 bits with sampling frequencies in the range of tens of kS/s to 20 MS/s. The input voltage ranges of these modules are hundreds of mV up to tens of V. Some modules also contain digital filters with adjustable cut-off frequencies in the range of 2 Hz to 20 kHz. Typical

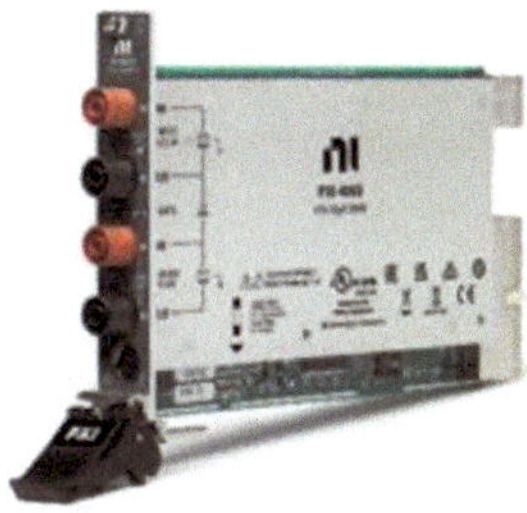
a) PXI-4070 multimeter

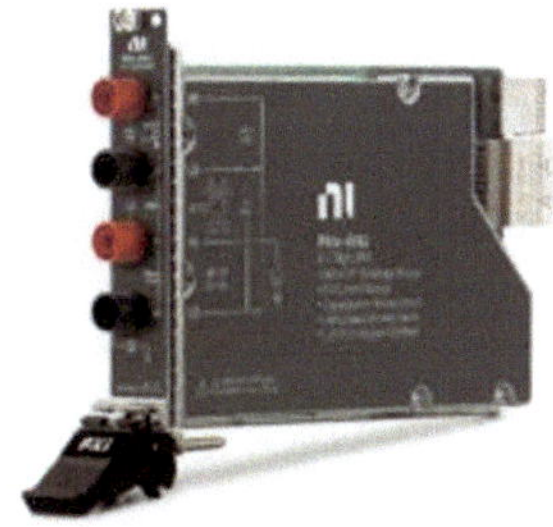
b) PXI-4082 multimeter

Figure 7.18 Digital multimeter modules.

a) PXI-4480 module

b) PXI-4481 module

Figure 7.19 Analog input modules.

analog input modules are, for example, the PXI-4480 and PXI-4481 modules from *National Instruments* (Figure 7.19).

In addition to 4–16 *multiplexed differential analog inputs*, *multifunctional I/O modules* also contain bidirectional numerical inputs and outputs, equipped with several counters. Analog signals are sampled by 12- to 16-bit A/D converters in the range of several hundred kS/s. Digit channels are sampled at sampling rates ranging from a few kS/s to a few MS/s. Typical representatives of these multifunctional I/O modules are, for example, the PXI-6115, PXI-6120, PXI-6122, PXI-6123, PX-I6133, and PXI-6143 modules from *National Instruments*.

Digital oscilloscope modules are produced in two- and four-channel designs with 8- to 10-bit A/D converters, enabling recording and displaying of voltage signals in the range of mV units to hundreds of V in the frequency range of 60 MHz to 5 GHz at sampling frequencies of 60 MS/s to 12.5 GS/s. The data memory capacity of these oscilloscopes ranges from 16 to 256 MB per channel. Typical digital oscilloscope modules are, for example, modules of the PXIe-5413, PXIe-5423, and PXIE-5433 series by *National Instruments* (Figure 7.20).

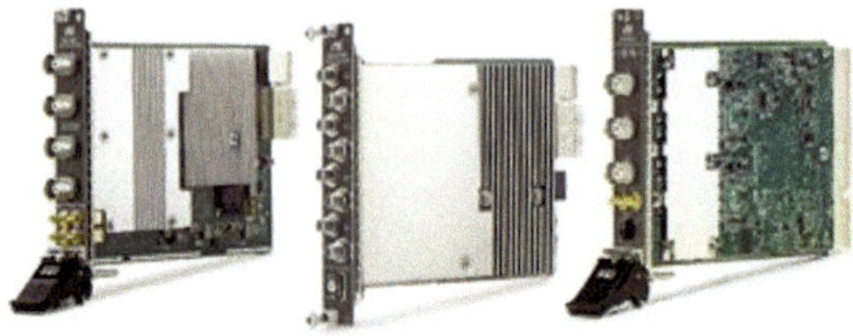

Figure 7.20 PXI modules of digital oscilloscopes.

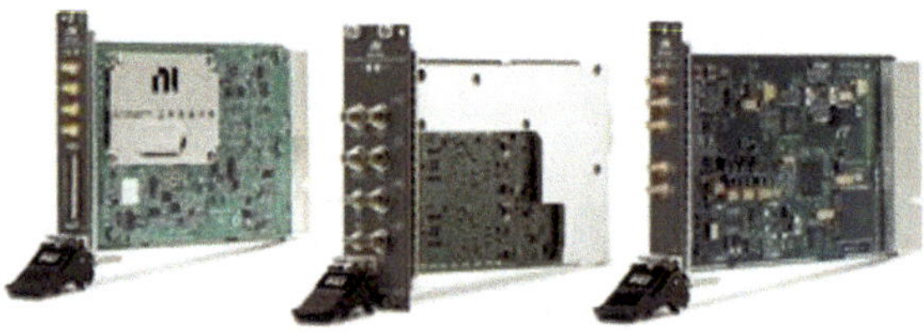

Figure 7.21 PXI function generator modules.

Functional generator modules can generate in 1–12 channels any harmonic and non-harmonic waveforms of voltage signals up to ±12 V in the frequency range of 20–80 MHz at sampling frequencies of 60–800 MS/s. Typical modules of these generators are, for example, modules of the PXIe-5413, PXIe-5423, and PXIE-5433 series by *National Instruments* (Figure 7.21).

Spectrum analyzer modules enable vector frequency analysis of signals in the frequency range of 50 MHz to 23 GHz, or in the shifted frequency range of 22.5–44 GHz with a signal-to-noise ratio of up to 55 dB at 100 MHz and retuning times in the range of 175–380 ms. These modules also contain a harmonic signal generator with a frequency range of 50 MHz to 23 GHz, which can be used to excite the measured object and subsequently determine its amplitude and frequency transmission characteristics, which can be presented in polar or Cartesian coordinates with the determination of their parameters. Typical modules of these analyzers are, for example, the PXIe-5830, PXIe-5841, and PXIe-5842 modules from *National Instruments* (Figure 7.22).

7.5 AXIe Modular Measuring Systems

The AXIe (*advanced extension for instrumentation*) modular data collection measurement system, introduced by *National Instruments* in 2009, is a more

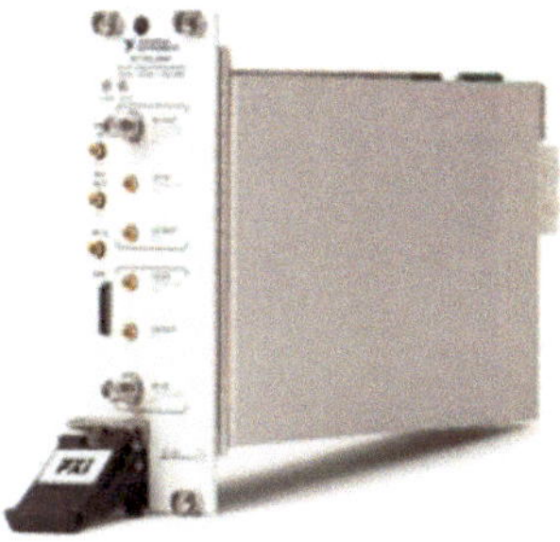

Figure 7.22 PXI module of the PXI-5842 spectrum analyzer.

powerful variant of the classic PXI system. The basis of the AXIe system is the industrially resistant frame of the ATCA (Advanced Telecom Computing Architecture) system, the power supply of which is dimensioned so that each slot of the system frame provides power of up to 200 W. It is possible to connect PCIe modules and a LAN bus (*local area network*).

In October 2017, the *AXIe Consortium* announced a new *optical data interface* OD specification, suitable for high-speed instrumentation systems, addressing challenging applications in 5G communication and advanced communication research. This new standard enables data transfer connections with transfer rates of up to 20 Gb/s (160 Gbit/s) between devices connected by *multimode optical fibers*.

7.6 LXI Modular System

Modular measuring systems LXI (*LAN extensions for instrumentation*) are made up of devices that communicate with each other via a LAN data network. The first specification of this system was presented in 2005. These systems can be used to connect not only individual measuring devices but also entire measuring systems, e.g., to measure the properties of microwave data links.

Figure 7.23 shows the LXI system 6-03 B modular system and the PXI system 40-923A modular system, which are housed in the same 13" *National Semiconductor* frames. An *Ethernet* interface with RJ45 connectors and a microwave signal switch are located in the leftmost slot of the LXI system.

7.7 Compact DAQ Modular Measurement System

In addition to the above-mentioned modular measuring systems, there are also a large number of compact modular measuring systems, enabling the realization of less demanding measuring applications. A typical example

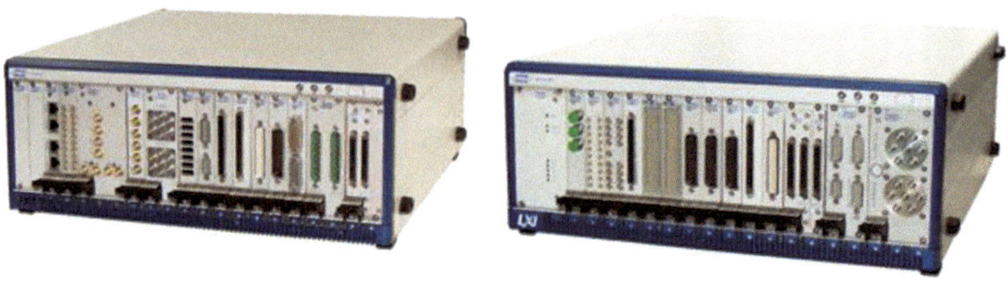

a) measuring system LXI b) measuring system PXInm

Figure 7.23 National Instruments LXI and PXI modular systems.

of these systems is the *CompactDAQ* data collection system from *National Instruments*, which is a cheaper variant of the PXI system from the same company.

CompactDAQ systems are controlled by a frame controller, which controls data transfer between I/O modules, the number of which is limited to 8. For small portable systems, a USB interface is used to transfer data to a computer, and in distributed systems, an Ethernet interface is used. A Wifi interface is used for wireless data transfer.

CompactDAQ system modules are represented by I/O modules of analog and digital signals, designed to modify and process the signals of typical sensors of physical quantities, including their galvanic isolation from the system's power source. This enables the use of these systems even in heavily disturbed industrial environments.

Figure 7.24 shows the CompactDAQ measuring system with *plug-in modules* designed to modify and process signals from sensors of typical physical quantities.

Figure 7.25 shows plug-in modules with voltage input, universal voltage modules, modules for processing signals of vibrometric, tensometric and temperature sensors of the CompactDAQ measuring system.

7.8 Software Equipment of Modular Measuring Systems

The standard *operating system* of the PXI and PXIe measuring systems is the Windows operating system of various versions. In some applications, other operating systems are also used, e.g., Linux. A key factor in choosing an operating system is that it can connect to the PXI bus and that its drivers are available. PXI (PXIe) system modules can be controlled using either registers or messages. In register-based control of a measurement system, the system controller has access to the registers in the modules that control their

Figure 7.24 Measuring system CompactDAQ from National Instruments.

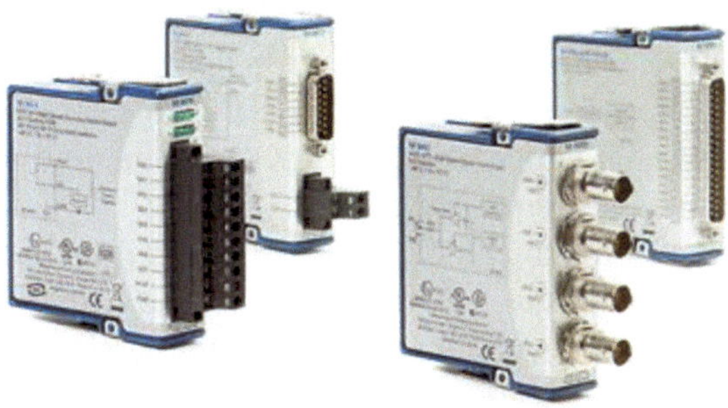

Figure 7.25 Plug-in I/O modules of the CompactDAQ measuring system.

operation. Message-based communication is based on the exchange of text messages between modules. Messages have a defined format and argument range. Message-based control is more widespread as it simplifies communication and control of PXI/PXIe peripherals.

Device drivers are used to simplify communication between the controller and system modules. These drivers allow not only interaction with the hardware interfaces of the modules but also easier communication between the operating system and application-specific software.

The PXI/PXIe standard is based on drivers of the VISA (*Virtual Instrument Software Architecture*) type, which defines the so-called software layer. This layer allows users of measurement systems to control the instruments without having to know the details of their operation. This so-called consistency method of access enables the control of devices with the same interfaces from different manufacturers.

Interchangeable virtual instrument (IVI) technology is a driver technology that provides a standard application program interface (API) for communicating with instruments. Because this software interface is built on top of *VISA drivers*, it allows users to potentially swap instrument controls without changing the source code.

Device drivers are also necessary in cases where users of measuring systems want to write their own programs in *Windows* operating systems, or *Linux*. Commonly used programming languages are, e.g., *Visual C++*, *Borland C++ Builder*, *LabWindows/CVI*, *LabVIEW*, *Borland Delphi*, *Visual Basic*, *VB.NET*, *C#*, *J#*, *Python*. *Linux* support for individual compiled program modules is included for easy distribution into *RedHat*, *Fedora*, *Suse*, *Ubuntu LTS*, or *Debian* software packages.

Controlling *multi-peripheral systems* requires system integration software. For this purpose, for example, the *LabWindows/CVI* or *MATLAB* programming language is used. A standard graphic programming language often used in industry is, for example, the *LabVIEW* language, which consists of several different libraries and several open *Virtual Instruments*, which allow controlling measuring systems, transferring their measured data to a computer, and subsequently processing them with more advanced methods of digital data processing in the time and frequency domain.

The functions of virtual instruments can be activated either directly in the *LabVIEW* environment or through an EXE command by the *Run Time subprogram*. The *LabVIEW* program is available either as a full version (*full development system*) or in a basic configuration (*base package*). The full configuration of the program contains, in addition to libraries for controlling the IEEE-488 (GPIB VI Library) and RS-232 (RS-232 VI Library) interfaces, also libraries for controlling and collecting data from plug-in cards (*Data Acquisition VI Library*) and mathematical analysis (*Analysis VI Library*), which includes digital processing of signals in the time and frequency domain, generation of signals, application of digital filtering, numerical and statistical analysis of signals.

The full configuration of the LabVIEW program also includes a tool (*Code Interface Developers Toolkit*) enabling the inclusion of externally created blocks in the program modules of the block diagram structure of the program. Program modules must be compiled with the *Watcom C* compiler.

Figure 7.26 shows the presentation of measurement results by the LabVIEW program.

References

[1] Oebelin. G. *Measurement system-application and design.* Mc. Graw Hill Publishing Company 1990, ISBN 0-07-017338-9

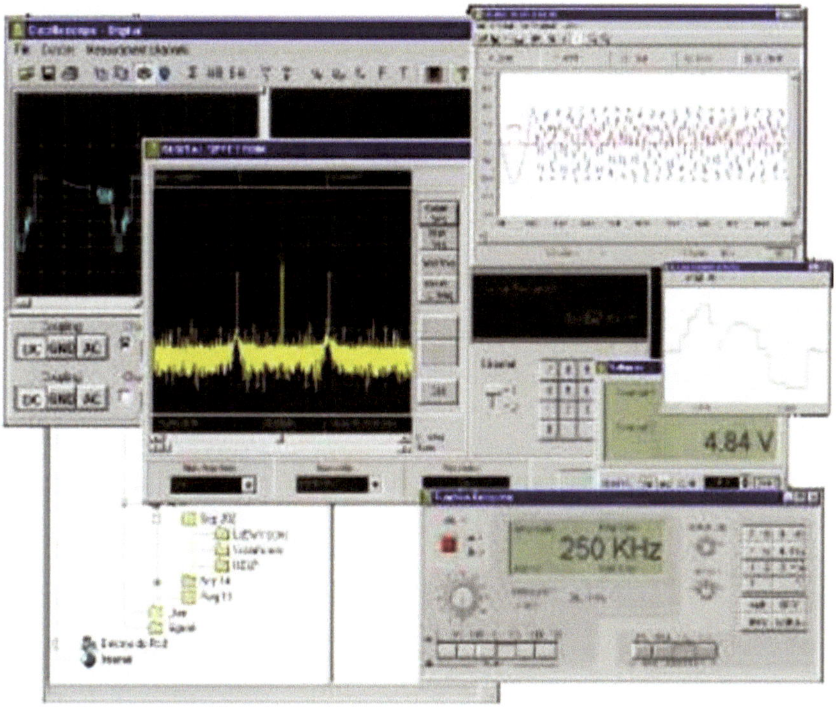

Figure 7.26 Example of presentation of measurement results by the *LabVIEW* program.

[2] Garrett, P. H., *Advanced Instrumentation and Computer I/O Design*: Real-Time System Computer Interface Engineering, New York: IEEE Press, 1994. ISBN 0-7803-0402-0
[3] Black, J. A. *The System engineer's handbook: a guide to building VMEbus and VXIbus systems.* Morgan Kaufmann. ISBN 978-0-12-102820-6.
[4] Boyes, W., (ed.), *Instrumentation Reference Book*, 3rd ed., Boston, MA: Butterworth-Heinemann, 2003. ISBN 9780080478531.
[5] Park, J., Mackey, S. *Practical Data Acquisition for Instrumentation and Control Systems,* Amsterdam, the Netherlands: Elsevier, 2003. ISBN 978-07506-5796-9.
[6] Nawrocki, W. *Measurement systems and sensors*. Second edition, Artec House 2016. ISBN 1-58053-945-9.
[7] *VMEbus Extensions for Instrumentation. VXIbus*. System Specification VXI-1. Revision 4.0, May 27, 2010.

[8] *Introduction to PXImc - Technology for High Performance Test, Measurement & Control Applications*. National Instruments. *www.ni.com*. Retrieved 2016-05-17.

[9] Rowe, M. *PXI expands to multiple processors. Test* & Measurement World. Archived from the original on 22 February 2012.

[10] www.ni.com.

[11] www.pickering.com.

[12] www.apex.com.

8

Measuring Systems Based on Data Networks

These measurement systems use computer networks to transfer data files. According to the available distances, data networks can be divided into LAN (*local area network*), designed, e.g., for data transmission in residential buildings and offices, urban MAN networks (*metropolitan area network*), designed for transmission in urban agglomerations, and networks for longer distances WAN (*wide area network*), designed for data transfer, e.g., between large agglomerations or countries.

Local and metropolitan LAN and MAN networks most often use a star or ring topology and transmit data via metallic or optical links. WANs usually have a line or tree structure and usually use fiber optic links to transmit data. The most widely used local area networks are *ARCNet* and *Ethernet.*

8.1 ARCNet Computer Network

The ARCNet (*Attached Resource Computing Network*) local computer network was launched in 1976 and subsequently standardized in 1992 by the ANSIA 878.7 standard, which was subsequently updated. This network was originally designed to transfer data between large computers. Its applications were developed in the DATABUS language, which is similar to the COBOL language. With the advent of personal computers, this network was gradually replaced by the Ethernet network. Currently, the ARCNet network is used in those applications where the majority requirement is cheap and reliable data transmission over shorter distances.

The transmission medium of this network is the RG-62/U coaxial cable, which with passive or active hubs HUB (host bus universal) enables the creation of line and star network topologies, of which the star topology allows for a simpler and cheaper implementation of this network. For data transmission, so-called dipulses are used, whose signal level log. 1 is expressed by a

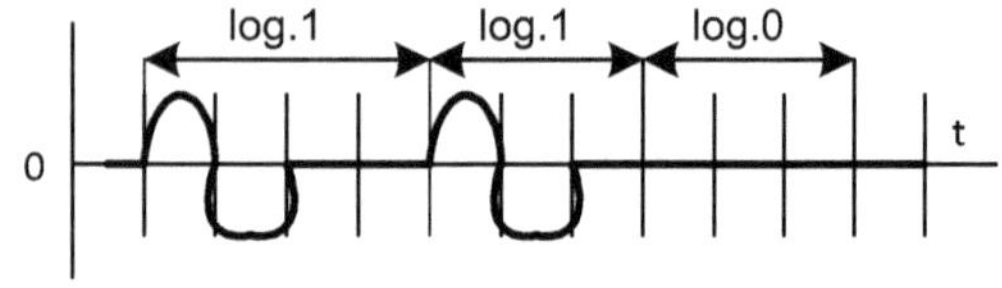

Figure 8.1 Time course of dipulses.

bipolar pulse with a length of 200 ms and a log level. 0 in the absence of this impulse (Figure 8.1).

The maximum length of this network is up to 1.8 km when using RG-58 coaxial cable.

In contrast to the Ethernet network, *deterministic selection* of the data recipient is used for data transmission. Each node of the network has a unique address (*MAC ID*) in the range of 1–255. Mediating the access of network units to its bus is handled by a special access data block (*Token*), which authorizes its holder to transmit data. If none of the network stations is active, i.e., none of them is transmitting, this block of data is passed from station to station, and none of the stations can immediately send a request to access the bus. If a certain station wishes to send a message, it waits for its turn, and after receiving an access message, sends its message to the network. After sending it, this station transmits the access message again to all stations in the network.

First, one of the stations sends an ITT (*invitation to transit*) message, and before sending a PAT (*packet*) data frame, an FBE (*free buffer enquiry*) query precedes it. Subsequently, either an ACK (*acknowledgement*) message is generated, confirming the release of the bus, or a NAK (*negative acknowledgment*) message, confirming the rejection of the bus release.

Since the topology of the ARCNet network has a distributed star nature, the access request cannot be sent from station to station in a circle. If a new node joins the network or an existing node disconnects from the network, a reconfiguration of the network occurs, which is generated by the RSU (*reconfiguration symbol unit*) command, after which each new network participant is assigned the address of the node to which the participant joined.

Other network messages are the SOH (*start of header*) messages, identifying the data packet 0x01 and the EOT (*end of transmission*) message, confirming the end of the transmission, and handing over the access data block to the next network node 0x04. One of the main advantages of this network over other networks is that the ARCNet network provides the sender of messages with confirmation of their receipt, which increases the reliability of data transmission and, at the same time, allows to significantly shorten the recovery of the network in the event of errors.

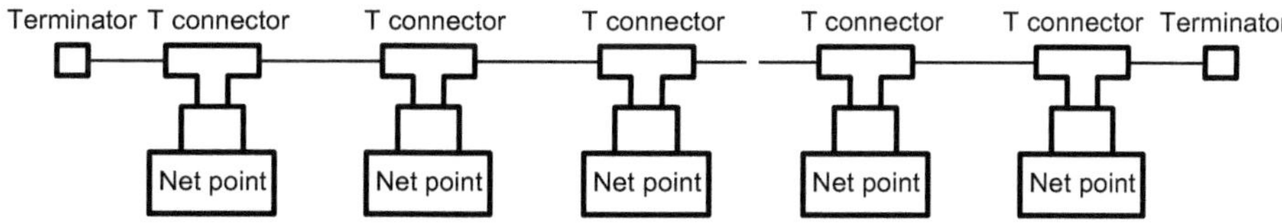

Figure 8.2 Ethernet 10Base-2 data network.

This network is suitable, for example, for controlling real-time RPC (*real process control*) processes, where reliability of data transmission is a major requirement.

8.2 Ethernet Computer Network

This currently most widespread computer network was created at the beginning of the seventies of the last century by *Xerox* and *Intel*. Its original variant had a transmission speed of 3 Mb/s. Later, this network was updated by the IEEE 802 standard, which was gradually expanded by the LLC (*logical link control*) and MAC (*medium access control*) standards, defining the Ethernet II network. These networks are used not only in companies, state administration, schools, and other institutions, including households. The leading manufacturers of measuring technology also equip their devices and systems with interfaces that enable the communication of these devices via the *Ethernet network*.

8.2.1 Physical layers of the Ethernet network

The earliest type of *Ethernet* in use was the *10Base-2* transmission system, which was standardized by the IEEE 802.3a standard. This system used a thin coaxial cable with a diameter of 5 mm, terminated with resistors (terminators) of 50 Ω (Figure 8.2).

The maximum length of each network segment was 200 m. Up to two repeaters could be used for this network. The individual nodes of the network were connected by a RG-B58 coaxial cable with a wave impedance of 50 Ω and T-type connectors, while the network driver was part of the network card located in the slot of the control computer. The maximum number of nodes in one network segment was limited to 30. This network is practically not used anymore.

The more modern *10Base-5* transmission system, standardized by the IEEE 802.3 standard, also used the same type of coaxial cable for data transmission. The maximum length of each of the network segments was 500 m and it was possible to extend it with up to two repeaters. The nodes of this network were connected through a separate driver MAU (*medium attachment unit*), which was connected to the network card of the computer through the AUI (*attachment unit interface*). The maximum number of nodes in one network segment was 100.

The *10Base-F* transmission system, standardized by the IEEE 802.3j standard, used a pair of *multi-mode optical fibers* in full duplex transmission for data transmission and reached transmission speeds of up to 10 Mb/s. The maximum length of a segment of this network was 2 km.

The *10Base-T* transmission system, standardized by the IEEE 802.3i standard, used the so-called structured cabling with a maximum segment length of 100 m for data transmission. A separate twisted shielded pair (*twisted pair*) was designated for each direction of communication, which made it possible to use duplex data in this sewing. The transmission speed of this network was also 10 Mb/sb.

A significant acceleration of the transmission speed up to 100 Mb/s was made possible by the *100BaseTX* (*fast Ethernet*) system, standardized by the IEEE 802.3 standard, which used two or four shielded twisted pairs of *Category 3* cable conductors with a wave impedance of 100 Ω for duplex data transmission. This network with a star topology and a central controller made it possible to transmit data up to a distance of 100 m.

The *100Base-FX* (*fast Ethernet*) transmission network, standardized by the IEEE 802.3u standard, used full-duplex data transmission by pairs of optical fibers with a maximum segment length of 400 m for multi-mode fibers and 2 km for single-mode fibers. In both of these cases, the achievable transmission speed was up to 100 Mb/s.

The currently most commonly used *1000BaseT* (*gigabit Ethernet*) network, standardized by the IEEE 802.3ab standard, enables full duplex data transmission via four twisted shielded pairs of wires with a resistance of 5, or 5e, achieve transmission speeds of up to 1 Gb/s. This network uses multi-bit coding to secure messages.

The optical variant of the previous *1000Base-SX* network uses multi-mode and single-mode optical fibers in a star topology to transmit data with a transmission speed of up to 1 Gb/s, while the maximum segment length, depending on the type of optical fiber, is up to 70 km.

So far, the fastest variant of the *Ethernet* transmission network is the 10GBaseT network, which uses four twisted shielded pairs of cable conductors with a resistance of 6 and a frequency bandwidth of 500 MHz to transmit data with a transmission speed of up to 10 Gb/s. With these cables, a data transmission distance of up to 100 m can be achieved. There is also a variant of this network, consisting of a pair of cables with single-mode optical fibers, enabling data transmission over distances of up to several km. Table 8.1 provides an overview of the typical properties of *Ethernet networks*.

Table 8.1 Properties of Ethernet networks.

Network	IEEE protocol	Transmission speed	Transmission distance	Transmission path
10Base2	IEEE-802.3a	10 Mb/s	200 m	Coaxial cable
10Base5	IEEE-802.3	10 Mb/s	500 m	Coaxial cable
10BaseF	IEEE-802.3j	10 Mb/s	32 km	Multi-mode optical line
10BaseT	IEEE-802.3i	10 Mb/s	100 m	Shielded double twisted line
100BaseTX	IEEE-802.3u	100 Mb/s	100 m	Shielded four twisted lines
100BaseFX	IEEE-802.3u	100 Mb/s	10 km	Single-mode optical line
100BaseT	IEEE-802.3u	100 Mb/s	100 m	A pair of twisted double lines
100BaseSX	IEEE-802.3u	100 Mb/s	500 m	Multi-mode optical line
1000BaseCX	IEEE-802.3z	1 Gb/s	25 m	Shielded double twisted line
1000BaseT	IEEE-802.3ab	1 Gb/s	100 m	Shielded four twisted lines
1000BaseSX	IEEE-802.3z	1 Gb/s	500 m	Multi-mode optical line
1000BaseFX	IEEE-802.3z	1 Gb/s	10 km	A pair of single-mode optical lines
10GBaseT	IEEE-802.3ae	10 Gb/s	100 m	Shielded four twisted lines
10GBaseFX	IEEE-802.3ae	10 Gb/s	10 km	Quadruple single-mode optical lines

Figure 8.3 shows a typical plug-in card with an *Ethernet* interface with a coaxial cable connector and an RJ-45 connector, used to connect shielded twisted pair and quad lines, and fiber optic cable connectors used in Ethernet networks.

Table 8.2 shows the assignment of the contacts of the RJ-45 connector when using twisted two-line and four-line cables to connect Ethernet data networks.

8.2.2 Data transmission in the Ethernet network

In connection with data transmission in the Ethernet data network, the so-called Ethernet frame is defined, the structure of which is shown in Figure 8.4, where the lengths of the individual parts of the frame are expressed by *hexadecimal code*.

Figure 8.3 Plug-in card with Ethernet interface and optical connectors.

Table 8.2 Assignment of RJ-45 network connector contacts.

Pin	10BaseT	100BaseTX	100BaseFX	1000BaseT
1	Tx+	Tx+	Tx_D1+	BI_D1+
2	Tx−	Tx+	Tx_D1−	BI_D1−
3	Rx+	Rx+	Rx_D2+	BI_D2+
4	NC	NC	BI_D3+	BI_D3+
5	NC	NC	BI_D3−	BI_D3−
6	Rx−	Rx−	Rx_D2−	BI_D2−
7	NC	NC	BI_D4+	BI_D4+
8	NC	NC	BI_D4−	BI_D4−

The frame begins with the so-called Ethernet header (*preamble*), containing the address of the receiving unit (*destination MAC address*), the address of the transmitting unit SMA (*source MAC address*), and additional data on the length of the data field LDF (*length data field*). The second part of the frame consists of its own PL (*payload*) data field and the frame control sequence FCS (*frame control check*).

A MAC (*media access control*) address is a unique identifier of a network device used by various data transfer protocols. This address is assigned to each network card and when it is manufactured for older cards, it is directly stored in the EEPROM memory, and is therefore also referred to as a *physical address*.

The MAC address consists of 48 bits and should be written as six two-digit hexadecimal numbers separated by dashes (e.g., 01-23-45-67-89-ab) according to the IEEE 802 standard. Often this address is also separated by colons (01:23:45:67:89:ab), sometimes also in the form of three groups of four hexadecimal numbers 0123.4567.89ab, which in binary form are expressed as a sequence

$$0000\ 0001\ 0010\ 0011.\ 0100\ 0101\ 0110\ 0111.1000\ 1001\ 1010\ 1011. \quad (8.1)$$

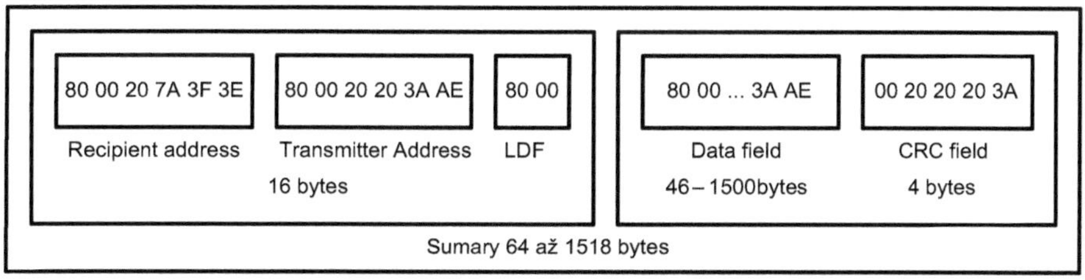

Figure 8.4 Standard Ethernet Frame Etherrnet II.

This 48-bit address space contains 248, or 281,474,976,710,656 possible MAC addresses. Therefore, the original MAC address standard assumed that the address would always be globally unique. In terms of address allocation, the MAC address is divided into two halves. The manufacturer must request the central administrator of the address space for the first half, and it is the same for all cards of a given manufacturer (or at least for large groups of cards). Major manufacturers have several values available for the first half of the address. The manufacturer then assigns a unique value of the second half of the address to each manufactured card or device. The unambiguity of the address facilitates the management of local networks, i.e., a new card or other device in the network can be uniquely identified.

However, in reality, the manufacturer's unique identification of the card may be violated. In the event that two devices with the same MAC address appear in the local network, this collision is resolved with modern network cards by a virtual change of their MAC addresses. However, the probability of this collision is very low and usually occurs only in large data networks.

Locally managed addresses are assigned by the network administrator and override the addresses specified by the manufacturer. Locally managed addresses do not contain an OUI label.

Universally managed addresses are distinguished by the second least significant bit of the most significant byte of the address. This bit is referred to as UL bit (*universal local bit*). If this bit is in the log state. 0, the address is not locally managed. If it is in state 1, the address is locally managed. For example, for the address 06-00-00-00-00-00, the most significant byte is 06(hex), which in binary is 0000 0110, where the second least significant bit is log. 1. Therefore, this address is locally administered.

The length of the data field depends on the type of transmission frame. The most widely used transmission frame is the Ethernet II or Ethernet Version 2 transmission system frame, or DIX-Ethernet. Other transmission frames used are the Nowell frames, standardized by the IEE 802.3 standard, and the LLC (*local link control*) frame and the SNAP (*subnetwork access protocol*) frame, which are standardized by the IEE 802.2 protocol.

Systems with transfer rates of 10 and 100 Mb/s allow the transmission of data up to 1500 8-bit words (bytes) in length. Systems with higher transmission rates of 1 Gb/s and 10 Gb/s support larger frames (*jumbo frames*) up to several tens of thousands of 8-bit words long.

The last part of the frame is reserved for the CRC (*cyclic redundancy check*) checksum, which is sent or stored together with data that could be transmitted or stored in error. After receiving the data, this sum is again compared in the receiving unit with the sum of the sent data. If this checksum is different from the transmitted or stored code, it is obvious that there was an error in the transmission or storage of the data. If this sum is the same, it can be assumed with a certain probability that there was no error when transferring data or writing it to the data memory of the receiver. When choosing a 16-bit reference code, the probability of detecting an error is 2^{-16}, i.e., approximately 0.999985.

8.2.3 Measuring systems with LAN/IEEE-488 converters

The emergence of these systems was conditioned by the possibility of incorporating measuring devices and systems with an IEEE-488 interface into the *Ethernet* data network. Connecting measuring devices with an IEEE-488 interface to an *Ethernet network* can be done either by connecting a computer with an IEE-488 interface to which these devices are connected and connecting that computer to the Ethernet network, or IEEE-488-*Ethernet* converters can be used to connect these devices.

The first method of connecting devices to the *Ethernet network* requires the use of a central control computer, to which slave computers are connected through the Ethernet network and HUB, which then communicate with the devices through their IEEE-488 interfaces. The disadvantage of this arrangement is the need for a larger number of slave computers (Figure 8.5).

Ethernet-IEEE-488 converters enable direct connection of devices with an IEEE-488 interface to an Ethernet network. For this purpose, *Tektronix* produced the AD007 converter, which allowed the data conversion connection between the 10BaseT network and the IEEE-488 interface. In 100BaseTx networks, it is possible to use, for example, converters from the same company GPIB-ETERNET/100, enabling data to be transferred with a transfer rate of up to 900 kB/s in the IEEE-488.1 transfer protocol, and with a transfer rate of up to 1.2 Mb/s in the HS-488 transfer protocol (Figure 8.6).

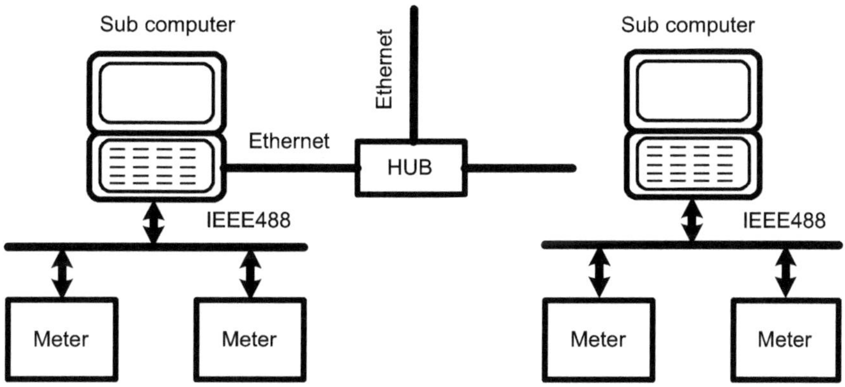

Figure 8.5 Connecting devices with an IEEE-488 interface to the Ethernet network via secondary computers.

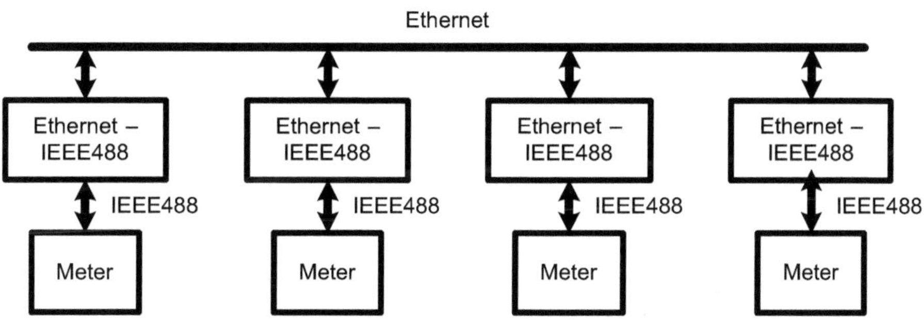

Figure 8.6 Using Ethernet-IEEE-488 converters to connect devices to an Ethernet network.

Table 8.3 shows the parameters of several typical Ethernet-IEEE-488 converters.

Measuring devices enabling their direct connection to the *Ethernet network* are, for example, digital oscilloscopes of the MDO3, MSO4/5/6 DPO/MSO70000 series, generators of the AFG31xxx series and spectrum analyzers of the RSAxxxx series from *Tektronix*, digital oscilloscopes of the EXRx00, MXRx00 series, signal generators of the VXRx00 series, and spectral *Keysight's* VXGx00 and PSGx00 series analyzers (originally *Hewlett Packard*, then *Agilent*), *LeCroy's* WaveRunner 6000 and WaveSurfer 400 oscilloscopes, and *Keithley's* measuring instruments, and *Rohde & Schwarz's* generators and spectrum analyzers. All of these devices are equipped with RJ-45 sockets, enabling the devices to be connected to *10BaseT* and 100BaseT networks, which enable their remote control and access to measured data via standard web browsers.

Table 8.3 Parameters of *Ethernet converters* – IEEE-488.

Convertor	Ethernet network type	Transmission speed	Computer operating system	Manufacturer
AD007	10BaseT	Not known	Windows 95	Tektronix
GPIB-ENET	10BaseT 100BaseTX	50 kB/s	Windows 95	National Instruments
GPIB-NET/100	100BaseTX	900 kB/s (IEEE-488) 1.2 Mb/s (HS-488)	Windows 2000, Windows ME, Windows XP	National Instruments
E2050B [2]	10BaseT	100 kB/s	Windows 98, Windows ME	Agilent Technologies
E5810A	10BaseT 100BaseTX	1 Mb/s	Windows 98, Windows 2000, Windows XP	Agilent Technologies

In addition to the cable connection, these devices can also be connected to this wireless network via GSM (*groupe special mobile*) data converters, allowing the operation of these devices to be controlled, including the transmission of data at virtually any distance.

8.2.4 Measuring systems with built-in web servers

These measuring systems use local web servers, built directly into these instruments, to transfer the data of the measuring instruments over the *Ethernet network*. Local web servers are produced either in the form of plug-in cards for computers, or in the form of separate modules. A block diagram of a typical Internet-based distributed metering system with local web servers connected to an *Ethernet network* is shown in Figure 8.7.

Data transfer over the *Ethernet network* is carried out using the TCP/IP protocol. Since data transmitted over an Ethernet network can follow different paths with different delays, it is not recommended to use this method of data transmission when measuring or controlling dynamic processes. Table 8.4 provides an overview of typical web servers used to transfer data over *Ethernet networks*.

A block diagram of a typical Intel WEB 6590 *Embedded Web Server*, enabling connections to *10BaseT* and *100BaseTX* networks, is shown in Figure 8.8.

This embedded server includes a *Celeron* 400/650 MHz microprocessor with Intel 515E(B) chipset, 16 MB RAM, 256 MB SDRAM data memory,

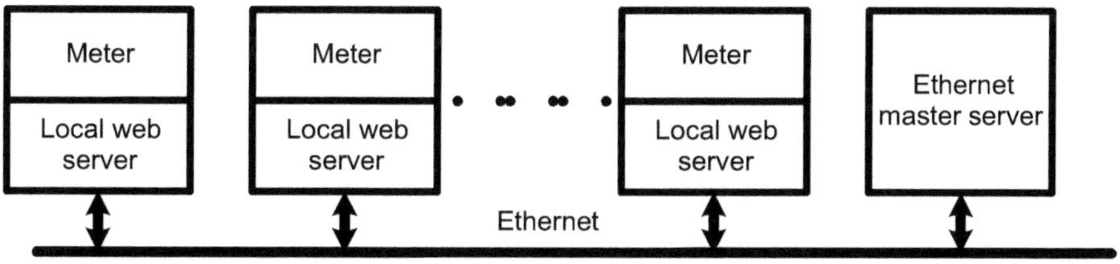

Figure 8.7 Distributed measurement system with web servers connected to the Internet.

Table 8.4 Embedded web servers.

Web server	Ethernet type	Interface	Memory capacity	Manufacturer
AVR 460	10BaseT	RS-232	256 kB	*Atmel*
PICDEM.NET	10BaseT	RS-232	32 kB	*Microchip*
C7520	10BaseT, 100BaseTX	RS-232, I²C, Fast SPI	8 MB	*IO Limited*
Series TINI (e.g., TBM390)	10BaseT	CAN, RS-232, 1-Wire	1 MB	*Maxim, Dallas Semiconductor*
Co-Box-Micro	10BaseT	RS-232	No data	*Lantronix*
WEB-6590	10BaseT, 100BaseTX	RS-232, USB, IrDA	256 MB	*Intel*

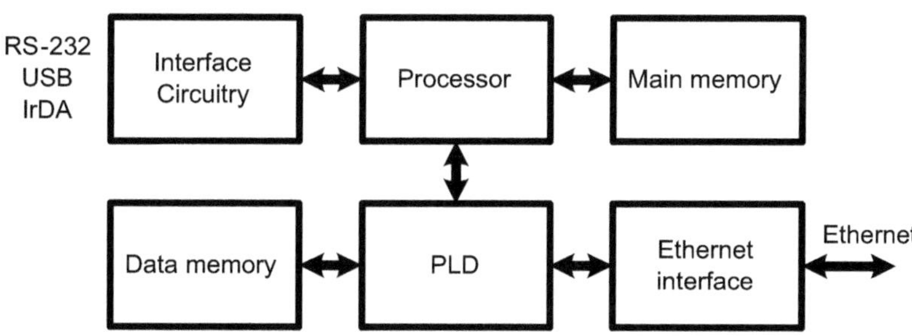

Figure 8.8 Block diagram of the WEB 6590 embedded web server.

integrated Intel 815(B) graphics controller, and RS 232, USB, IrDA interfaces. The *Intel* 82 560ET circuit is used to connect to *Ethernet 10BaseT* and *100BaseTX networks*.

Devices with built-in web servers enable their control and data transmission over practically unlimited distances while simultaneously ensuring a high level of transmission reliability.

References

[1] Boyes, W. (ed.), *Instrumentation Reference Book*, 3rd ed., Boston: Butterworth-Heinemann, 2003. ISBN 978-008040-47853-1.

[2] Corea, C., Kozierok, C. M. *Automotive Ethernet. The Definitive Guide*, Intrepid Control Systems 2014, ISBN 978-09905-3880-6.

[3] IEEE-802.3 *Standards, Computer Local and Metropolitan Networks,* 2004.

[4] Wilder. F. *A Guide to the TCP/IP Protocol Suite*, 2nd ed., Norwood, MA: Artech House, 1998. ISBN 089-0066-930.

[5] Park, J., Mackay, S. *Practical Data Acquisition for Instrumentation and Control Systems*, Amsterdam, the Netherlands: Elsevier, 2003. ISBN 978-075065-7969.

[6] ARCNET and the ATA History. Retrieved 2022-10-14.

[7] IEEE Std. 802-1990: IEEE standards for Local and Metropolitan Networks: Overview and Architecture New York. 1990.

[8] IEEE802.1 - *www.ieee802.org*. Retrieved 2022-11-27.

[9] IEEE802.3. *Data Communincation Standards and Protocols. EE Herald*. Retrieved 2012-01-25.

[10] The fate of 100 Mbps Ethernet now definitely two-fold. *FDDI News. Boston. Information Gatekeepers, Inc. 4 (7): 1–2. July 1993.* Retrieved 2013-11-21.

[11] IEEE 802.15 WPAN Task Group 2 (TG2). *IEEE Standards Association. May 12, 2004*. Retrieved June 30, 2011.

9

Wireless Measuring Systems

Wireless measuring systems enable the collection and transfer of data to a computer from remote, hard-to-reach, and moving objects, such as vehicles, aircraft, and satellites. Compared to a fixed data connection, these systems are often more economically advantageous, as they do not require the construction of metallic or optical routes. Their disadvantage is generally lower transmission speed and less resistance to interference.

Wireless data transmission is always *serial*, even with multi-channel measuring systems.

Measuring systems using wireless data transmission are used

- mobile phone networks,
- data transmission via radio modems,
- short distance data transmission systems.

The first two types of measurement systems use communication systems of worldwide coverage and can therefore also include very distant objects. Unlike systems based on mobile phone networks, designed in principle for audio transmission, or video signals, systems with dedicated radio channels are intended exclusively for data transmission. In addition to the infrared light beam IrDA connection, short-range wireless measurement systems also use high-frequency transmission systems such as *Bluetooth, Wifi, ZigBee*, and *LoRAN*.

In addition to these transmission systems, the license-free ISM (industry, science, and medicine) band is also used for wireless data transmission. These systems, originally intended mainly for industry, science, and medicine, have dedicated frequency bands of 2.4–2.4835 GHz, 5.72–5.85 GHz, and 902–928 GHz.

9.1 Measurement Systems Based on Mobile Phone Networks

The first GSM (*Groupe Special Mobile*) mobile phone network, referred to as the 1G network, was created in 1982 on the basis of the work group of the same name. This was intended only for voice transmission and worked in the 450 and 900 MHz frequency bands with a transmission speed of up to 9.6 kb/s in one channel. Typical representatives of these networks were, for example, the NMT (*Nordic Mobile Telephone*) systems, which were used mainly in Scandinavia and other countries of Central and Eastern Europe. In the USA, the AMPS (*advanced mobile phone system*) mobile network became this standard.

The *2G generation*, operating in the 900 MHz and 1.8 GHz frequency bands, was able to transmit not only voice messages but also short SMS (*short message system*) text messages, and/or short data messages. 2G network services were referred to as PCS (*personal communications service*) services in the USA. The first commercial launch of a 2G network was in 1991 in Finland. Gradually, this network began to be used in the USA as well. In Japan and other Asian countries, this network was called JDC (*Japan Digital Cellular*). This network used the 900 MHz and 1.8 GHz frequency bands. This generation was supplemented by the GPRS (*general packet radio service*) to increase the data transfer speed and is referred to as the 2.5G generation.

The *3G generation*, standardized as ITU-IMS-2000 (*Universal Mobile Telecommunications 2000*), is currently the most widespread telecommunications wireless network. This network operates in the 1.95 and 2.15 GHz frequency bands and allows data to be transferred at a speed of up to 2 Mb/s. This system also allows the use of public switched telephone networks PSTN (*public system telecommunication network*), including BISDN (*broadband integrated services digital network*). Later editions of the 3G standard, including the HSPDA and HSUPA protocols, allow the transmission speed to be increased up to 384 kbit/s.

The *4G generation*, operating in the 1.95 and 2.15 GHz frequency bands, allows data to be transferred at a transfer rate of up to 1 Gb/s for stationary or slow-moving devices and up to 100 Mb/s for fast-moving devices.

The *5G generation* operating in the frequency bands 410 MHz to 7.125 GHz and 2.425 GHz to 52.6 GHz with a maximum transmission speed of up to 20 Gb/s and a response time of less than 1 ms enables its users, for example, to control production processes online.

Table 9.1 shows the basic parameters of mobile networks.

The GSM system consists of mobile base stations and a GSM control center (Figure 9.1).

Mobile stations can be not only *mobile phones*, enabling the transmission of not only voice but also data signals, but also separate *GSM modules* that can be placed, for example, in means of transport in monitoring systems (Figure 9.2).

Table 9.1 Basic parameters of mobile networks.

Generation	Frequency bands	Transmission speed	Application	Year of introduction
1G	450 MHz, 900 MHz	9.6 kb/s	Voice services only	1982
2G	900 MHz, 1.8 GHz	80 kb/s	Voice services, SMS messages	1991
3G	1 950 MHz, 2 150 MHz	384 kb/s	Voice services, internet, video telephony	2001
4G	1 950 MHz, 2 150 MHz	100 Mb/s to 1 Gb/s	Internet, IP telephony, HD television, 3D television	2009
5G	410 MHz to 7.125 GHz 2.425 GHz to 52.6 GHz	20 Gb/s	Voice services, internet, transfer of large data files	2020

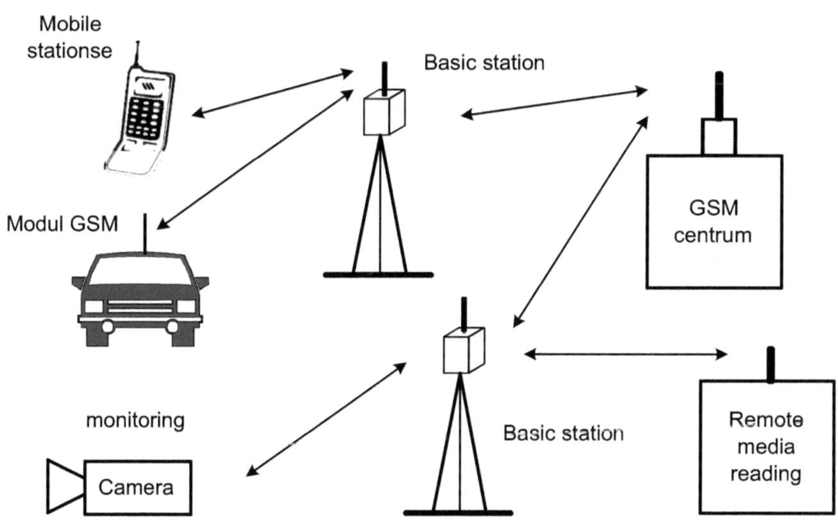

Figure 9.1 GSM network structure.

9.1.1 Mobile phones and stations

In order for data transmission to be possible in the GSM system, all its parts, including mobile stations (*telephones*), must be able to transmit this data. Figure 9.3 shows a block diagram of a typical mobile phone, operating in the 900 MHz frequency band, which enables the transmission of data messages in addition to audio signal transmission.

a) contactless mobile phone

b) GSM module

Figure 9.2 Mobile station.

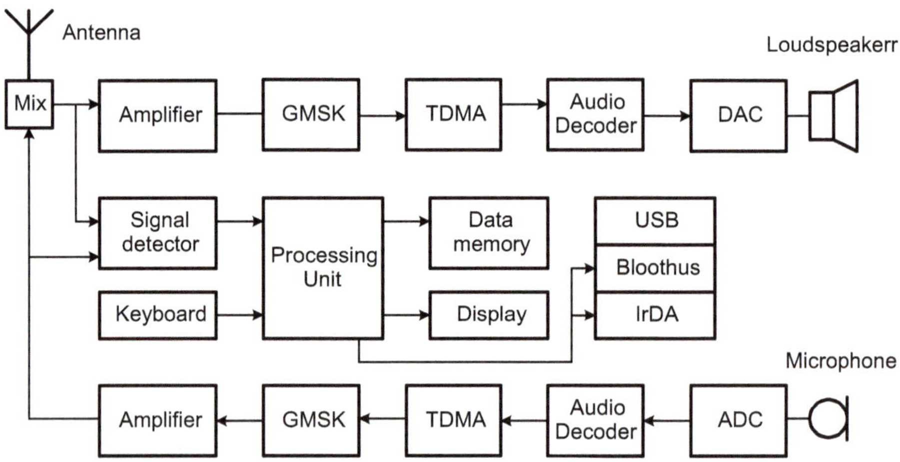

Figure 9.3 Data mobile phone block diagram.

The *telephone part* of the data mobile phone consists of a piezoelectric microphone, the signal of which is digitized by a 13-bit A/D converter with a sampling frequency of 8 kS/s, which is fully sufficient to digitize the audio signal in the frequency range up to 3.2 kHz. The 104 kb/s (13 bit × 8 kS/s) data stream created in this way is processed by an audio decoder, in which the data stream is compressed to a speed of 22.8 kb/s. After GMSK modulation (*Gaussian minimum shift keying*) and amplification, the mobile phone antenna is excited by this signal. The next part of the telephone line is made up of a GMSK demodulator, a TDMA decoder, and, after processing by an audio encoder, converted into an audio signal by a 13-bit B/A converter, which drives the speaker.

Figure 9.4 Mobile phone chip card.

The *data part* of the mobile phone is made up of a signal detector, which through the control unit enables data transfer to one of the common serial interfaces, e.g., USB, Bluetooth, and IrDA. For systems with a frequency band of 900 MHz, the maximum transmission rate is 270.833 kb/s, and the duration of the transmission of 1 bit is 3.692 ms, which corresponds to approximately 3300 signal periods at this signal carrier frequency.

In the *processing unit*, the data is organized into packets and prepared for TDMA transmission (i.e., transmission within a single time slot). The carrier frequency is keyed by the digital data signal in the GMSK modulator. The amplified digital signal with a data rate of 270.833 kb/s is then fed to the antenna combiner and then transmitted by the antenna. The transmission duration of 1 bit is 3692 μs, which corresponds to approximately 3300 signal periods at a signal carrier frequency of 900 MHz.

Each mobile station contains a mobile terminal and a subscriber identity module. This module has a unique number, referred to as IME (*international mobile equipment identity*), which enables the unique identification of a mobile station in a global wireless network. This number is stored in the SIM chip card, which also contains the local operating system of the mobile station, including security passwords, referred to as PIN1, PIN2, or PUK1 or PUK2 (Figure 9.4).

The BSSD (*Base Station System*) consists of a series of base stations located on poles or towers. Each base station is equipped with an antenna, a receiver, a transmitter of radio waves, and a signal processing unit. The base stations are connected to each other either by metallic or optical data links, and/or wireless connection.

The *GSM center* controls the activity of the base stations, consisting in the choice of communicating base stations with mobile stations so that each mobile station is connected to at least three base stations. This solution allows not only to increase the reliability of data transmission but also to identify the location of the mobile station.

Data transmission in the GSM system takes place in *duplex mode*, which requires the use of a double (duplex) transmission channel. In GSM systems using the 900 MHz frequency band, separate frequency bands are allocated for transmission channels between mobile and base stations and between base and mobile stations, referred to as *Uplink* channels in the 890–915 MHz frequency band and *Downlink* channels in the frequency band 935–960 MHz. The number of these channels in these bands is 124, which corresponds to a

Table 9.2 Transmission frequency bands and channels in the GSM network.

GSM type	Frequency range	Uplink band	*Downlink* band	Number of frequency bands	Number of transmission channels
GSM 900	900 MHz	890 MHz 915 MHz	935 MHz 960 MHz	124	992
GSM 1800	1800 MHz	1710 MH 1785 MHz	1805 MHz 1880 MHz	374	2992
GSM 1900	1900 MHz	1850 MHz 1910 MHz	1930 MHz 1990 MHz	299	2396

width of each channel band of 200 kHz. The channel-free 100 kHz bands are at the borders of these bands.

In addition to frequency ranges, the so-called time slots are defined, which make up the transmission channel of the transmission system. The 2.5G network and higher variants use TDMA (*time division multiple access*) data transmission, which works with multiplexed transmission channels, in order to increase the transmission speed.

The GSM 1900 system has an up-band frequency range of 1850–1910 MHz and a down-band frequency range of 1930–1990 MHz.

The parameters of the GSM 900, GSM 1800, and GSM 1900 systems are listed in Table 9.2.

9.1.2 Data transmission in mobile networks

When using TDMA (time division multiple access) data transmission technology, the transmission bands are divided into eight time slots with a period of 4.615 ms, each of which lasts 577 μs. When using 124 frequency channels with eight time slots, 992 duplex transmission channels are created simultaneously. When using the 1.8 GHz frequency band, the number of frequency bands can be increased to 374, thus increasing the number of transmission channels to 2992. In this case, the frequency range of the transmitting bands is 1710–1785 MHz, and the receiving bands are in the range of 1805–1880 MHz.

Mobile phones are divided into three groups according to their data transmission capabilities: MT0, MT1, and MT2.

The MT0 (*mobile terminal 0*) group includes mobile phones without an external data interface. In its simplest form, the MT0 phone only transmits voice messages and SMS (*short message service*) messages. It cannot transmit any data coming from another digital device. An example of such a phone

Figure 9.5 Nokia Card Phone 2.0 mobile.

is, for example, a mobile phone integrated with a palmtop. As it is a source of digital data in itself, it does not need any interface with external devices.

TMT1 phones communicate with the ISDN (*integrated services digital network*) through their interfaces but require a separate device called a TA (terminal adapter) to communicate with a computer. The task of the adapter is the two-way adaptation of standard computer signals to the mobile phone, e.g., USB signals. The terminal adapter function (TAF) is divided between the mobile phone and the adapter. This can be, for example, a PCMCIA card functioning as a GSM modem. An example of such a PCMCIA card is the *Cellular Data Card* for the Nokia 2110 MT1 phone.

MT2 phones can transfer digital data from a computer either via *USB*, wireless *IrDA*, or *Bluetooth* radio.

Another option is to transfer data via a PCMCIA card designed for laptops, which works as a mobile phone in itself. An example of such a PCMCIA card is the *Nokia Card Phone 2.0* card with a built-in antenna. After installation in a notebook, this card functions as a dual-band mobile phone, capable of working in the GSM 900 or GSM 1800 frequency bands (Figure 9.5).

Newer products such as *Sony-Ericsson Card Phone GC 19*E or *Nokia Card Phone D211 PMCIA cards* can be used for both HSCSD and GPRS transmission. These cards have an additional IEEE 802.11b radio interface and can work in many personal computer operating systems such as *Windows ME*, *Windows XP*, *Windows 8*, and *Windows 10*. For example, the *Nokia Card Phone 2.0* card with the dimensions of a PC card type II, 124 × 54 × 10 mm and weighing 58 g, allows the transmission of SMS messages and digital data with a maximum transmission speed of 43.2 kb/s for receiving data and 28.8 kb/s for their transmission.

Another type not covered by the above classification includes devices such as the *GENERIC GPRS Laptop* with a mobile phone module or the

Motorola Accompli 09 palmtop. Their development is a step forward in the integration of GSM-based data transmission devices.

The GSM system offers a number of digital data transmission services:

- Short message service or SMS (*short message*) is the transmission of alphanumeric messages of up to 160 characters.
- Multimedia messaging service or MMS (*Multimedia Messaging Service*) is the transfer of text, graphics, audio, and video files from storage to forwarding.
- CSD data transmission (*circuit switched data*) is a switched transmission of digital data at a speed of up to 9.6 kb/s via a service channel; this type of transfer is also referred to as SDT (switched data transfer).
- High-speed data transmission with circuit switching HSCSD (*high-speed circuit switched data*) is a transmission in which data is transmitted usually in four time slots with a total transmission speed of 57.6 kb/s.
- ECSD (*enhanced circuit switched data*) transmission is an extended data transmission protocol, using EDGE (*enhanced data rates for GSM evolution*) technology to speed up transmission in individual time slots. The maximum transfer rate for this transfer is 59.2 kbit/s per one slot, which is 236 kb/s using four slots.
- GPRS transmission (*general packet radio service*), packet data transmission mode.

The types and basic properties of digital data transmissions in mobile telecommunication networks are listed in Table 9.3.

9.1.3 Transmission of SMS and MMS messages

The SMS *short message service* enables the transmission of short alphanumeric messages of up to 160 characters to GSM or email users. Messages can also be sent without interruption during a phone call. SMS is a *person-to-person* service, which means that the user receives the message directly to the mobile station without checking their inbox.

The SMS represents a very low channel load and is the cheapest way to transfer data. Optionally, a delivery message can be sent from the receiver to the sender to verify that the message was delivered. If communication with the recipient is not established at the time of sending the message, the

Table 9.3 Types of digital data transmission in mobile telecommunication networks.

Transmission type	Transmission characteristics	Max. transmission speed	Network type
CSD	Standard digital data transmission	9.6 kb/s	GSM
CSD	Digital data transmission with its compression	14.4 kb/s	GSM
HSCSD	Fast digital data transfer	57.6 kb/s	GSM + slight adjustments
GPRS	Packet data transfer	115 kb/s	GSM + GPRS basic adjustments
EDGE	Extended data transmission in a modified GSM system	384 kb/s	GSM + substantial modifications
UMTS	Universal mobile data transmission communication system	1960 kb/s	UMTS global network

message is stored in the SMS center and is forwarded to the recipient after connecting. The typical delivery time of a message is a few seconds from the moment it is sent, if the communication is not disrupted.

However, the delay in message delivery can be much longer. The message may be delivered after a few hours or days, and in rare cases, it may remain undelivered (which was the reason for creating the delivery confirmation option). All SMS messages are transmitted via the short text message center (SMS-C), which is part of the GSM switching system. Because SMS-C features are not standardized, different operators may offer different SMS services.

For measurement purposes, the SMS service can be used, for example, when monitoring objects for the remote transmission of small volumes of data. The maximum number of characters in an SMS message can actually be six times greater than the nominal number and can be equal to $6 \times 160 = 960$ characters. In this case, the messages are divided into shorter messages with a maximum length of 160 characters, which are sent and billed separately. The message delivery delay is then in the order of tens of seconds, which may be unacceptable for some measuring systems.

MMS (multimedia message service) is the transfer of multimedia files from one subscriber to another or between GSM devices. The MMS service allows you to transfer:

- plain text (unformatted or formatted with formatting instructions);

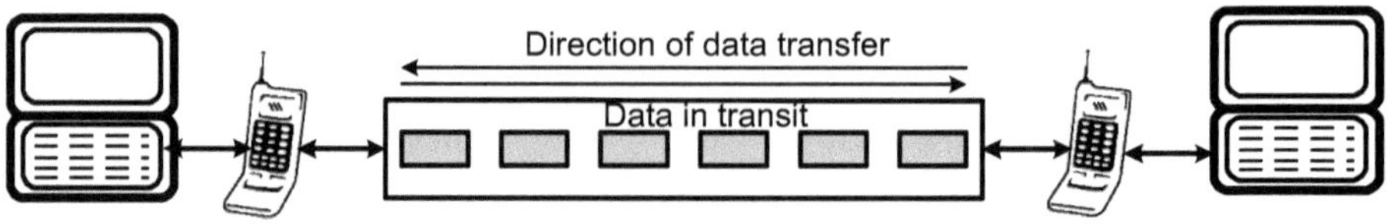

Figure 9.6 Block diagram of CSD data transfer in the GSM system.

- graphics such as JPEG, GIF, and *Portable Network Graphic* (PNG) files;
- sound (basic files, MP3, or WAV);
- video (MPEG files).

An important feature of MMS messages is the ability to transfer files from the Internet. Each MMS message, like an Internet packet, consists of an envelope (instructions needed to deliver and interpret the message) and message content. The MMS standard uses WAP (*wireless application protocol*) as its transport protocol. Therefore, the MMS standard uses improved transmission modes (HSCSD, GPRS, and EDGE). MMS messages are not used very often in measurement systems, but they can be used, for example, in traffic or water level monitoring, as well as in monitoring and alarm systems in industrial production. All these applications involve the transfer of image or video files.

Data compression by switching CSD (*compression switched data*) has been used since the very beginning of GSM. However, its use in measurement systems is limited due to low data transfer rates, which reach a maximum of 9.6 kb/s. The cost-effectiveness of CSD transmission is affected by the fact that this type of transmission occupies the entire traffic channel, while the data connection is billed as a telephone call. The solution to this problem is the HSCSD transfer mode.

Figure 9.6 shows the block diagram of CSD data transmission in the GSM system.

High-speed circuit switched data (HSCSD) uses several (up to four) traffic channels within one physical channel to increase the transmission speed. Figure 9.7 shows the block diagram of HSCSD data transmission in the GSM system, which uses four operational channels for data transmission.

When using four traffic channels, the data transfer rate can reach up to 57.6 kb/s. However, this requires GSM base stations to be suitably modified and users to have suitable mobile phones. In both CSD and HSCSD transmission modes, data is transmitted either between two GSM network users or between a GSM network user and a PSTN (*public switched telephone network*) stationary telecommunication network user.

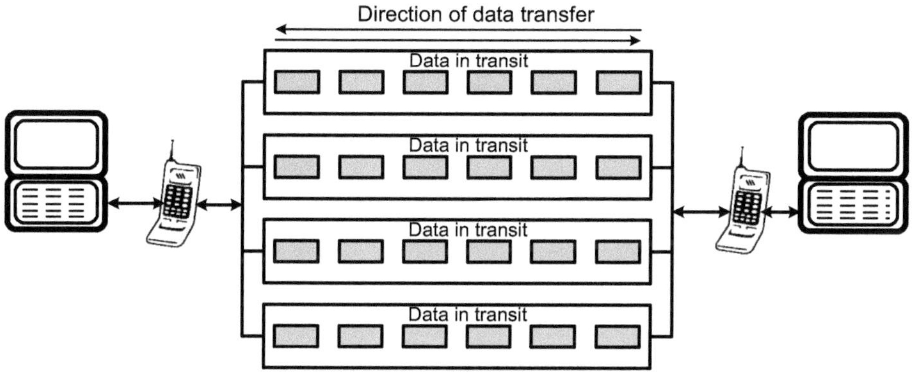

Figure 9.7 Block diagram of HSCSD data transfer in the GSM system.

9.1.4 General packet radio service

The general packet radio service GPRS (*general packet radio service*) enables the transmission of packet data over the Internet without switching transmission channels. While the transmission modes discussed above are based on channel switching, GPRS uses packet switching instead. A GPRS session can be activated in "always connected" mode and data can be transferred without interference during phone calls. Each packet or set of transmitted data is an integrated whole and can be transmitted independently of other packets, with the destination Internet address specified in the packet header, as shown in Figure 9.8.

In addition to permanent access to the web, the advantage of GPRS is high data transfer speed. This is caused by the use of up to eight time slots, allocated when setting the transmission channel. The capacity of one time slot of 14.4 kb/s multiplied by the number of allocated time slots gives the maximum speed of GPRS data transfer, i.e., $8 \times 14.4 = 115.2$ kb/s. Transfer speeds can theoretically be increased up to 170 kb/s with suitable data encoding methods. The implementation of GPRS requires network operators to modify the base stations even more than in the case of the HSCSD mode. However, these modifications are limited enough to qualify both GPRS and HSCSD transport modes as complementary services to the GSM system (i.e., hardware and software to be added to the core GSM network).

GPRS is considered the first usable digital data transfer technology between GSM and UMTS. The maximum speed of data transfer in GPRS mode depends on the quality of the GSM network and also on the parameters of the mobile phone, which should meet GPRS standards. For cost-effectiveness reasons, GPRS mode is recommended for non-periodic data transfer,

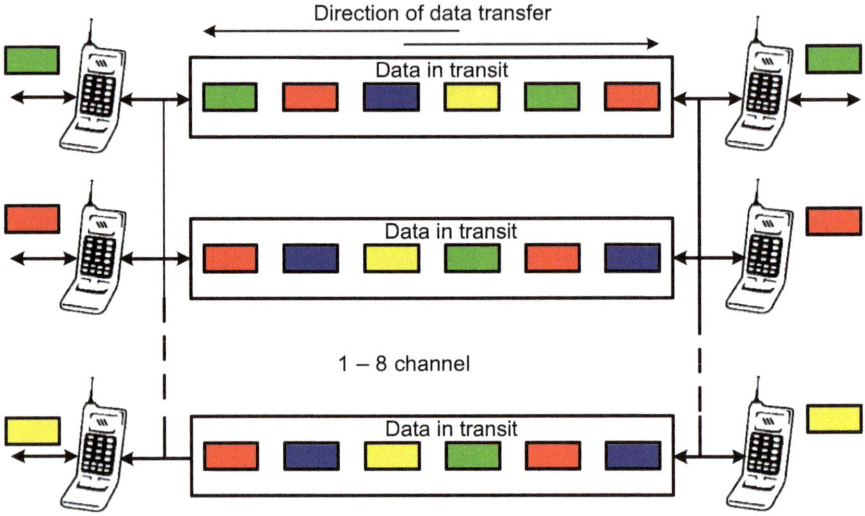

Figure 9.8 Block diagram of GRPS transmission in the GSM system.

Table 9.4 Transmission rates when using GRPS code schemes.

Code diagram GRPS	CS-1	CS-2	CS-3	CS-4
Transmission speed	8.0 kb/s	12.0 kb/s	14.4 kb/s	20 kb/s

frequent small-volume transfers of less than 500 octets of data each, performed several times per minute, and larger data transfers performed at a frequency of several times per hour or less.

Maximum transmission speeds specified for certain applications may differ from those guaranteed by network operators. The GPRS Internet transmission mode involves the use of the internet protocol (IP), whose procedures reduce the maximum single-channel compressed data transfer rate from 14.4 to 13.4 kb/s. Based on the standardized GPRS protocols, the GSM network operator can provide a number of additional services, such as data reading from the Internet, remote meter reading, credit card authentication, and many others. However, these services require special equipment for mobile stations.

Table 9.4 shows the transmission rates for individual GRPS code schemes.

9.1.5 EDGE data transfer

This type of transmission, which has been used since the 3G generation, enables transmission speeds of up to 473.6 kb/s; so it can also be used to

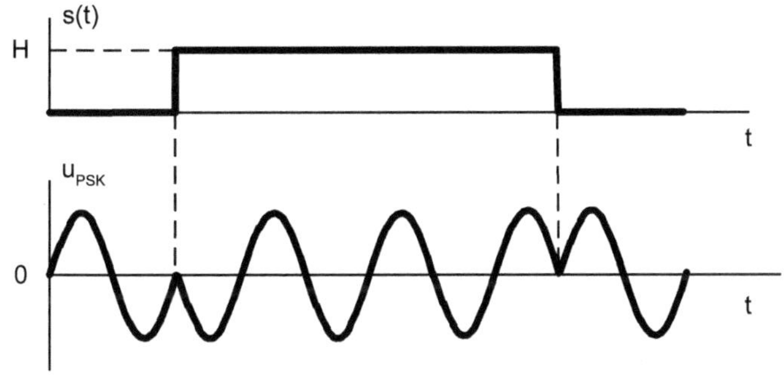

Figure 9.9 Time course of the PSK signal.

transmit moving images. GSM frequency bands and radio channels with a spacing of 200 kHz are reserved for EDGE transmission. This solution facilitates the cooperation between classic GSM and EDGE/GSM systems and thus allows achieving a significant reduction in the costs of its implementation.

The high transmission speed of EDGE transmission is due to the use of two modulation methods, which are phase pulse modulation PSK (*phase shift keying*) and quadrature phase pulse modulation QPSK (*quadrature phase shift keying*).

Phase pulse modulation modulates the phase of the carrier harmonic signal with a modulating two-valued signal $s(t)$, while its lower level L corresponds to the zero phase shift of the carrier signal and the higher level H corresponds to a phase shift of 180° (Figure 9.9).

QPSK (*quadrature phase shift keying*) modulates two harmonic (quadrature) signals that are phase-shifted by 90° with two-value or multilevel modulation signals

$$u_{C1}(t) = U_c \cdot \sin\omega_c t, \qquad u_{C2}(t) = U_c \cdot \cos\omega_c t, \qquad (9.1\text{ a, b})$$

If these signals are modulated by a one-bit (binary) modulation signal $s_1(t)$, it is 2-QPSK modulation. When quadrature signals are modulated by a two-bit modulation signal $s_1(t)$ and $s_2(t)$, it is a 4-QPSK modulation (Figure 9.10).

In EDGE data transmission, 8-QPSK and 16-QPSK quadrature modulation with three-bit and four-bit modulation signals are used to modulate the carrier signals.

It follows from the character of these modulations that by increasing the number of bits of the coding signal, the requirement for their higher signal-to-noise phase separation increases, which, for example, in the case of 16-QPSK modulation should be at least twice as small as 360/16 = 22.5°.

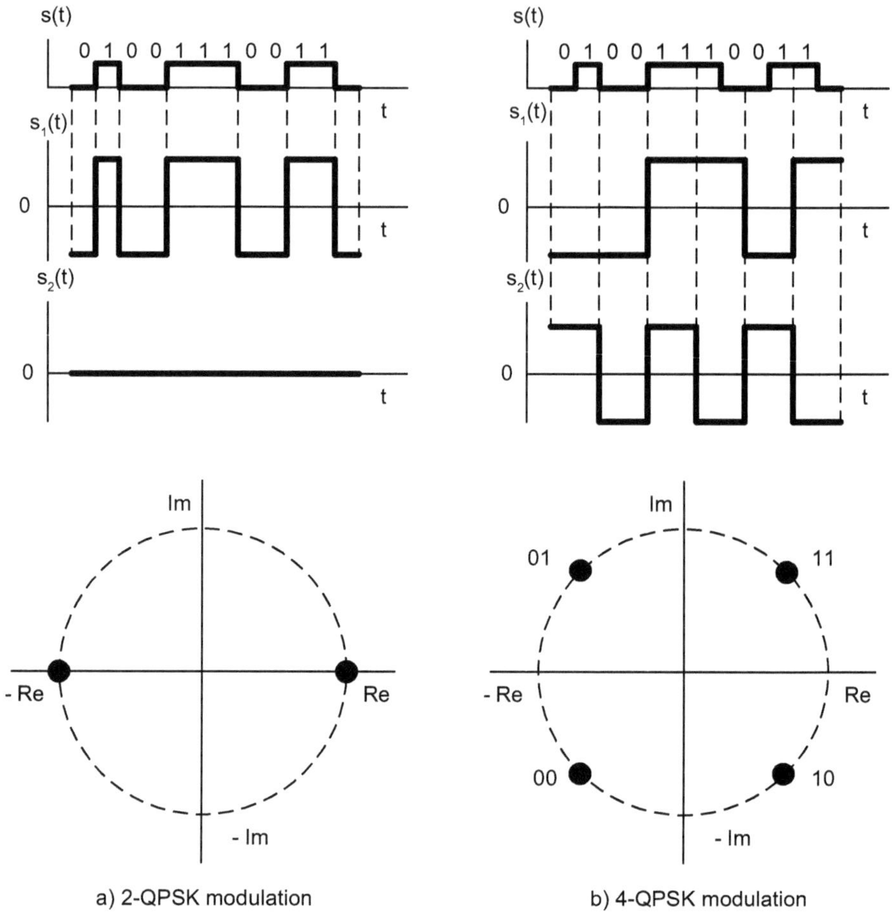

Figure 9.10 2-QPSK and 4 QPSK-quadrature modulation.

Therefore, the type of QPSK modulation is chosen based on the distance and quality of the data link between the base stations.

Depending on the quality of the signal, GPSK modulation with transmission rates of 8.4 kb/s is used for less demanding requirements and QPSK-8 modulation for high requirements, with which transmission rates of up to 59.2 kb/s can be achieved using one data channel (slot). When using eight time slots, the theoretical maximum data rate ranges from 8 × 8.4 = 67.2 kb/s to 8 × 59.2 = 473.6 kb/s. The recommended data rate for moving image transmission is 384 kb/s, which is well below the EDGE upper limit recommended by *the International Telecommunication Union* (ITU) for moving image transmission.

Table 9.5 Transmission rates for one channel when using EDGE code schemes.

Code scheme EDGE	Speed rate	Modulation
MCS-1	8.8 kb/s	GPSK
MCS-2	11.2 kb/s	GPSK
MCS-3	14.8 kb/s	GPSK
MCS-4	17.6 kb/s	G/PSK
MCS-5	22.4 kb/s	8-QPSK
MCS-6	29.6 kb/s	8-QPSK
MCS-7	43.8 kb/s	8-QPSK
MCS-8	54.4 kb/s	8-QPSK
MCS-9	59.2 kb/s	8-QPSK

Table 9.5 shows the transmission rates for individual code and modulation schemes of the EDGE transmission technology.

9.1.6 5G mobile network

This network is currently the most powerful wireless transmission network. It uses the 26 GHz (24.25–27.5 GHz) and 42 GHz (40.5–43.5 GHz) frequency bands. The 66–71 GHz bands are intended for local wireless interconnection of devices and sensors within a single operation.

Since the requirements of different technologies can be different (for example, in mobile phones), connection speed is more important, while for autonomous vehicles, the shortest possible response is key, a method called network slicing is being developed, which allows dividing the 5G network into subnets with different parameters, which require appropriate levels of technology. Another benefit of the new standard is the partial decentralization of the network, enabling direct communication between individual devices without the operator, which enables higher reliability and speed of data transfer and lower reaction times with reduced costs for transferring large datasets.

9.1.7 GSM-based distributed measurement systems

In addition to the system controller of this GSM network, distributed measurement systems using the GSM network are made up of GSM modules, which can also be data mobile phones. These modules can be connected either directly to individual measuring devices via their interfaces (*USB*, *Bluetooth*, *IrDR*), or these devices can be equipped with GSM modules, or PCMIA plug-in interface cards, which are equipped with GSM modules, can be used to connect measuring devices (Figure 9.11).

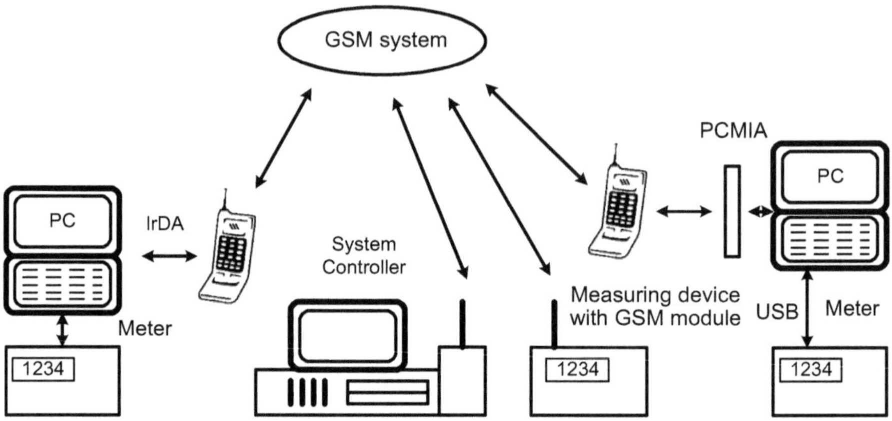

Figure 9.11 Block diagram of a distributed measurement system with data transmission via GSM network.

Data is transmitted to the system controller either through the fixed telephone network or over the Internet. In the second case, data can also be received by a mobile phone that allows access to the Internet via the WAP protocol.

The SMS message path from the computer in the measuring station to the SMS center can be divided into two sections with different transmission protocols. The first part is from the terminal adapter (computer) to the mobile phone. Since there is no standard protocol for transferring digital data between such devices, there is a lot of freedom in the choice of protocol, and phone manufacturers are not always willing to disclose the protocols used in their mobile phone models. The second part of the message path is from the mobile phone to SMS-C. In this part, the SMS data is transmitted within the PDU (*protocol data unit*).

There are two types of PDU protocols:

- SMS-SUBMIT, transmission of an SMS message from a mobile phone to SMS-C;
- SMS-DELIVER, sending an SMS message from SMS-C to a mobile phone.

The formats of both PDU types are shown in Table 9.6.

The gray UD fields contain the text of the SMS message with a size from 0 bytes (invalid message) to 140 bytes. Other PDU fields include:

- PDU type, specifying whether the PDU is of type DELIVER or SUBMIT, whether the transmission of the message is paid for by the

Table 9.6 Data formats of SMS messages.

Format of the received message SMS							
1 byte	**1 byte**	**2–12 bytes**	**1 byte**	**1 byte**	**1–7 bytes**	**1 byte**	**0–140 bytes**
Type PDU	MR	DA	PID	DCS	VP	UDL	UD
Format of the message sent							
1 byte	**2–12 bytes**	**1 byte**	**1 byte**	**1–7 bytes**	**1 byte**	**1 byte**	**0–140 bytes**
Type PDU	OA	PID	DCS	SCTS	VP	UDL	UD

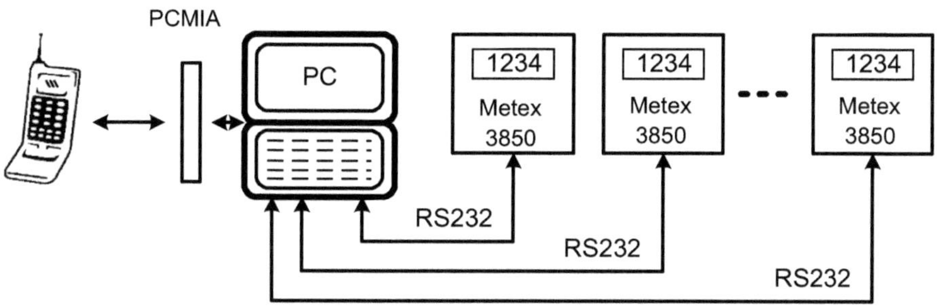

Figure 9.12 Block diagram of distributed measurement system based on GSM.

sender of the message or the recipient of the message and whether a confirmation of delivery is required or not,

- *message reference number* MR (0–255),
- *destination address* (DA), phone number of recipient,
- *originator's address* (OA), sender's phone number,
- *protocol identifier* (PID), specifying the type of data (text, telex, or fax),
- *used data encoding scheme* (DCS),
- *validity period* (VP), the maximum time for saving a message in SMS-C in case of delivery failure,
- *user data length* (UDL), message length,
- the time of receipt of the SMS message (*SCTS – service center time stamp*).

A block diagram of a simple wireless distributed measurement system with GSM-based Metex 3850 multimeters with data transfer from the measuring instruments to a computer via a serial RS-232 interface equipped with a PCMCIA card is shown in Figure 9.12.

Data transfer from these multimeters to a computer via the RS-232 serial interface enables transfer with a maximum transfer rate of 1.2 kb/s without checking data transfer by parity. The data frame sent by the multimeter contains 14 SDU characters and contains one measurement result. The 14th character is CR, closing the transmission of the frame. A PCMCIA GSM card is used to transfer data from a computer to the GMS network. The disadvantage of this system lies in the limited transmission speed and the absence of data transmission reliability security. Therefore, such a system is not suitable, for example, for transferring larger volumes of data, e.g., from digital oscilloscopes or spectrum analyzers.

9.1.8 Universal mobile telecommunication system

The UMTS transmission system (*universal mobile telecommunication system*) is the standard under which the development of GSM mobile networks takes place, led by the 3GPP association. UMTS meets the ITU IMT-2000 requirements for third generation (3G) mobile cellular networks. The UMTS transmission system was first introduced in 2001 in Japan, has been operating in SNR since June 2004, and was gradually introduced in other European countries. Typical representatives of mobile phones with the possibility of connecting to UMTS networks are, for example, mobile phones from the companies *Samsung*, *LG*, and *Iridium*.

- The UMT transmission system itself does not bring any benefits to users, but it is a framework that provides space for the deployment of other technologies, of which HSDPA in particular brings users a substantial increase in data transfer speed.

They are:

- 2G GSM: 9.6 kbit/s (CSD) or multiples when using HSCSD,
- 2.5G GPRS: up to 107 kbit/s (with the possibility of increasing to EDGE-type transmission),
- 2.75G EDGE: up to 384 kbit/s,
- 3G UMTS: up to 384 kbit/s (with the option of upgrading to HSDPA),
- 3.5G and HSDPA: up to 14 Mbit/s (with the possibility of increasing to HSDPA up to 84.4 Mbit/s),
- 3.9G LTE: up to 326 Mbit/s,
- 4G LTE advanced up to 1 Gbit/s,

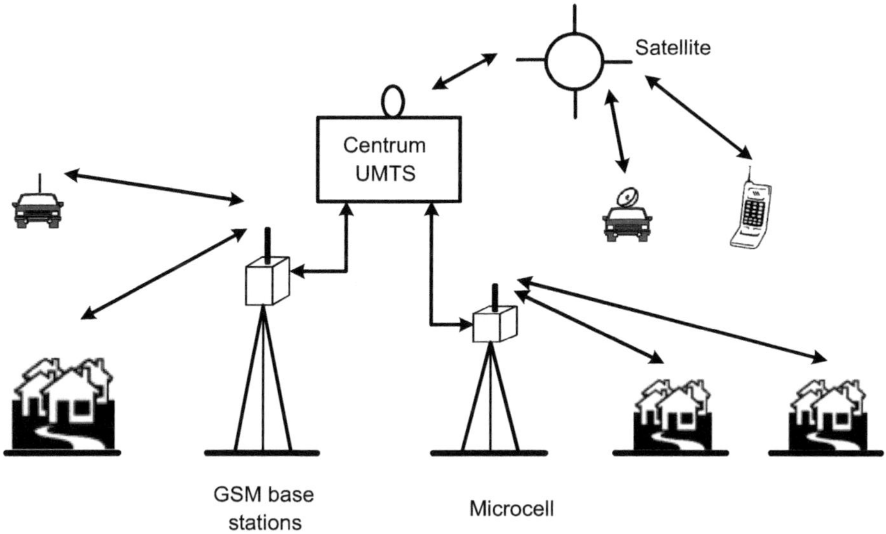

Figure 9.13 Diagram of the UMTS 3G wireless communication system.

- 5G up to 80 Gb/s.

The advantage of the UMTS system is that it combines TDMA transmission on the usual GSM bands with WCDMA (*wideband code division multiple access*) transmission to share the access channel in the 1.9–2.2 GHz frequency bands. Another benefit is the possibility of cooperation with fixed networks, including the B-ISDN network (*broadband integrated services digital network*). A subscriber in the GSM network who has an activated data plan and a phone with UMTS support is automatically transferred to the UMTS system in places where a UMTS signal is available. It goes without saying that UMTS networks are also used to transmit telephone calls.

The characteristic features of the UMTS system are the preservation of one individual user number in any telephone network, an individual set of transmission services provided anywhere by any operator and, in particular, global access to their networks, including satellite networks.

Uninhabited areas such as oceans or deserts, as well as regions with very low population density, can thus be covered by large-area macrocells via the SCS (*satellite communication system*) organized within UMTS. In other parts of the world, the UMTS service is based on terrestrial networks, e.g., GSM, or microcells of this system are built in densely populated regions (Figure 9.13).

Table 9.7 UMTS data transmission services.

Service	Transmission speed	Error rate
Telephone transmission	8–32 kb/s	10^{-4}
Video transmission	64–384 kb/s	10^{-7}
Audio transmission of data	2.4–64 kb/s	10^{-6}
SMS messages and calls	1.2–9.6 kb/s	10^{-6}
Digital data transmission	64–1920 kb/s	10^{-6}
Database access	2.4–768 kb/s	10^{-6}
Remote control	1.2–9.6 kb/s	10^{-6}

Table 9.8 Transfer speeds and times required to transfer 1 MB of data.

Transmission type	Transmission speed	Transmission time
CSD	9.6 kb/s	14 minutes
GRPS	115 kb/s	90 s
EDGE	384 kb/s	20.8 s
UMTS	1.92 Mb/s	4.6 s

Landline organization (i.e., system hierarchy and cell sizes) is operator dependent. *Satellite communication* is based on a group of 50–500 geostationary satellites moving in low Earth orbits, on the order of 1000 km above it, or on a group of approximately 100 satellites moving in a medium Earth orbit at a distance of the order of 10,000 km from the Earth.

Digital data transmission in UMTS networks enables the achievement of transmission speeds listed in Table 9.7 within the 3G generation.

To illustrate the properties of individual methods of data transfer, Table 9.8 shows the transfer speeds and times required to transfer a data file with a capacity of 1 MB.

The accepted UMTS standards also include the function of determining the location of the mobile station (or mobile phone). Currently, there are several autonomous systems for determining the location of objects, the most widespread of which is the *American Global Positioning System* (GPS) with a resolution for commercial use of 10 m (1 m for special use) around the globe. For example, positioning systems are used for the following purposes:

- sea and land navigation;
- monitoring the location of trucks by transport bases;
- identification of the location of cars, e.g., when they are stolen;
- optimization of car and truck routes;

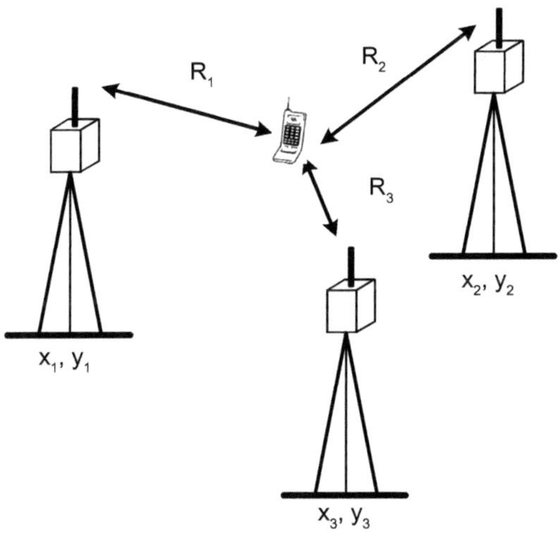

Figure 9.14 Mobile station positioning.

- setting out points and routes in engineering geodesy;
- monitoring the movement of suspicious persons;
- determining the location of injured persons even in otherwise inaccessible locations.

The methods of determining the location of UMTS mobile stations are defined by system standards. Two independent positioning methods are used:

- Measuring the signal level and the time required for the signal to get from the mobile station to the base station (*uplink signal*) and back from the base station to the mobile station (*downlink signal*). This method is referred to as RTT (*round trip time*).
- Determining the position of the arrival time difference of OTDOA, using the measurement of the differences between the arrival times of the signals at the base mobile stations, while using at least three nearby stations, managed by the same controller of these stations.

The choice of positioning method is entirely up to the operator. None of these methods are intended to disrupt ongoing connections or degrade transmission quality. The first of the above-mentioned techniques is difficult to implement, especially in large cells of the system, and therefore the second method, based on knowledge of the coordinates of ground stations, is most often used (Figure 9.14).

If the coordinates of the mobile stations are x_i, y_i, $i = 1, 2, 3$, then the distance difference between them and the mobile phone is determined by the following equations:

$$R_1 - R_2 = c \cdot \Delta T_{12}\left[\sqrt{(x_1 - x)^2 + (y_1 - y)^2} - \sqrt{(x_2 - x)^2 + (y_2 - y)^2}\right], \quad (9.2\text{ a})$$

$$R_1 - R_3 = c \cdot \Delta T_{13}\left[\sqrt{(x_1 - x)^2 + (y_1 - y)^2} - \sqrt{(x_{32} - x)^2 + (y_{32} - y)^2}\right], \quad (9.2\text{ b})$$

$$R_2 - R_3 = c \cdot \Delta T_{23}\left[\sqrt{(x_2 - x)^2 + (y_2 - y)^2} - \sqrt{(x_3 - x)^2 + (y_3 - y)^2}\right], \quad (9.2\text{ c})$$

where c is the speed of propagation of electromagnetic waves in air, which is approximately equal to the propagation of these waves in a vacuum of $3 \cdot 10^8$ m/s, and ΔT_{ij}, $i, j = 1, 2, 3$ determine the time differences between the arrival of signals from the mobile to the base stations.

The achievable accuracy of determining the location of mobile stations depends on the weather conditions and reaches several units of m to hundreds of m.

9.2 Radio Modem Measurement Systems

Radio communication is also used to transmit measured data over distances of up to hundreds of km, which is an alternative to communication systems based on either wired PSTN or wireless GSM telephone networks through dedicated non-telephone channels. Radio modems are used for this method of wireless connection.

9.2.1 Radio modem measurement system

Measuring systems with radio modems consist of a system center in which data is collected and processed, and a number of measuring stations communicating with this center (Figure 9.15).

A radio modem with a suitable directional antenna is installed in each measuring station, which enables the reduction of transmission power, the sensitivity of receivers to interference, and also signal transmission in an unwanted direction. Signals transmitted within the measurement system can

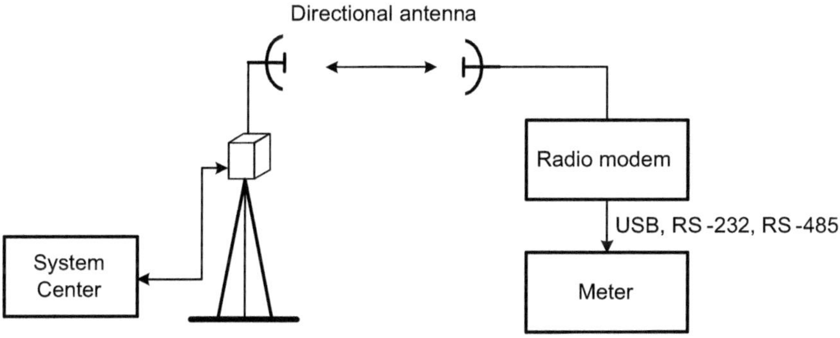

Figure 9.15 Measuring system with radio modem.

be interfered with by other transmitters. Directional and encrypted signal transmission also reduces the risk of unauthorized data interception. Data transmission over a non-telephone radio channel offers significantly higher security than transmission over a public telephone network.

However, the use of radio modems can involve problems such as unstable connection quality or even its interruption. Therefore, it is recommended to use radio modems with the additional function of a radio telephone, which enables simple acoustic control of the radio channel. The reason for unstable radio transmission is mainly changes in weather conditions affecting the propagation of radio waves. An improvement in reception can be achieved by increasing the transmit power or the height of the antenna. However, both of these parameters are limited by license and license fee. The problem of poor connection quality and limited area of use of the radio connection can also be solved by using some measuring stations also working as repeaters (retranslators) (Figure 9.16).

In this system, one of the measuring stations is also used for radio signal retransmission. In order to enable the distinction between reception, transmission, and retransmission of a radio signal, a modified retransmission communication protocol must be used for this purpose. For signal retransmission, e.g., *Satel 2ASxE*, *3AS EPIC*, and *3AS(d)*, radio modems can be used.

9.2.2 Radio modems and radio channels

In addition to the serial interface, radio modems also contain modems, modulators, demodulators, transmitters, receivers, and high-frequency signal antenna combiners. Common serial interfaces include the USB interface, or RS-232. Transmitted signals are converted in modems to modulated signals,

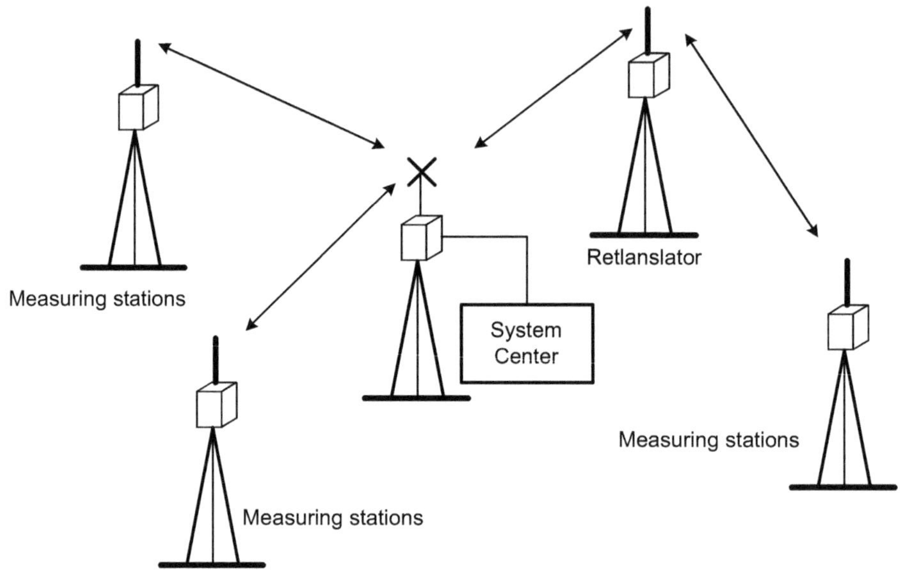

Figure 9.16 Measuring system based on a radio modem with measuring stations and a repeater.

which after amplification are fed through the antenna combiner to the antenna of the radio modem (Figure 9.17).

When receiving signals, the signal is demodulated and converted to a serial interface signal by the modem.

The *frequency bands* reserved for radio communication are determined by the relevant radio communication authority in each country. Radio frequency bands are allocated by these authorities in the range indicated in Table 9.9.

In order to legally use a radio frequency band for data transmission, a license must be obtained from the appropriate government agency, which issues this license for a fee.

No license is required to use radio modems with transmitters with antenna power less than 20 mW in frequency ranges less than 800 MHz. In addition to frequency bands and maximum power, the type of modulation of the carrier high-frequency signal can be specified in each license.

9.2.3 Modulation of signals in radio modems

In addition to frequency modulation FM (*frequency modulation*), frequency pulse modulation FSK (*frequency shift keying*), also modulation with a spread

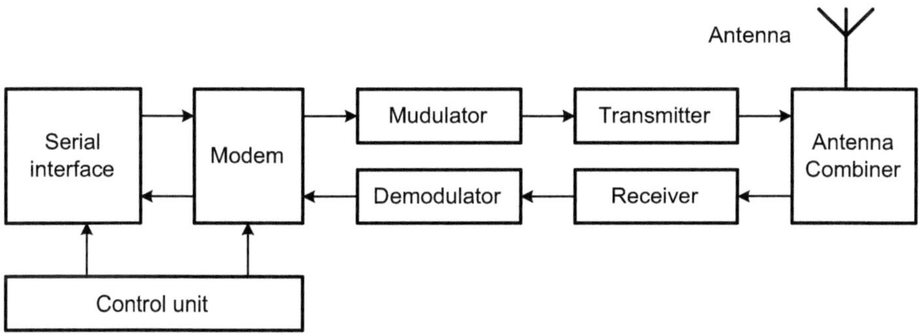

Figure 9.17 Block diagram of a radio modem.

Table 9.9 Radio communication frequency bands in the Czech Republic.

Frequency band	Bandwidth	Antenna length	Station type
77.5–79 MHz	12.5 kHz	≈ 1 m	Mobile
159.4–160 MHz	12.5 kHz	≈ 0.5 m	Mobile
299–309 MHz	12.5–25 kHz	≈ 0.5–5 m	Mobile or stationary
335–345 MHz	12.5–25 kHz	≈ 0.5–3 m	Mobile or stationary
452–457 MHz	12.5–25 kHz	≈ 0.5–3 m	Mobile or stationary
890–919 MHz	12.5–25 kHz	≈ 0.5–2 m	Mobile

frequency spectrum SSFM (*spread spectrum frequency modulation*) is used in the radio modem connection.

In *FM frequency modulation*, the frequency of the modulated (*carrier*) harmonic signal

$$u_c(t) = U_c \sin(\omega_c \cdot t) \tag{9.3}$$

modulated by a harmonic modulation signal

$$u_m(t) = U_m \sin(\omega_m \cdot t). \tag{9.4}$$

It then applies to a signal modulated in this way

$$u_{\mathrm{FM}}(t) = U_c \sin\left[\omega_c \cdot t + \frac{\Delta\omega}{\omega_m} \cdot \sin(\omega_m \cdot t)\right], \tag{9.5}$$

where
$$\omega_c = \omega_{c0} + U_m \cdot \Delta\omega \cdot \cos(\omega_m t) \tag{9.6}$$

and $\Delta\omega = 2\pi \cdot (1/T_{\min} - 1/T_{\max})$ is the frequency shift of the *modulated signal* (Figure 9.18).

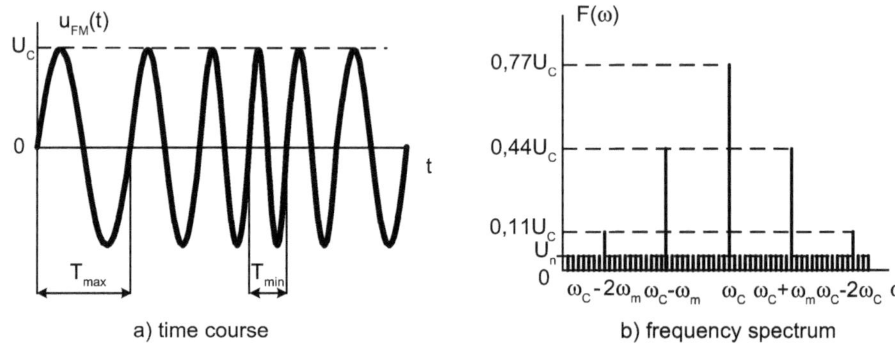

Figure 9.18 Frequency modulated signal $m_{FM} = 1$.

When introducing the *modulation index* of the frequency-modulated signal $m_{FM} = \Delta\omega/\omega_m$, expressing the relative range of changes in the circular frequency of the modulated signal, the frequency-modulated signal can be expressed in the form

$$u_{FM}(t) = U_c \cdot \sin\begin{bmatrix} \sin(\omega_c \cdot t) \cdot \cos(m_{FM} \cdot \sin(\omega_m \cdot t)) \\ +\cos(\omega_c \cdot t) \cdot \sin(m_{FM} \cdot \sin\omega_m) \end{bmatrix}. \tag{9.7}$$

The amplitudes of the symmetrical spectral components of the frequency spectrum of the FM signal are determined by *Bessel functions* of the first order. For small changes in the modulation frequency, the properties of the frequency-modulated signal approach those of the amplitude-modulated signal.

Table 9.10 shows these amplitudes for modulation indices in the range of 0.25–5.

Frequency pulse modulation FSK modulates the frequency of the carrier harmonic signal with a modulating two-valued signal $s(t)$, while its lower level L corresponds to the lower frequency $f_L = 1/T_L$ of the modulated signal and vice versa (Figure 9.19).

Modulation with a spread frequency spectrum uses DSS (direct sequence system) systems to modulate signals, either with linearly changing modulation frequencies of carrier signals LFM (*linear variable frequency modulation*) or with step changes of modulation frequencies SFM (*step frequency modulation*), in which the frequencies generated by PRBFM (*pseudo-random binary frequency modulation*) pseudo-random binary signals. Figure 9.20 shows a block diagram of a system with DSS modulation.

In the primary modulator of the transmitter, the input signal is modulated, which was previously modulated by PCM pulse code modulation.

Table 9.10 Amplitudes of the spectral components of the frequency spectrum of an FM signal.

m_{FM}	ω_c	$\omega_c \pm \omega_m$	$\omega_c \pm 2\omega_m$	$\omega_c \pm 3\omega_m$	$\omega_c \pm 4\omega_m$
0.25	0.98	0.12	<0.01	<0.01	<0.01
0.50	0.94	0.24	0.03	<0.01	<0.01
1.00	0.77	0.44	0.11	0.02	<0.01
1.50	0.51	0.56	0.23	0.06	<0.01
2.00	0.22	0.58	0.35	0.13	0.03
2.50	0.05	0.50	0.45	0.22	0.07
3.00	0.26	0.34	0.49	0.31	0.13

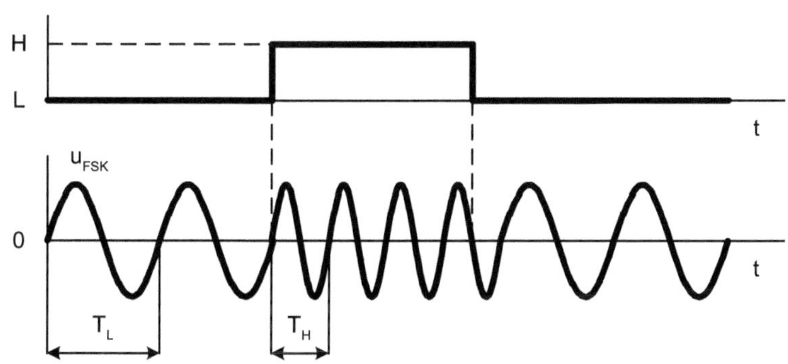

Figure 9.19 Time course of the FSK signal.

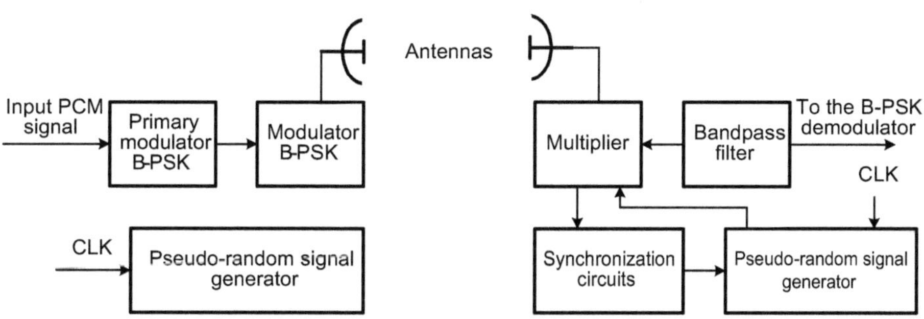

Figure 9.20 Block diagram of a pseudorandom code sequence frequency modulated spread spectrum communication system.

Two-state phase modulation with binary phase shift BPSK (*binary phase shift modulation*) is most often used for this modulation.

Figure 9.21 shows the time courses of signals during two-state phase modulation.

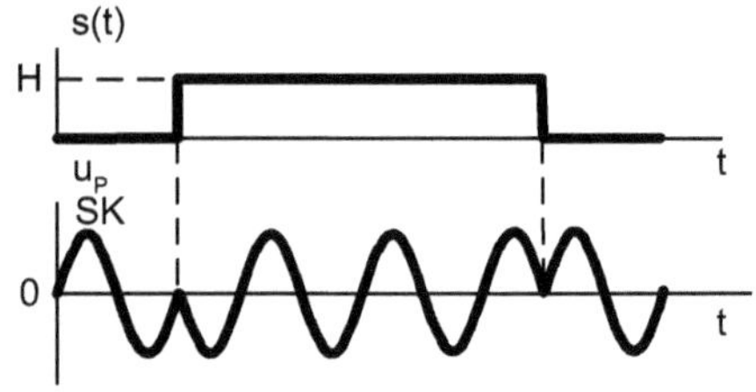

Figure 9.21 Signal courses during two-state phase modulation.

The output signal of the primary modulator occupies a frequency band with a width of FB (*frequency bandwidth*), equal to twice the bit rate (BR) of the PCM modulation signal. The signal is then fed to another B-PSK modulator, where it is again modulated by a binary signal, but in this case, it has a pseudo-random character, i.e., its bits change pseudo-randomly with the same probability of log occurrence. 1 and log. 0.

The bit rate of the pseudo-random signal BR_{pn} is chosen to be several orders of magnitude higher than the bit rate of the modulation signal BR_i. As a result, the bandwidth of the output signal of the second modulator B_{vj} is also significantly larger than the bandwidth of B. In practice, bit rates are chosen in the relation $BR_{pn} = (10^3 \text{ to } 10^5) \cdot BR_i$, so that the bandwidths

$$B_i \approx 2 \cdot BR_i, \qquad B_{vf} \approx 2 \cdot BR_{pn} = (10^3 \text{ to } 10^5) B_i). \qquad (9.8\text{ a, b})$$

The effect of the pseudo-random signal thus results in a significant spread of the frequency spectrum of the transmitted signal.

On the receiving side, a signal with a spread spectrum enters the multiplier, which with the following bandpass performs the role of a *correlato*r. The second input of the multiplier is supplied with a coding signal from a pseudo-random signal generator, generating a pseudo-random sequence of code words that are identical to the sequence generated by the transmitter's *pseudo-random generator*. This achieves that the output of the multiplier is a signal from which the pseudo-random sequence of the modulation signal is completely removed. Therefore, the signal demodulated in this way already has a narrowband character and can be demodulated after frequency filtering in a BPSK demodulator. This results in a baseband PCM signal that, apart from some distortion and noise, matches the modulation signal coming into the transmitter.

A *pseudo-random signal generator* with linear feedback contains a shift register, composed of *N* flip-flops, some of whose output signals are summed *modulo 2* and then fed to the register input. According to the arrangement

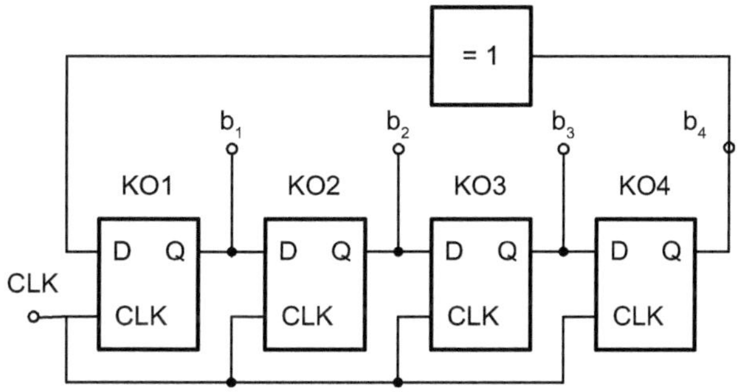

Figure 9.22 Block diagram of a four-bit pseudorandom generator.

of the feedback, the period and other properties of the generated sequence change, while its maximum achievable length is

$$L = 2^N - 1. \tag{9.9}$$

Figure 9.22 is a block diagram of a four-bit pseudo-random generator with linear feedback.

When initially setting the flip-flops to log states. 1, quadruplets of code words with a repeatability of 15 will be gradually generated at their outputs

$$111101110011000110000100001010011100011010110101101011011110, \tag{9.10}$$

whose decimal equivalents are

$$15\ 7\ 3\ 1\ 8\ 4\ 2\ 912\ 6\ 11\ 5\ 10\ 13\ 14. \tag{9.11}$$

The frequency of the generation of pseudo-random sequences of code words is determined by the frequency of the clock pulse generator CLK.

Modulation with a spread frequency spectrum is also used in DSSS (direct sequence spread spectrum, frequency hopping systems) systems that use multi-state keying of the frequency shift of modulation signals (Figure 9.23).

The transmitter in these systems is a fast *broadband frequency synthesizer*, the frequency of the output signal of which changes in equidistant time segments in equidistant graduated frequencies, evenly distributed in the B_{vf} frequency band. The size of these frequencies is determined by a pseudo-random binary signal, generated by a *pseudo-random number generator*, with a bit rate BR_h, which is approximately one order of magnitude higher than the bit rate of the information signal BR_i.

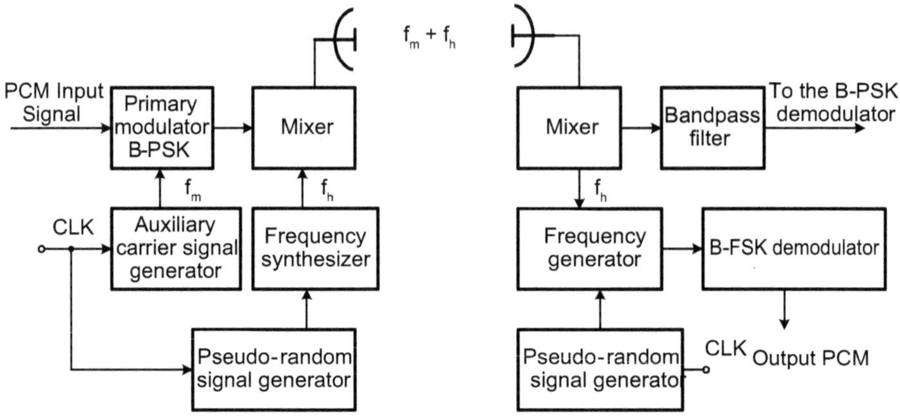

Figure 9.23 Spread-spectrum system with step-change carrier frequency.

In order for this system to be effective, the synthesizer must generate a large number of discrete frequencies in the range 10^3–10^5. The signal obtained in this way is mixed in the mixer of the transmitter with an auxiliary carrier signal of frequency f_m, which was previously modulated in the primary modulator by two-state phase modulation B-FSK transmitted signals. The medium frequency $(f_h + f_m)$ modulated signal in this way is transmitted by a directional antenna after amplification.

On the receiving side of this system, there is an identical frequency synthesizer and pseudo-random code word generator, which is frequency and phase synchronized with the received signal. The frequency range of the input signal of the receiver is reduced in the mixer to the intermediate frequency band f_m, in which the frequency of the carrier signal is already removed. The signal obtained in this way is demodulated with a pseudo-random signal and subsequently demodulated with a B-FSK modulator.

Again, a shift register with linear feedback is used to generate a pseudo-random coding signal. A certain frequency of the output signal of the synthesizer is thus assigned to each of the pseudo-randomly generated code combinations.

In addition to the two basic types of spread spectrum communication systems described above, systems with linear frequency modulation LFMS (*linear frequency modulation system*) are used in practice, in which pulse signals are used to transmit data, the frequency of which increases or decreases linearly, resulting in the desired spread of their frequency spectrum. These systems were first used in radiolocation, but they are also used in the wireless data transmission of radio modems.

The essence of the operation of the system based on the method of time jumps SFM (*step frequency modulation*) is that a modulation signal is fed

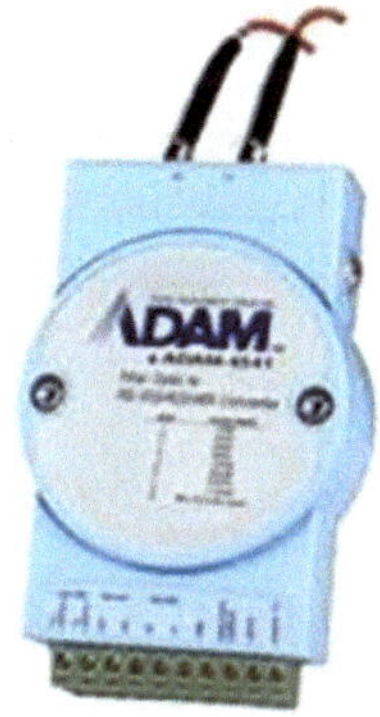

a) ADAM-4550 b) WISE 4660

Figure 9.24 Radio modems.

into the modulator of the transmitter, which implements a certain type of discrete (pulse) modulation. In this case, however, the modulator does not work continuously in time but is switched on and off by a pseudo-random coded signal. This creates an apparently irregular transmission after short periods of time, between which the transmitter is silenced. This method ensures high resistance against unauthorized interception of messages.

Mixed systems with a *spread frequency spectrum* are created by combining the basic modulation variants described above. In practice, a combination of a system with phase modulation and code sequence modulation and a system with a change in the frequency of the carrier signal is most often used, as well as a system with pulse-time modulation and a system with a change in the frequency of the carrier signal and a system with pulse-time modulation. Compared to the modulation techniques mentioned above, combined modulation systems with a spread frequency spectrum enable better transmission properties, including increased resistance to the occurrence of errors.

Distributed measurement systems with radio transmission are used, for example, for wireless measurement of water, gas, and thermal energy consumption in residential buildings and in energy networks. In order to ensure the reliability of data transmission, it is recommended to supplement wireless data transmission with cable transmission systems in exposed applications.

A typical modern radio modem is, for example, the ADAM-4550 radio modem from *National Instruments* with a power of 100 mW, operating in the 2.442 GHz frequency band with selectable transmission rates in the range of 3.2–115.2 kb/s. This modem uses the *direct sequence spread spectrum* (DSSS) method for data transmission. The modem can be connected to the RS-232 and RS-485 interfaces (Figure 9.24).

9.3 Data Transmission Over Short Distances

In addition to the *IrDA infrared beam* connection, high-frequency *Bluetooth* radio connection, *Wifi connection*, and other types of connections defined by the IEEE 802.11b standard are used for wireless data transmission over shorter distances.

9.3.1 IrDA connection

The IrDA modulated infrared wireless connection was introduced in the early 1980s by the *Infrared Data Association*. Since then, this method of control has spread not only in everyday life but is widely used in office, laboratory, and industrial environments. Its advantage is not only the exclusion of cabling but also the isolation of individual devices from the power supply network and from each other. In measuring technology, this connection is used to transfer data not only from touchable but also from moving objects with minimal investment costs.

The achievable connection distance of IrDa versions 1.0 and 1.1 is approximately 1 m, the transmission speed ranges from 2.4 kb/s up to 16 Mb/s and can work in normal daylight up to 10,000 lx. In addition to regular IrDa, there is also a *low-power version* of this connection (low-power IrDa), enabling data to be transmitted with a lower transmission speed of up to 115.2 kb/s.

Data transfer in an IrDa connection is declared by the following protocols:

- IrLAP (*IrDA infrared link access protocol*) is an HDLC protocol modified for IrDA communication. This protocol defines the way of communication between devices, their numbering, and sets the default transmission speed of the connection to 9.6 kb/s. It also defines transmission synchronization and creates communication channels between individual devices.
- IrLMP *(IrDA infrared link management protocol*) defines how to detect a newly connected device.
- TinyTP (*IrDA transport protocols*) defines the creation of a virtual channel between devices, the correction of transmission errors, and the division of data into data packets (similar to TP).
- IrOBEX (*IrDA object exchange protocol*) is a protocol with PUT and GET commands, allowing binary data to be transferred between devices. There is also the extension protocol, Extensions to IrOBEX for Ir Mobile Communications, intended for mobile devices, e.g., mobile

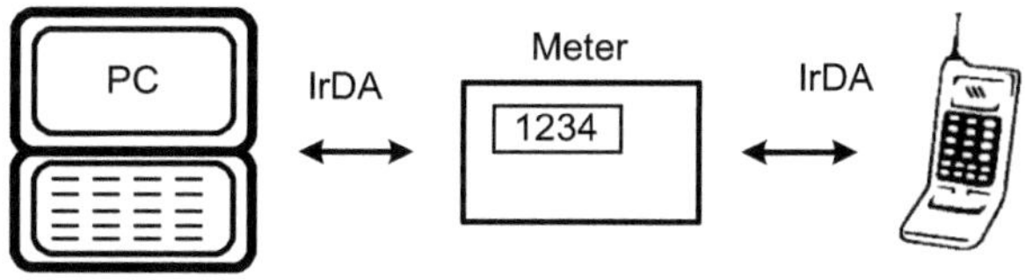

Figure 9.25 Measuring system with IrDa connection.

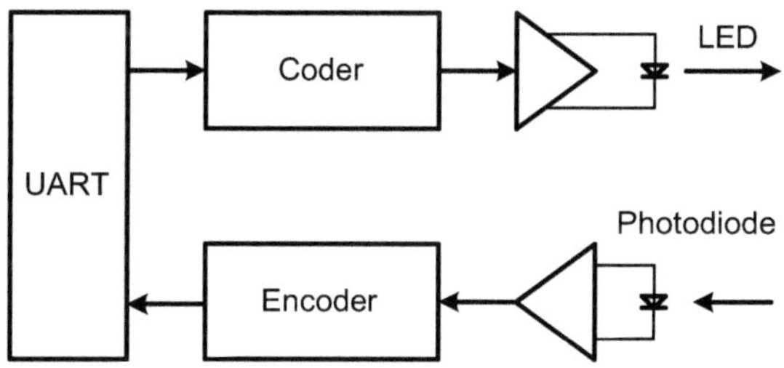

Figure 9.26 Block diagram of an IrDa transmission device.

phones, which defines data transfer by dialing a subscriber number or SMS message.

- IrTran-P (*infrared transfer picture*) is a specification designed to transfer images from digital cameras via an IrDa connection.

An example of the connection of a measuring system consisting of a computer, a measuring device, and a mobile phone with an IrDa connection is shown in Figure 9.25.

The circuit of the infrared connection consists of a light emitting diode LED, generating a modulated infrared light beam. This beam is converted into a sequence of pulses by a fast photodiode and, after decoding, transmitted by a universal asynchronous transmitter-receiver UART (*universal asynchronous receiver-transmitter*) to the appropriate device (Figure 9.26).

For data transmission, pulses with a width equal to 3/16 of their duration according to Figure 9.27 are used in the modulation of transmitted signals, while the level of log. 0 of the transmitted data signal corresponds to either one or several current pulses that drive the LEDs or that are generated by the receiving photodiodes.

Table 9.11 shows the typical parameters of an IrDA line according to the IrDa-1.3 standard.

Table 9.12 shows the types of modulation, the parameters of current pulses, the intensity of the infrared beam, and the sensitivity of the IrDA receiver to the transmission speed of IrDA transmission.

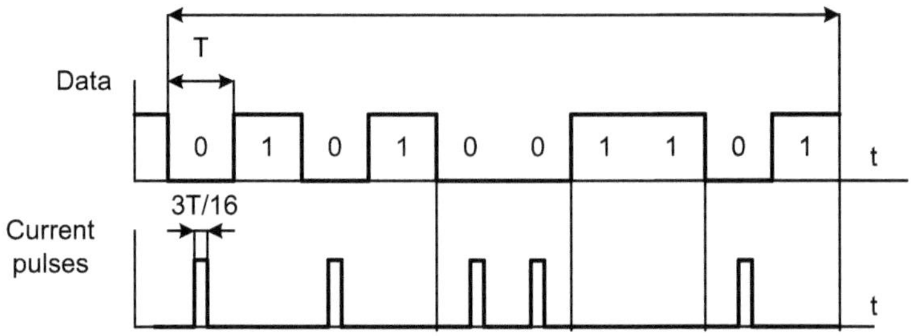

Figure 9.27 Time course of the pulse signal in IrDA transmission.

Table 9.11 IrDa line parameters according to the IrDA-1.3 standard.

Parameter	Typical value	Limit value
Range	1 m	3 m
Coverage angle	±15°	±30°
Transmission speed	2.4 kb/s to 4 Mb/s	16 Mb/s
Error rate	10^{-8}	–
Wavelength	850–900 nm	900 nm

Table 9.12 IrDa transmission parameters according to the IrDA-1.3 standard.

Transmission speed	Modulation type	Minimum pulse duration	Nominal pulse duration	Beam intensity	Receiver sensitivity
19.2 kb/s	RZI	1.41 μs	9.77 μs	40 mW/SR	4 μW/cm^2
38.4 kb/s	RZI	1.41 μs	4.88 μs	40 mW/SR	4 μW/cm^2
57.6 kb/s	RZI	1.41 μs	3.26 μs	40 mW/ SR	4 μW/cm^2
115.2 kb/s	RZI	1.41 μs	1.63 μs	40 mW/ SR	4 μW/cm^2
576 kb/s	RZI	295.2 ns	434 ns	100 mW/SR	10 μW/cm^2
1.152 Mb/s	RZI	147.6 ns	217 ns	100 mW/SR	10 μW/cm^2
4 Mb/s	4PPM	115.0 ns	125 ns	100 mW/SR	10 μW/cm^2
16 Mb/ps	HHH (1,13)	38.3 ns	41.7 ns	100 mW/SR	10 μW/cm^2

Note: The intensity of the light beam is given in units of *W/SR*, where *SR* denotes the solid angle.

RZI (return zero inverted), four-pulse modulation 4PPM (*four pulse position modulation*), and its fast variant, known as HPPM modulation (*high pulse position modulation*), are used to transmit data through the IrDa interface. With these modulations, transfer speeds of up to 16 Mb/s can be achieved. The IrDa 1.4 standard makes it possible to increase the transmission speed up to 100 Mb/s.

9.3.2 Bluetooth connection

This low-energy radio connection, operating in the 2.45 GHz frequency band, enables data transmission with a transfer rate of up to 1 Mb/s over a distance of approximately 100 m. This type of connection was originally designed by Ericsson and subsequently became a low-capacity computer wireless peripheral for positioning devices and controllers (mice, joysticks), printers, scanners, memory readers, bar, and two-dimensional matrix codes, but also mobile phones.

The difference between this connection and the *IrDA connection* is that the *Bluetooth connection* allows the connection of up to eight devices, unlike the IrDa connection, which allows the connection of only two devices. Also, the range of radio waves using the Bluetooth connection is significantly higher compared to IrDA transmission. In addition, the spherical nature of radio wave propagation enables a transmission range of 10–100 m, depending on the climatic conditions, because radio waves with a length of 13 cm are propagated much better with this connection than infrared radiation waves with a length of 900 nm. Another, no less significant advantage of the Bluetooth connection is that its use is not subject to a license.

Table 9.13 shows the frequency ranges of the license-free Bluetooth radio connection.

Bluetooth data transmission technology is standardized by the IEEE 802.15.1 standard. It falls under the PAN (personal area network) category. It comes in several versions. The *Bluetooth* 2.0 specification released on the market in 2020 is implemented in most devices such as mobile phones, laptops, and televisions. The *Bluetooth* 2.0 EDR (*enhanced data rate*) specification introduced the 4-DQPSK modulation technique, increasing the transmission data throughput to three times that of Bluetooth 1.2, which is 2.1 Mb/s. In 2011, the *Bluetooth* 4.0 interface was developed for a longer range, up to 100 m. This was followed by the *Bluetooth* 4.1 version and, from 2014, the *Bluetooth* 4.2 version. Since 2016, there is a version of *Bluetooth* 5.0 with a transmission speed of up to 2 Mb/s. This version allows four choices of transfer speeds ranging from 2 Mb/s to 125 Kb/s.

Broadcasting was not very popular when using *Bluetooth* 4.2 due to the small message size, limited to 31 bytes. The Bluetooth 5 version enables the transmission of messages of up to 255 bytes in size, which makes it useful, for example, in beacons used for the advertising market or navigation, in shopping centers and similar spaces. This version is backward compatible with previous versions and is also less power intensive. It also has significantly greater resistance to electromagnetic interference.

Table 9.13 Frequency ranges of the *Bluetooth radio connection.*

Frequency range	Transmitted power	Note
902–928 MHz	<10 mW	US only
2400–2483.5 MHz	<10 mW	In Europe, Asia, and the USA
5725–5875 MHz	<25 mW	In Europe, Asia, and the USA

The *Bluetooth* 5 version works in the same way as *Wifi* transmission, in the 2.4 GHz frequency band and uses the FHSS (*frequency hopping spread spectrum*) method for data modulation, during which 1600 jumps (retuning) between 79 frequencies with a frequency spacing of 1 MHz are performed within one second. This method of modulation increases the resistance of the connection to interference on the same frequency. Several power levels are defined (1, 10, and 100 mW), with which communication is possible up to a distance of hundreds of m. However, the given values are only valid in free space. If there are obstacles between the communicating devices (typically for example walls, the user's body), the transmission range decreases rapidly. In most cases, however, there is no sudden loss of connection, but the number of erroneously transmitted packets gradually increases.

The transmission speed is around 720 kbit/s (90 kB/s) and it is possible to create a symmetric or asymmetric data link, where the transmission speed during reception (*downlink*) is higher than when sending (*uplink).*

Individual devices are identified using their BD_ADDR (*BlueTooth Device Address*), similar to the Ethernet MAC address.

The *Bluetooth connection supports* both two-point and multi-point communication. If several stations are connected in a so-called piconet, one radio station acts as a master and can simultaneously serve up to seven subordinate devices (*slaves*). All devices in the piconetwork synchronize to the clock of the control station with discrete frequency changes. The specification allows the simultaneous use of up to 10 piconets on an area with a diameter of 10 meters and the further grouping of these piconets into scattered networks (*scatternets*) (Figure 9.28).

In Table 9.14, the transfer rates and maximum throughputs of the versions of the *Bluetooth transmission* system are listed.

In the *Bluetooth interface* system, data is transmitted in packets consisting of three parts, i.e., access code, packet header, and own data (Figure 9.29).

The access code consists of four start bits (*start bits*), designed to open the code, a 64-bit synchronization word (*sync bits*) and four end bits (*end bits*), designed to close the code. Information about the pseudo-random sequence of signal frequency changes in the transmission channel is contained in the synchronization word. This information is encoded into a new

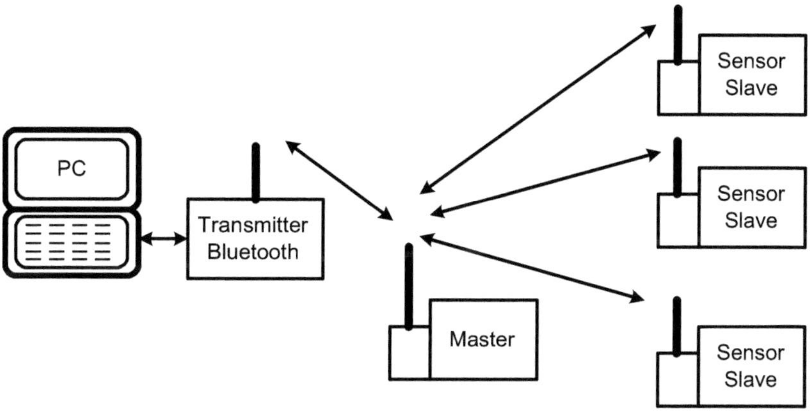

Figure 9.28 Measuring system with Bluetooth connection.

Table 9.14 *Bluetooth* system parameters.

Version	Power	Range	Transmission speed	Max. throughput
1.2	100 mW	100 m	1 Mbit/s	0.7 Mbit/s
2.0 + EDR	2.5 mW	10 m	2.1 Mbit/s	1.4 Mbit/s
3.0 + HS	1 mW	1 m	24 Mbit/s	16 Mb/s

64-bit word, starting with the 40 most significant bits overwritten from the sync word, the remaining 24 bits being the result of a modulo 2 logic operation performed on at least 24 bits. The 54-bit header contains a three-bit address of the active member of the piconet AMA (*active member address*), a 4-bit packet type specification, one flow bit, one ARQ (*automatic repeat re quest*) bit, and an error check of the HEC header. As a standard, up to 2745 bits can be transmitted.

Figure 9.30 shows a measurement system that uses *Bluetooth transmission* from the tensometric sensor of the mechanical stress of the rotating shaft of the turbogenerator to the computer.

9.3.3 Wifi connection

The high-frequency interface Wifi (*wireless fidelity*), or WLAN (*wireless local area network*), standardized by the IEE 802.11 standard, is primarily intended for connecting infrastructure wireless computer networks without using a network base or network center. This system requires at least one access port that can be connected to either a wired LAN or a GSM wireless network.

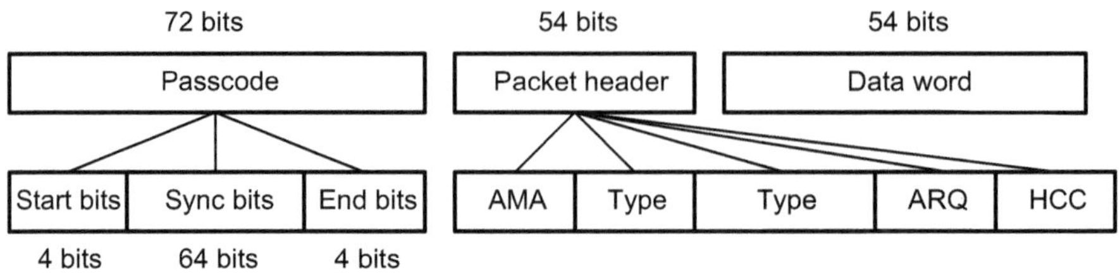

Figure 9.29 Data packet format during Bluetooth transmission.

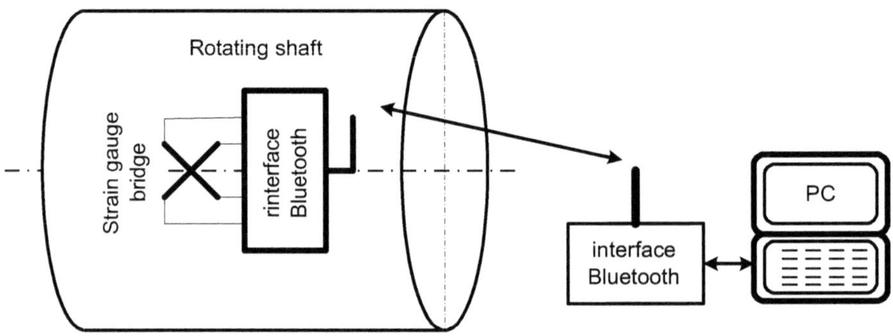

Figure 9.30 Transfer of data from the strainometric load sensor of the rotating shaft via the *Bluetooth interface.*

The 2.45 and 5.4 GHz frequency bands are reserved for the transmission of signals in the Wifi network. Two ways of encoding the carrier signal are used in radio transmission:

- DSSS (*direct sequence spread spectrum*) with a step change in the frequency of the carrier signal.
- RSSS (*random sequence spread spectrum*) with a pseudo-random change in the frequency of the carrier signal.

Regardless of the physical medium of the transmission channel, access to the computer interface is controlled by the *distributed foundation wireless medium access control* (DFWMAC) protocol, which is designed for local area networks and is part of the IEEE 802.11 standard.

The IEEE 802.11b version provides a maximum transfer rate of 11.5 Mb/s. The hardware of this interface includes a transmission moderation system and adaptation of the transmission speed to the level of interference. When the level of interference is high, the transmission speed will slow down. The system selects one of the following values for the maximum data transfer rate in the range of 1–11.5 Mb/s.

A more modern version of the Wifi interface, referred to as HIPERLAN 2 (*high performance radio local area network* 2), is designed to provide wireless connections to local area networks (LANs). The use of this interface is determined by its range, up to 150 m. The interface can therefore work within one or more buildings. Different modulation schemes are used in the HIPERLAN 2 interface depending on the expected transmission rates. This is mainly binary phase modulation BPSK (*binary phase shift keying*), with a transmission speed of up to 9 Mb/s, quadrature phase modulation QPSK (*quadrature phase shift keying*) with a transmission speed of up to 18 Mb/s, and quadrature eight-channel phase modulation 8-QAM (*quadrature amplitude modulation*) with a transmission speed in the range of 27–54 Mb/s.

9.3.4 ZigBee connection

The concept of this wireless communication, created in 1998, was standardized in 2004 by the IEEE 802.15.4 standard. The *ZigBee radio interface*, operating in the license-free frequency bands of 868 MHz, 902–928 MHz, and the 2.4 GHz band, enables data transmission up to a distance of several tens of meters, with transmission rates of 20, 40, and 250 kb/s.

Since the transmission power of this interface is approximately 1 mW, and in addition it is possible to put this interface into *sleep mode*, up to two years of operation can be achieved when using battery power. For less demanding applications, this interface uses an internal RAM data memory with a capacity of 8 kb, and for more demanding applications, a memory with a capacity of up to 32 kb is reserved.

This interface makes it possible to create even complex star networks, allowing to optimize the data path in terms of minimum time and energy consumption of the connection. Individual parts of the network are identified by an end point and identified by so-called clusters. The *ZigBee interface* is often used for wireless data collection, e.g., from water, gas, electricity, and heat consumption meters in residential buildings.

9.3.5 LoRaWAN connection

The LoRaWAN (*long-range wide area network*) radio communication protocol is a communication protocol operating in Europe in the 868 MHz frequency band and in the USA in the 915 MHz band.

This connection uses frequency modulation with a spread frequency spectrum for data transmission, based on a continuous change in the frequency of the modulation signal CSSM (*chirp spread spectrum modulation*).

Table 9.15 Measured signal-to-noise distances of narrowband 2-SK and eight-channel LoRa modulation with switchable frequency channels.

Distance	2-FSK (4.8 kb/s)	LoRa (22 kb/s)
80 m	97 dBm	91 dBm
150 m	100 dBm	102 dBm
280 m	112 dBm	114 dBm
330 m	115 dBm	124 dBm
480 m	118 dBm	120 dBm
560 m	–	121 dBm
1350 m	–	126 dBm
1750 m	–	127 dBm
2150 m	–	129 dBm

Similar to the *ZigBee connection*, the emphasis is placed on minimum consumption with this interface.

The transmission protocol allows you to adaptively change the transmission speed of the connection depending on the volume of transmitted data and the frequency of errors during its transmission. If two different files with a frequency difference of less than 6 dB are simultaneously transmitted, i.e., in a 1:2 ratio on the same channel, then a disproportionate increase in the transmission error rate can occur. In this case, the system will choose a lower transmission speed, which will increase the signal-to-noise ratio of the transmitted signal. For this reason, eight frequency ranges of modulation signals are distinguished in the range of 125–500 kHz, of which a bandwidth of 500 kHz allows data to be transmitted at a transmission rate of up to 22 kb/s.

In Table 9.15, the measured signal-to-noise and distances between narrow-band binary keyed 2-FSK phase modulation with a transmission rate of 4.8 kb/s and sequential LoRa modulation with a spread frequency spectrum in eight frequency ranges with a transmission frequency of 22 kb/s are compared, from which the priority of sequential modulation during the transmission of signals in a heavily disturbed urban environment at distances greater than 500 m.

Due to the asynchronous method of data transmission, the influence of interfering signals is also significantly suppressed in this *frequency sequence modulation*, which can be optimally set by the number of frequency channels used. Compared to, for example, amplitude or frequency modulation, sequential frequency modulation can achieve an increase in the signal-to-noise ratio by approximately 50 dB, and compared to 2-PSK modulation by approximately 20 dB.

Table 9.16 shows the parameters of typical wireless interfaces over short distances.

Table 9.16 Parameters of the most common short-range wireless interfaces.

Interface type	Frequency band	Transmission speed	Range	Coverage angle	Transmission power
IrDA	3.3 THz	16 Mb/s	1 m	30°	10–100 mW
Area InfraRed	3.3 THz	4 Mb/s	8 m	120°	80 mW
HomeRF	2.45 GHz	2 Mb/s	50 m	360°	75 mW
Bluetooth	2.45 GHz	≤54 Mb/s	150 m	360°	1–100 mW
Wifi	2.45 GHz	54 Mb/s	50 m	360°	10–50 mW
ZigBee	2.45 GHz	250 kb/s	300 m	360°	1–10 mW
LoRAN	868 MHz	≤kb/s	500 m	360°	1–500 mW

9.4 Comparison of Properties of Wireless Connections

In order to compare the features of these measurement systems, their coverage range, maximum data transfer rate, connection setup response time, data security level, number of measurement stations, communication system construction costs, and expected operating costs must be taken into account.

The range of coverage of *Ethernet* and *GSM connections* in urbanized areas is practically one hundred percent. In addition, GSM networks enable practically unlimited range of connections around the world when using satellite communication. The range of radio modem networks is limited by the maximum allowed transmission power, height, length of antennas, and geographical and weather conditions for the propagation of radio waves. The range of common radio modem systems is also conditioned by the number of their stations, the distance of which usually does not exceed tens of km.

Transmission speeds of *Ethernet networks* reach units of Gb/s when using metallic lines and up to 10 Gb/s when using optical lines. Transmission speeds of GSM 3G networks reach up to 2 Mb/s and 5G networks up to 20 Gb/s. *Radio modem transmission* systems achieve significantly lower transmission rates in the range of 1.2–19.6 kb/s.

The reaction response in *Ethernet networks* ranges from several tens to hundreds of ms. In radio modem data transmission, there are no switching delays, neither in the transmission of data from the measuring station to the system center nor in the transmission of commands from the system center to the measuring station, and each receiver is continuously available.

In addition, the possibility of directional transmission of the radio signal reduces the risk of unauthorized interception of the signal. In contrast to the above-mentioned characteristics of radio transmission, the signals transmitted in the *GSM network* are freely available, and the used transmission protocols and channel frequencies are standardized and publicly known. The best security is data transmission in the *Ethernet system*, especially when using optical data lines. When using *GSM data transmission*, the frequency of occurrence of errors is up to 10^{-6}, with the Ethernet system up to 10^{-16}.

The cost of building an *Ethernet communication* system consists of building a data network, including channel switches and servers. An integral part of these costs are network usage fees, depending on the required speed and volume of data transferred. The cost of acquiring end devices depends on their number and quality. The implementation of the radio communication network involves the construction of radio modem stations including receiving and transmitting units. The costs of using the *GSM network* only include the acquisition of mobile data units, e.g., mobile phones, and fees that depend on the required speed and volume of transmitted data.

Ethernet networks are therefore used to transfer large volumes of data with high demands on transmission speed, or *GSM wireless network*, also enabling satellite connection. Radio modem connections, including short-distance connections, are suitable for local wireless data transfers of smaller volumes at lower transmission speeds.

References

[1] Le Bidic, G. *Mobile Messaging Technologies and Services* SMS, EMS a MMS, New York: John Wiley & Sons, 2003. ISBN 978-047085-8035.
[2] Wessoowski, K. *Mobile Communication Systems*, New York. John Wiley & Sons, 2002. ISBN 978-047149-8377.
[3] Nawrocki, W. *Measurement systems and sensors.* Second edition, Artec House 2016. ISBN 1-58053-945-9.
[4] Bekkers, R. *Mobile telecommunication Standards UMTS, GSM, TETRA and ERMES.* Norwood, MA: Artech House, 2001. ISBN 1-58056-0.
[5] El-Rabbany, A. *Introduction to GPS: The Global Positioning System.* Norwood, MA: Artech House, 2002. ISBN 9781580535472.
[6] Miller, B. A., Biskidian, A. K. *Bluetooth Revealed: The Insider's Guide to an Open Specification for Global Wireless Communications*, 2nd ed. Upper Saddle River, NJ: Prentice Hall, 2003. ISBN 978-0130672377.
[7] Castro, J. P. *The UMTS Network and Radio Access Technology*, New York. John Wiley & Sons, 2001. ISBN ISBN 0-471-81375-3.
[8] Cheah, J. *Practical Wireless Data Modem Design.* Norwood, MA. Artech House 1999. ISBN 978-1-580853-047-7.
[9] *A set of AT commands for GSM mobile devices.* European Telecommunications Standard Institute. GSM 07.07 version 7.4.0, 1998.
[10] Frequency allocation of GSM, DCS, UMTS and LTE in the Czech Republic. GSM web.cz.

10

Digitizing and Reconstruction of Continuous Signals

In all measuring systems, which are intended for the collection and processing of time-varying signals, their digitization takes place, including time and amplitude discretization of the signals. Conversely, during the generation of time-varying signals, their reconstruction and subsequent filtering takes place. In addition to A/D and D/A amplifiers, the peripheral means of these systems are converters, voltage reference sources, filters, and frequency synthesizers including phase synchronization circuits.

10.1 Digitization of Continuous Signals

When digitizing time-varying signals, they are sampled, i.e., *time* and amplitude *discretization*, of the signal samples taken during their sampling, and coding, i.e., the conversion of the number of quanta corresponding to the taken signal sample into a usually binary signal.

10.1.1 Sampling of continuous signals

Signal sampling is characterized by taking samples of the signal at certain points in time. Real-time equidistant sampling and non-equivalent (stroboscopic) sampling are used in measurement systems.

In RTS (*real-time sampling*) equidistant sampling, the sampled signal $x_s(t)$ can be expressed as the product of the sampled signal $x(t)$ with the sampled signal $s(t)$, which is given by *Dirac pulses* of amplitude 1 and period $T_s = 2 \cdot \pi/\omega_s$ (Figure 10.1)

$$x_s(t) = x(t) \cdot s(t) = \sum_{n=-\infty}^{\infty} x(n) \cdot T_s \cdot \delta(t - n \cdot T_s). \qquad (10.1)$$

The *frequency spectrum* of the sampled signal is determined based on the *discrete Fourier transform* (DFT) by the *convolutional product*, marked with

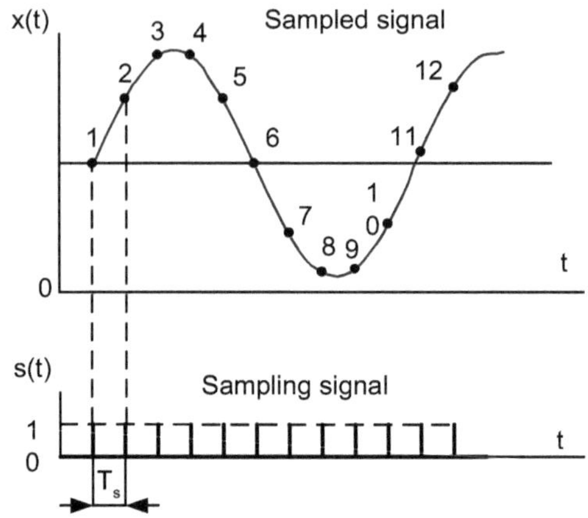

Figure 10.1 Equistatic real time signal sampling.

the symbol *, of the frequency spectrum of the sampled signal $\boldsymbol{X}(\omega)$ with the frequency spectrum of the sampled signal $\boldsymbol{S}(\omega)$

$$\boldsymbol{X}_s(\omega) = \boldsymbol{X}(\omega) * \boldsymbol{S}(\omega). \tag{10.2}$$

After performing the *convolutional product*, the frequency spectrum of the sampled signal can be expressed as the sum of the spectra of the sampled signal, mirror-spread around integral multiples of the circular sampling frequency ω_s

$$\boldsymbol{X}_s(\omega) = \frac{1}{T_s}\sum_{n=-\omega}^{\omega}\int_{\alpha=-\omega}^{\omega} \boldsymbol{X}(\alpha)\cdot\delta(\omega - k\cdot\omega_s - \alpha)d\alpha = \frac{1}{T_s}\sum_{k=-\omega}^{\omega}\delta(\omega + k\cdot\omega_s). \tag{10.3}$$

At the limiting angular frequency of the sampled signal $\omega_m < \omega_s/2$, the copies of its frequency spectra do not intersect and the signal after sampling is therefore fully reconstructable (Figure 10.2).

In the opposite case, when $\omega_m > \omega_s/2$, *undersampling* of the signal occurs, during which copies of the frequency spectra overlap, which causes the ambiguity of signal reconstruction with the appearance of reverberant signals (*aliasing error*).

This fact is expressed by the *Shannon–Kotelnik sampling theorem*, and the frequency $\omega_s/2$ is usually referred to as the *Nyquist cutoff frequency.*

Undersampling signals cause the dynamic range of the sampled signal to be reduced by an amount

$$\Delta X = \frac{\omega_m - \frac{\omega_s}{2}}{\omega_m - \omega_0}, \tag{10.4}$$

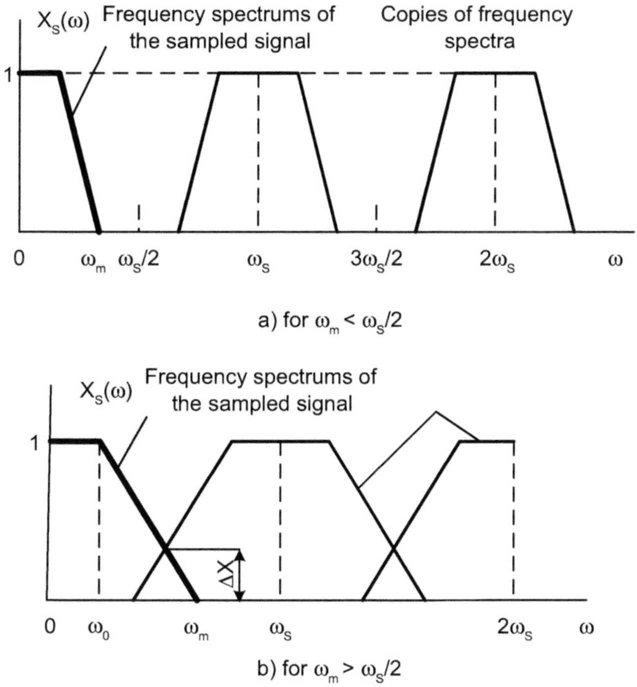

Figure 10.2 Frequency spectra of the sampled signal.

where ω_0 is the circular frequency at which the amplitude frequency characteristic of the sampled signal decreases, e.g., by 3 dB compared to the amplitude of this signal at frequencies $\omega << \omega_0$. For example, when sampling a harmonic signal with an angular frequency $\omega = 2 \cdot \pi/T$ and a sampling signal with an angular frequency $\omega_s = 2 \cdot \pi/T_s$, reverberant sine signals with angular frequencies arise during its reconstruction

$$\omega_a = \omega - \omega_s \cdot \text{ROUND}\left(\frac{\omega}{\omega_s}\right), \tag{10.5}$$

where *ROUND* means rounding the ratio of circular frequencies ω/ω_s down to the nearest whole number. Figure 10.3 shows the creation of a reverberation signal with a circular frequency $\omega_a = 2 \cdot \pi/T_a$, created by sampling a harmonic signal with a circular frequency $\omega = 2 \cdot \pi/\text{T}$ with a sampling frequency $\omega_s = 2 \cdot \pi/T_s$.

To limit the frequency range of the digitized signal, filters (*anti-aliasing filters*) of the lowpass- or bandpass-type are used, which limit the frequency range of the digitized signal to less than half the sampling frequency.

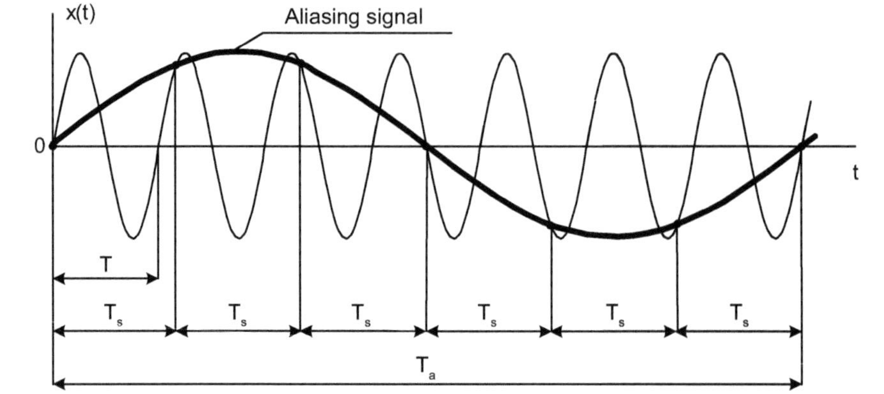

Figure 10.3 Reverberation signal generation by subsampling.

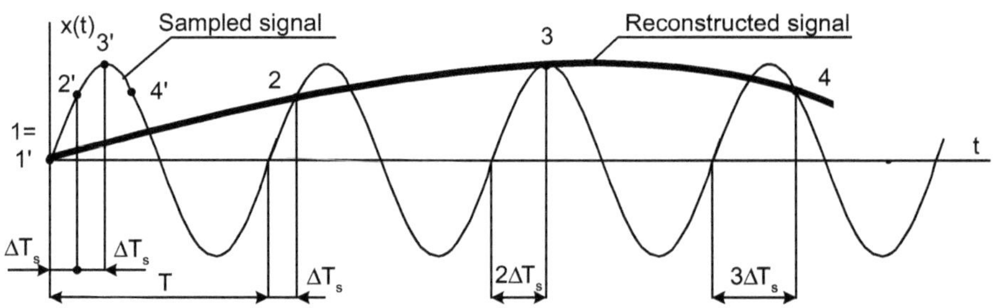

Figure 10.4 Equivalent sampling.

Equivalent sampling uses non-equidistant sampling of a periodically sampled signal at time-graded moments, always extended by a constant time ΔT_s after each signal sampling (Figure 10.4).

Assuming that the sampled signal is *periodic*, an *equivalent circular sampling* rate can be defined in relation to this sampling

$$\omega_{se} = \frac{1}{2 \cdot \pi \cdot \Delta T_s}, \tag{10.6}$$

which determines the cutoff frequency of the sampled periodic signal $\omega_{ms} < \omega_s/2$.

A prerequisite for the unequivocal reconstruction of a signal sampled in this way is its strict periodicity. If this condition is not met, even with this method of sampling, reverberation signals occur, which make it impossible to unambiguously reconstruct the signal.

For example, when sampling a signal in real time with a period of T_s = 10 ns of the sampling signal, i.e., with a sampling frequency f_s = 100 MS/s, it

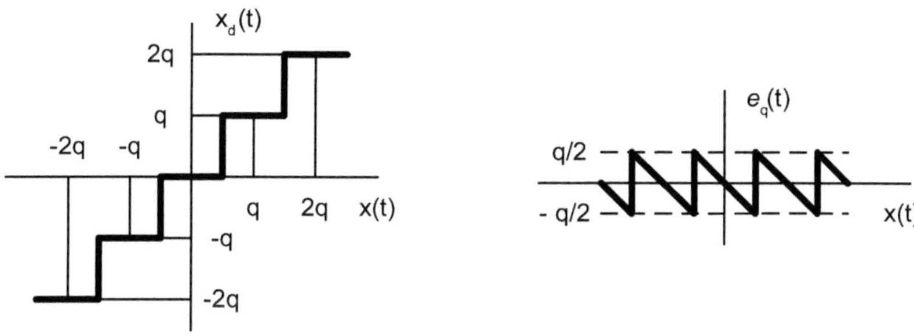

a) transfer characteristic

b) quantization error

Figure 10.5 A quantizer with uniform quantization.

is theoretically possible to sample signals up to a frequency range of 40 MHz. With equivalent sampling with $\Delta T_s = 1$ ns ($f_{se} = 1$ Gs/s), the frequency range of the sampled signal can be increased up to 1.8 GHz.

10.1.2 Quantization of signals

When quantizing signals, their amplitude is discretized. The conversion characteristic of the A/D converter has a step-like course during uniform quantization, causing its *quantization error* $e_q(t)$ (Figure 10.5).

It then holds for the *effective value* of the *quantization error*

$$E_q = \sqrt{\frac{1}{q}\int_{-q/2}^{q/2} e_q^2(t)\cdot dt} = \frac{2^{-n}}{\sqrt{12}}, \tag{10.7}$$

where 2^{-n} is the *resolution of the quantizer.*

If the quantizer is excited by a *sinusoidal signal* whose amplitude is equal to the range of the quantizer, then the *signal-to-noise ratio* (SNR) of the quantized signal is determined by the ratio of the effective value of the exciting sinusoidal signal $U_{\sin}$ to the effective value of the quantization error U_q

$$\text{SNR} = \frac{U_{\sin}}{U_q} = \frac{1}{2\cdot\sqrt{2}}\frac{\sqrt{12}}{2^{-n}} = 2^{-n}\sqrt{3/2}. \tag{10.8}$$

The *signal-to-noise ratio* is usually expressed in dB

$$\text{SNR}[\text{dB}] = 20\cdot\log(\text{SNR}) = 6{,}02\cdot n + 1.76. \tag{10.9}$$

If, for example, a 12-bit A/D converter is used to digitize a time-varying signal, then when it is fully excited, the signal-to-noise ratio is 74 dB.

Table 10.1 shows the signal-to-noise ratio for 8- to 24-bit A/D converters.

Table 10.1 Parameters of ideal n-bit A/D converters.

n	2^n	SNR [dB]	n	2^n	SNR [dB]
8	256	49.9	16	65,536	98.1
10	1024	61.6	18	262,144	110.1
12	4096	74.0	20	1,048,576	122.2
14	16,384	86.1	24	16,777,216	146.2

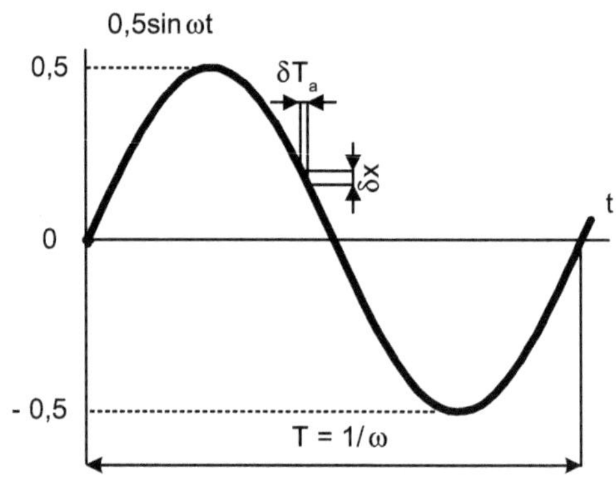

Figure 10.6 Determining the sampling uncertainty of a sinusoidal signal.

In real A/D converters, the theoretical *SNR* value is replaced by the *signal-to-noise ratio* and the SINAD (*signal to noise and distortion*), which is related to their *effective number of bits* (ENOB)

$$\text{SINAD}[\text{dB}] = 6.02 \cdot \text{ENOB} + 1.76. \tag{10.10}$$

For example, for fast 12-bit A/D converters used in multiplex measurement modules, the ENOB ranges from 10.5 to 11.2 bits, depending on the frequency of the digitized signal.

When sampling a time-varying signal, amplitude uncertainty δx. occurs due to the temporal uncertainty of the sampling time δT_a (*apertute jitter*). When sampling and quantizing a sinusoidal signal, this uncertainty has a maximum value when this signal passes through the zero value (Figure 10.6)

$$\delta x_{\max} = \frac{dx}{dt}\delta T_a = \omega \cdot \delta T_a.$$

If we equate this uncertainty with the resolution of an n-bit A/D converter $q = 2^{-n}$, then the maximum frequency of a sinusoidal signal that can be sampled and quantized in this way with its full resolution is determined by the equation

$$f_m \le \frac{2^{-n}}{\pi \cdot \delta T_a}. \tag{10.11}$$

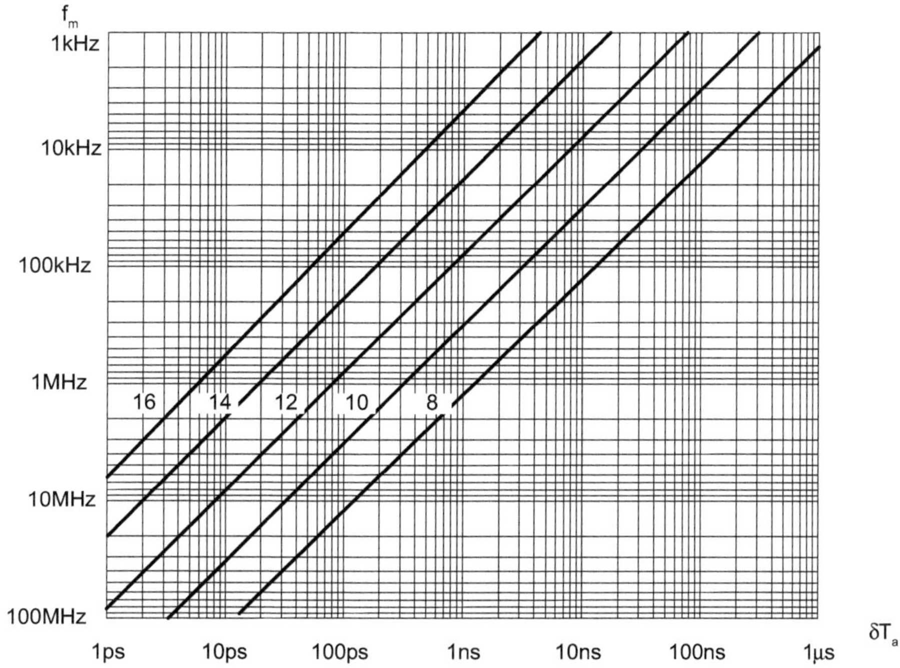

Figure 10.7 The relationship between the maximum f_m frequency of an *n* bit A/D converter and the sampling time uncertainty δ Ta.

Figure 10.7 shows the relationship between the maximum frequency f_m of an *n*-bit A/D converter and the sampling time uncertainty δT_a. For example, with sampling uncertainty $\delta T_a = 100$ ps, the maximum usable frequency range of 12-bit A/C is approximately $f_m = 1.24$ MHz.

From the previous analysis, it is clear that by reducing the required resolution and sampling time uncertainty, the usable frequency range of the digitized signal increases. For example, when the resolution is reduced by 1 bit, the frequency range of the digitized signal will double. The same situation occurs when a harmonic signal with an amplitude equal to half the range of the A/D converter is digitized.

10.1.3 Types of A/N converters

To digitize time-varying continuous signals, measuring systems use A/D converters with resolution in the range of 8–24 bits with sampling frequencies in the range of tens of GS/s to tens of kS/s.

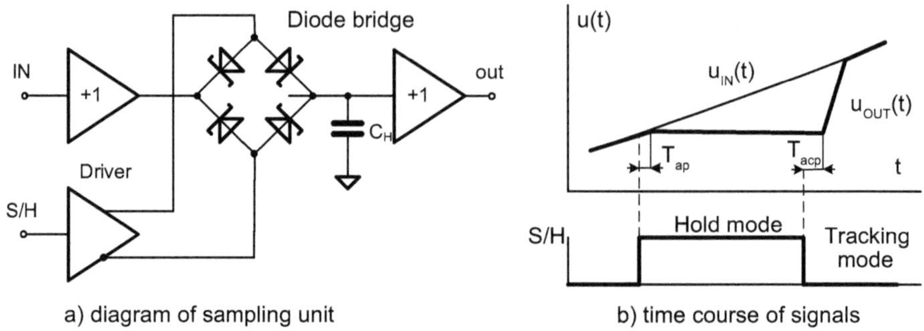

Figure 10.8 Sampling unit with switched diode bridge.

10.1.3.1 Parallel and cascaded A/D converters

To digitize fast signals, *parallel A/D converters* (*flash ADC*) with fast *sampling units* (*sample hold unit*) are used. A block diagram of a fast sampling unit, consisting of an input amplifier, a diode switching bridge, a memory capacitor, and an output amplifier is shown in Figure 10.8.

In the *tracking mode*, in which the switch is closed, the output voltage of the sampling unit copies the time course of its input voltage. Upon arrival of the *S/H* sampling signal, which is synchronized with the SYNC trigger signal, the switch is opened and the voltage sample taken at that time is stored in the storage capacitor. This voltage sample is subsequently converted into a corresponding binary signal in the A/D converter.

By using fast switching *Schottky diodes*, sampling times T_a *(aperture time)* of the order of tens of ps (10^{-12} s) with a time uncertainty of hundreds of ps and collection times T_{acq} (acquisition time) of the order of units to tens of ns (10^{-9}) can be achieved.

Fast voltage followers with transfer 1 are used to impedance-separate the memory capacitor from the signal source and the A/D converter.

Parallel A/D converters are made up of a set of fast differential voltage comparators, whose digital signals are decoded from a *positional code* expressing its value by the number of logs. 1 to binary code (Figure 10.9).

In order to avoid concurrent hazards when coding these signals, which occur when more than 1 bit in the code word changes, *Gray's codes* with a change of 1 bit in their neighboring code words are used to decode the output signals of the comparators (Table 10.2).

Since *n*-bit parallel A/D converters contain $2^n - 1$ comparators and the decoding of their digital outputs would require a complex decoder, the decoding of signals is divided into several stages, while these A/D converters use the principle of *time sharing*. This consists in time-gradual sampling and

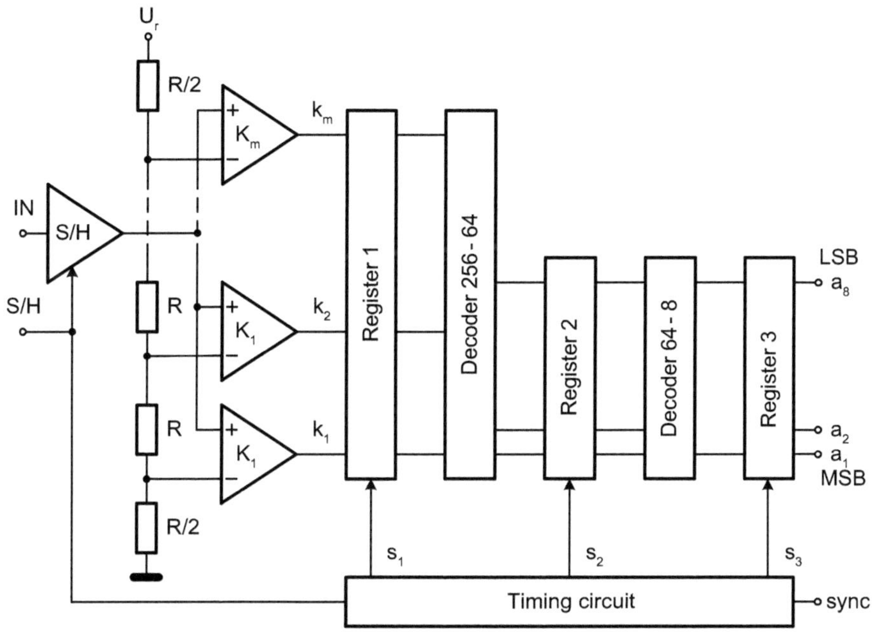

Figure 10.9 Parallel 8bit A/D converter.

Table 10.2 Coding of the output signals of the comparators of the 8-bit A/N converter.

U_{IN}/U_r	**Position code** $k_1\,k_2\,k_1\,k_2\,....\,k_1\,k_1\,k_2\,k_2$	**Gray's code** MSB LSB	**Binary code** MSB LSB
0	1 0 0 0 ...... 0 0 0 0	0 0 0 0 0 0 0 0 0	0 0 0 0 0 0 0 0 0
1/256	1 1 0 0 ...... 0 0 0 0	0 0 0 0 0 0 0 01	0 0 0 0 0 0 0 0 1
2/256	1 1 1 0 ...... 0 0 0 0	0 0 0 0 0 0 0 1 1	0 0 0 0 0 0 0 1 0
....	....		
254/256	1 1 1 1 ...... 0 0 1 0	1 1 1 1 1 1 1 0	1 1 1 1 1 1 1 0
1	1 1 1 1 ...... 0 0 1 1	1 1 1 1 1 1 1 1	1 1 1 1 1 1 1 1

quantization of the signal using several simpler digital decoders connected in succession, whose operation is controlled by multiphase sampling signals. If, for example, an 8-bit A/D converter contains 255 comparators and two decoders with an organization of 256/64 and 64/8 are used for decoding, then the sampling frequency can be increased to twice the value compared to the sampling frequency of an A/D converter with 1 decoder out of 25 to 8 bits.

To digitize fast signals with higher resolution, cascaded A/D converters (*propagation ADC*) are used, which are made up of several sampling units, parallel A/D and D/A converters, and differential amplifiers, each of which

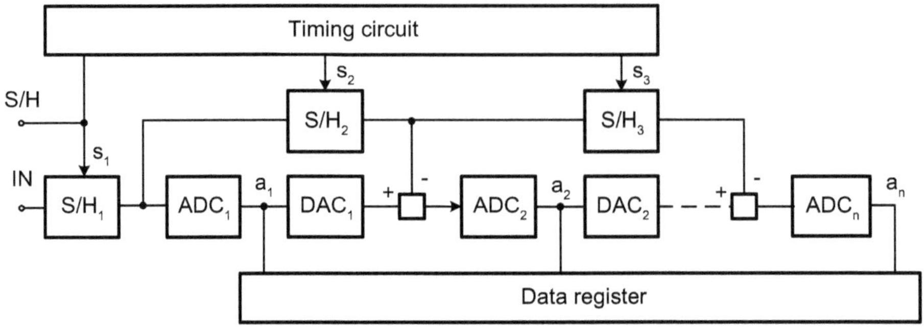

Figure 10.10 Cascade A/D converter.

always converts a sample of the difference signal of the previous ADC, reconstructed by its DAC (Figure 10.10).

For example, a 12-bit A/C converter made up of three sections of 4-bit ADC and DAC converters thus allows to achieve sampling frequencies given by the conversion time of each of the three A/C and C/A converters. Low-voltage differential data transmission LVDS (*low voltage differential signaling*) is used to transfer data from A/D converters to FIFO (*first in, first out*) memories, where the data is gradually refreshed, in order to achieve high transfer rates.

Another option to increase the frequency range of sampled signals is *interleaved sampling*, in which the sampled signal is digitized by several parallel connected board inputs (Figure 10.11).

10.1.3.2 Approximate A/C converters

Approximation A/D converters SAR (*successive approximation ADC*) work on the principle of successively connecting weight-graded charges to the input of a comparator, whose output signal decides whether the difference between the charge corresponding to the sample of the converted signal is greater or less than the accumulated charge of the charge A/D converter (Figure 10.12).

A condition for the correct determination of the digital equivalent of a voltage sample is its constant value during the determination of all bits of the digital signal. Therefore, all *approximation A/D converters* include an S/H (*sample hold*) sampling circuit, consisting of a switch and a memory capacitor CIN. In *tracking mode*, this capacitor is clamped to the input of the converter and its charge changes as the input voltage changes. In the state of memorization (*hold mode*), the charge corresponding to the sample of the input voltage at the moment of connecting the memory capacitor to the input

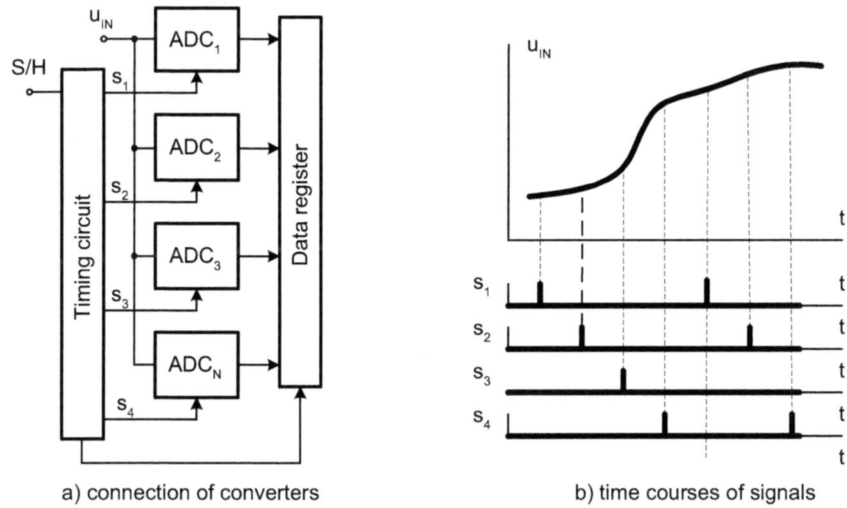

Figure 10.11 Interleaved sampling.

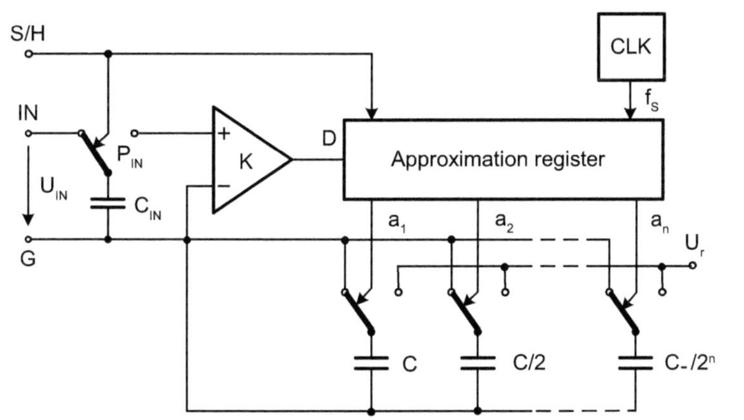

Figure 10.12 Block diagram of a charge-coupled approximation A/D converter.

of the comparator is compared with the weighted charges of the D/A converter, formed by capacitors with weighted capacitances.

The time sequence of stepwise approximation is controlled by the *approximation register*. If the sum of charges at the input of the comparator is positive, log level 0 is written to the log register. If it is negative, log level 1 is written into it. The conversion organized in this way begins with the determination of the MSB (*most significance bit*) and ends with the LSB (*least*

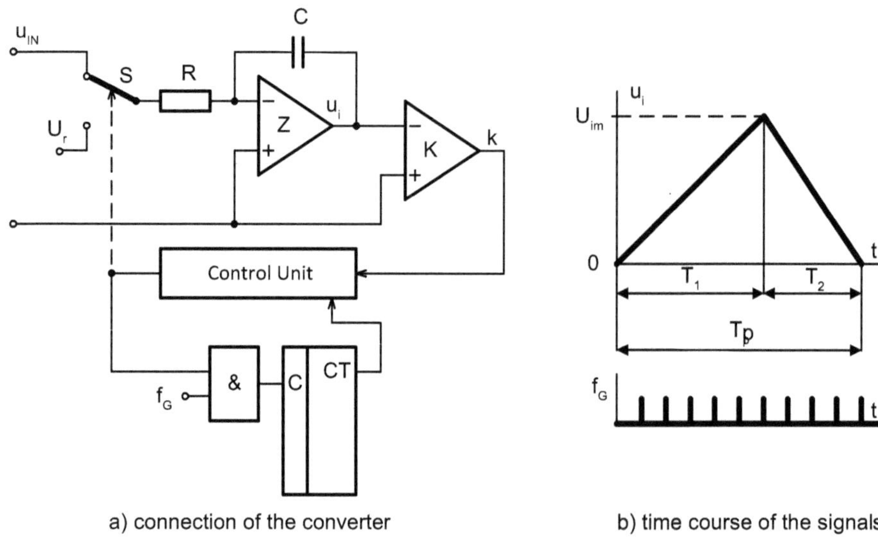

Figure 10.13 Integration converter with two-step integration.

significance bit) determination. This means that it takes *n* iterations to convert the voltage sample into an *n*-bit binary signal.

10.1.3.3 Integration A/D converters

These converters are used in measurement modules to measure voltage or current with high accuracy and effective suppression of DC and AC interference signals. The default method of these converters is the method of *dual slope integration* (*dual slope ADC*), in which the measured voltage is connected to the input of the inverting integration amplifier in the first cycle for a constant conversion time T_1 (Figure 10.13).

At the end of this time, the output voltage of the integrator is given by the *electrolytic mean value* of the input voltage

$$u_i = \frac{1}{RC}\int_0^{T_i} u_{\mathrm{IN}}(t) \cdot dt. \tag{10.12}$$

The time T_1 is determined by the capacity of the counter N_1 and the frequency of the generator f_g

$$T_1 = \frac{N_1}{f_g}. \tag{10.13}$$

In the second cycle of the conversion, the reference voltage U_r of the opposite polarity to that of the input voltage is integrated. The charge on the integrating

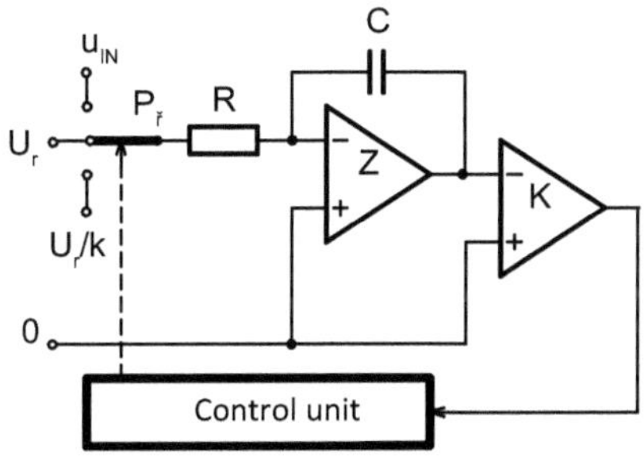

a) connection of the converter

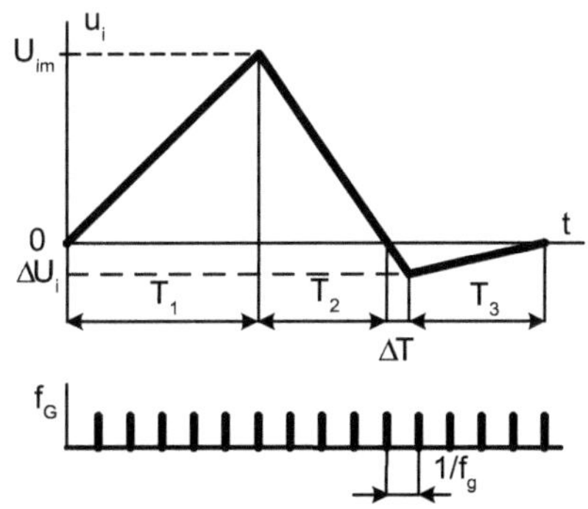

b) time course of the signals

Figure 10.14 Three-cycle integration A/D converter.

capacitor C decreases during T_2 to a zero value. It then applies to periods T_1, T_2

$$\frac{U_{im}}{RC}T_1 = \frac{U_r}{RC}T_2. \tag{10.14}$$

At the end of time T_2, the state of the counter $N_2 = f_g/T_2$ determines the *electrolytic mean value* of the input voltage

$$\frac{1}{T_i}\int_0^{T_i} u_i(t)\cdot dt = U_r\frac{T_2}{T_1} = U_r\frac{N_2}{N_1}. \tag{10.15}$$

since the accuracy of the conversion is not affected by the time constant of the integrator and the frequency of the generator, high accuracy, and stability of the conversion can be achieved with this method, which is determined only by the stability of the reference voltage.

By inserting additional integration clocks, integration errors of measured and reference voltages can be eliminated and resolution and accuracy of integration A/D converters can be increased.

The connection of the integration A/D converter with *three-step integration* (*three-slope ADC*) is shown in Figure 10.14.

In the first cycle of the conversion, the input voltage u_{IN} is integrated for T_1. At the end of this time, the switch connects the input of the integrator to a reference voltage U_r of the opposite polarity to the input voltage and this voltage is integrated for a time T_2, determined by reaching zero output voltage of the integrator

$$T_2 = \frac{N_2}{f_g} \pm \frac{1}{2\cdot f_g}. \tag{10.16}$$

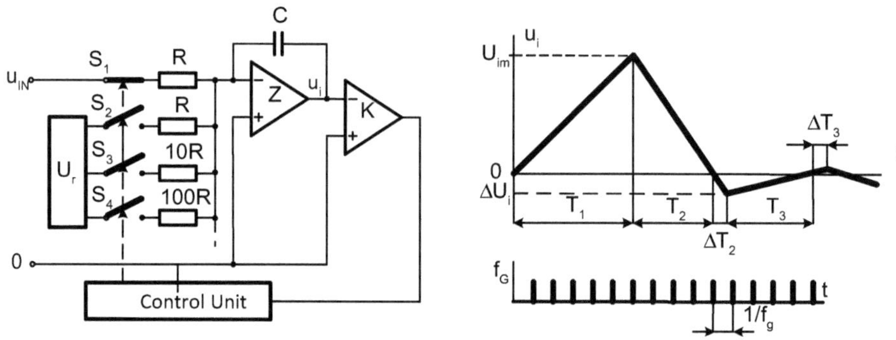

a) connection of the converter b) time course of the signals

Figure 10.15 Multi-step integration A/D converter.

From the equality of the output voltage of the integrator at the end of the integration time T_1

$$\frac{\Delta T}{RC} U_r = \frac{U_r}{k} \cdot \frac{T_3}{RC}, \tag{10.17}$$

the time T_3 can be determined

$$T_3 = k \cdot \Delta T = \frac{N_3}{f_g}, \tag{10.18}$$

where k is a *binary scaled ratio*, which is usually 1:16. This method partially suppresses the effect of time ΔT on the accuracy of determining the integration time T_2 of the reference voltage, which is caused by the asynchrony of the arrival of the edges of the generator pulses and the moment when the output voltage of the integrator passes its zero comparison level.

The *electrolytic mean value* of the input voltage of the converter is then defined by the equation

$$U_{IN} = \frac{N_2 - \dfrac{N_3}{k_1}}{N_1} \cdot U_r. \tag{10.19}$$

The integration A/D converter with *multi-step integration* (*multi-slope ADC*) in Figure 10.15 makes it possible to eliminate the influence of times T_i, i = 2, 3,..., which are caused, similarly to the previous case, by the asynchrony of the arrival of the edges of the generator pulses and the time moments when the output voltage of the integrator passes through the zero voltage level.

Multiple integration of the reference voltage not only increases the resolution of the A/D converter but also under the assumption of uncorrelated

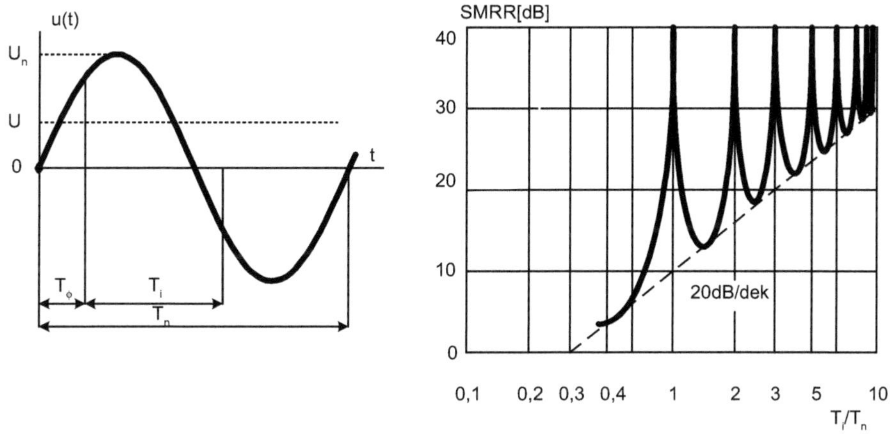

a) time course of the integrated signal b) dependence of the *SMRR* suppression factor on T_i/T_n

Figure 10.16 Suppression of AC interference by an integration A/D converter.

values of times ΔT_i reduces the mean square value of determining the integration time of the reference voltage in the ratio $1/\sqrt{N}$.

A distinctive feature of integration A/D converters with a constant integration time is the suppression of series AC interference voltage superimposed on the measured voltage.

If the disturbing voltage has a sinusoidal waveform with amplitude U_n and frequency f_n, then the output voltage of the integrator is

$$u_i(t) = u_a(t) + U_n \cdot \cos(2 \cdot \pi \cdot t \cdot f_n + \varphi), \tag{10.20}$$

where u_a is the measured voltage and φ is the phase shift between the beginning of the integration and the passage of the alternating disturbance voltage through zero with a positive derivative (Figure 10.16).

The mean value of the output voltage of the integrator is $T_n = 1/f_n$ during the disturbance voltage period

$$U_s = \frac{1}{T_i}\int_0^{T_i} u_i(t)dt = U_{as} + U_n \cdot \frac{T_i}{2 \cdot \pi \cdot T_n}\left[\sin\left(2 \cdot \pi \cdot \frac{T_i}{2 \cdot \pi \cdot T_n} + \varphi\right) - \sin\varphi\right], \tag{10.21}$$

which consists of the mean value of the measured voltage U_{as} and the disturbing component

$$\Delta U_n = U_n \cdot \frac{T_i}{2 \cdot \pi \cdot T_n}\left[\sin\left(2 \cdot \pi \frac{T_i}{2 \cdot \pi T_n} + \varphi\right) - \sin\varphi\right], \tag{10.22}$$

whose value is zero at $T_i/T_n = 1, 2, ...$

The series mode rejection ratio (*SMRR*), also referred to as NMRR (*normal mode rejection ratio*), is determined by the ratio $U_n/\Delta U_n$,

$$\text{SMRR}[\text{dB}] = 20 \cdot \log_{10} \max\left(\frac{U_n}{\Delta U_n}\right) = 20 \cdot \log_{10} \frac{\pi \cdot \dfrac{T_i}{T_n}}{\sin\left(\pi \cdot \dfrac{T_i}{T_n}\right)} \tag{10.23}$$

in phase shift

$$\varphi = -\pi \cdot \frac{T_i}{T_n}. \tag{10.24}$$

From the above analysis, it follows that by integrating the measured voltage over the entire period of the interfering AC signal, the influence of this interference can be completely eliminated. The integrator thus implements a selective comb filter, the transmission of which is limited to zero at $T_i/T_n = n$, where n is an integer.

For example, at the ratio of integration time to disturbance voltage period $T_i/T_n = n = 1.1$, *SMRR* = 19 dB, at $T_i/T_n = n = 1.01$, SMRR = 40 dB, and at $T_i/T_n = n = 1.001$, *SMRR* = 66 dB. By synchronizing the integration times with the period of the expected AC interference voltage, *SMRR* suppression of up to 160 dB can be achieved.

10.1.3.4 Delta sigma A/S converters

Delta sigma A/D converters (ΔΣ ADC) consist of several cascaded ΔΣ integrators whose signals are filtered by a digital filter with a *finite impulse response* (FIR) to the digital signal. An alternate schematic of a third-order delta sigma modulator is shown in Figure 10.17.

The DS modulators themselves are made up of integrators, integrating the differences between the input signals of the modulators and the output signals of the D/A converters. The input digital signals of the D/A converters are the sampled signals of the A/D converters, which digitize the output voltage of the integrators (Figure 10.18).

Digital signals of the modulators obtained in this way by multi-bit quantized feedback are filtered by digital filters with finite response FIR (*finite impulse response*) with *binary graded weighting coefficients* for the digital signal, the samples of which correspond to the weighted moving average of samples of the sampled signal.

Another effective means of increasing the resolution of ΔΣ converters is oversampling the digitized signal with a sampling frequency f_s, which is

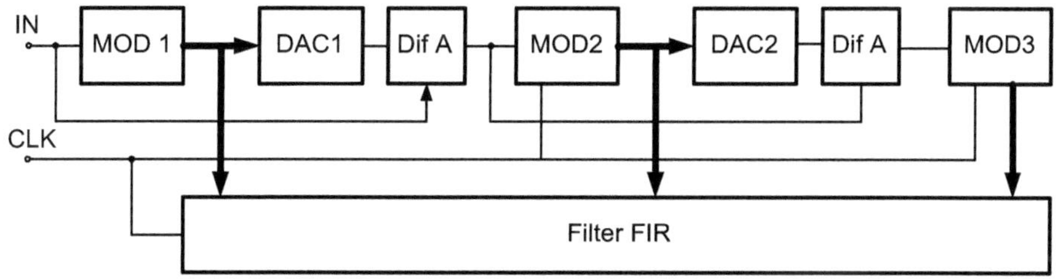

Figure 10.17 Replacement diagram of a 3rd order delta sigma modulator.

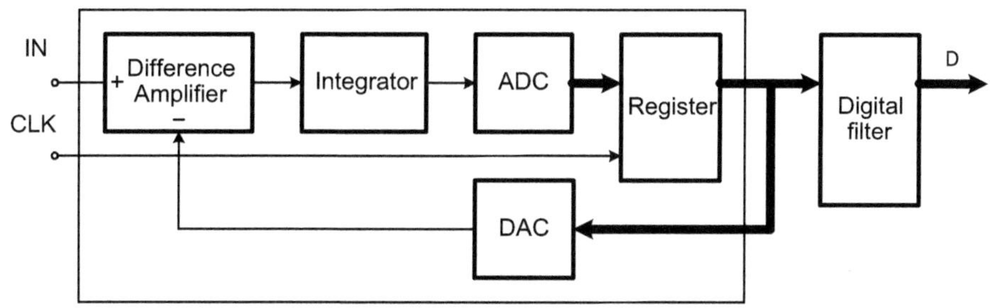

Figure 10.18 Delta sigma modulator with digital filter.

significantly higher than the sampling frequency f_N, determined by the *sampling theorem*, which determines the minimum sampling frequency as twice the cutoff frequency of the sampled signal. This resampling of signals and its subsequent filtering reduces the effective value of the quantization error of n-bit modulators from values of $2^{-n}/\sqrt{12}$ to values k times smaller

$$U_{qk} = \frac{2^{-n}}{k \cdot \sqrt{12}}, \tag{10.25}$$

where $k = f_s / f_N$ is the *oversampling coefficient*.

The signal-to-noise ratio of an unsampled signal, determined by the ratio of the rms value of the digitized sinusoidal voltage waveform U_{sin} to the rms value of the quantization error U_q of the n-bit A/D converter, whose voltage range is equal to the amplitude of the sinusoidal voltage waveform

$$\mathrm{SNR} = 20 \cdot \log\frac{U_{\sin}}{U_q} = 20 \cdot \log\left(\frac{1}{\sqrt{2}} \cdot \frac{\sqrt{12}}{2^{-n}}\right) = 6.02 \cdot n + 1.76[\mathrm{dB}] \tag{10.26}$$

will be increased to a value during resampling

$$\mathrm{SNR}_k = 6.02 \cdot n + 1.76 + 20 \cdot \log\sqrt{k} = 6.02 \cdot n + 1.76 + 20 \cdot \log k\,[\mathrm{dB}]. \tag{10.27}$$

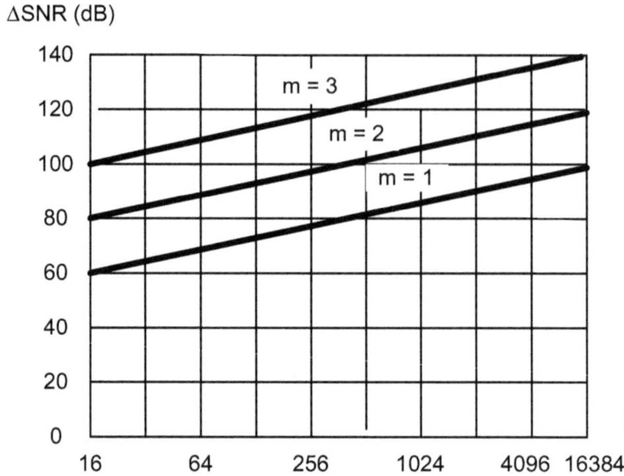

Figure 10.19 Dependence of the increase in the SNR on the resampling coefficient.

This increase in the signal-to-noise ratio is reflected in the increased resolution of the ΔΣ A/C converter

$$\Delta n_k = \log_2 \sqrt{k} = \frac{1}{2} \cdot \log_2 k . \tag{10.28}$$

Increasing the resolution of ΔΣ converters by 1 bit thus corresponds to increasing the signal-to-noise ratio by ΔSNR = 6 dB.

For example, with the oversampling coefficient $k = 256$, the resolution of the ΔΣ converter can theoretically be increased by 4 bits, which corresponds to an increase in the *SNR* by 24 dB. A further increase in the signal-to-noise ratio allows the use of multiple DS cascaded modulators, each of which increases the signal-to-noise ratio by approximately 20 dB.

Figure 10.19 shows an increase in the signal-to-noise ratio and a corresponding increase in the resolution of ΔΣ converters depending on the number m of cascaded modulators.

For example, when using three cascaded ΔΣ modulators, each of which has 8-bit quantization feedback, and choosing a oversampling coefficient of $k = 16{,}384$, the *SNR* of the ΔΣ converter can be reached up to 145 dB, which corresponds to a resolution of 24 bits.

These converters use FPGA (*field programmable gate array*) programmable gate arrays for signal processing, whose sampling frequency reaches several GS/s. These converters enable the digitization of signals with 16-bit resolution with sampling frequencies up to 200 kS/s and with 24-bit resolution with sampling frequencies of 50 kS/s (NI 9234).

However, in practical applications, especially with 24-bit converters, it is not possible to count on the full use of this resolution, as AC interfering signals can also be present in the entire measurement chain, including their preamplifiers and reference sources. Therefore, the so-called effective resolution is specified for these digitizers, given by their ENOB (*effective number of bits*), which for 24-bit digitizers ranges from 19 to 21 bits depending on the frequency of the digitized signal, which corresponds to the actual signal-to-noise ratio 116–128 dB.

From the frequency spectrum of digitized signals, in addition to the basic spectral component, corresponding to the frequency of the excitation signal, also spectral components whose frequencies are integral multiples of the frequency of the excitation signal f_1 determine the distortion by higher harmonics *THD* (*total harmonic distortion*), which is given by the ratio of their rms values to the rms value of the excitation harmonic signal

$$\mathrm{THD}[\mathrm{dB}] = 20 \cdot \log \frac{\sqrt{\sum_{i=2}^{k} U_i^2}}{U_1}, i = 1,2,\ldots,k. \tag{10.29}$$

Spectral components whose frequencies are not integral multiples of the frequency of the excitation signal f_1 determine the distortion by non-harmonic components *SNHD* (*signal non-harmonic distortion*), which is given by the ratio of their effective values to the effective value of the excitation harmonic signal

$$\mathrm{SNHD}[\mathrm{dB}] = 20 \cdot \log \frac{\sqrt{\sum_{i=2}^{k} U_i^2}}{U_1}, i \neq 1,2,\ldots,k. \tag{10.30}$$

Distortion by inharmonic components determines the so-called *noise floor* of the frequency spectrum of the digitizer. Since both of these spectral components are statistically independent, the signal-to-noise ratio and *SINAD* can be determined by the equation,

$$\mathrm{SINAD}[\mathrm{dB}] = 20 \cdot \log \sqrt{\mathrm{THD}^2 + \mathrm{SNHD}^2}. \tag{10.31}$$

Knowing the distance between signal noise and distortion, the *number of effective bits ENOB* of the digitizer can then be determined

$$\mathrm{ENOB} = \frac{\mathrm{SINAD}[\mathrm{dB}] - 1.76}{6.02}. \tag{10.32}$$

Figure 10.20 shows the amplitude spectrum of the NI 9234 24-bit digitizer from National Instruments, which uses a DS A/C converter with a sampling frequency of up to 52 kS/s to digitize the ΔΣ signal.

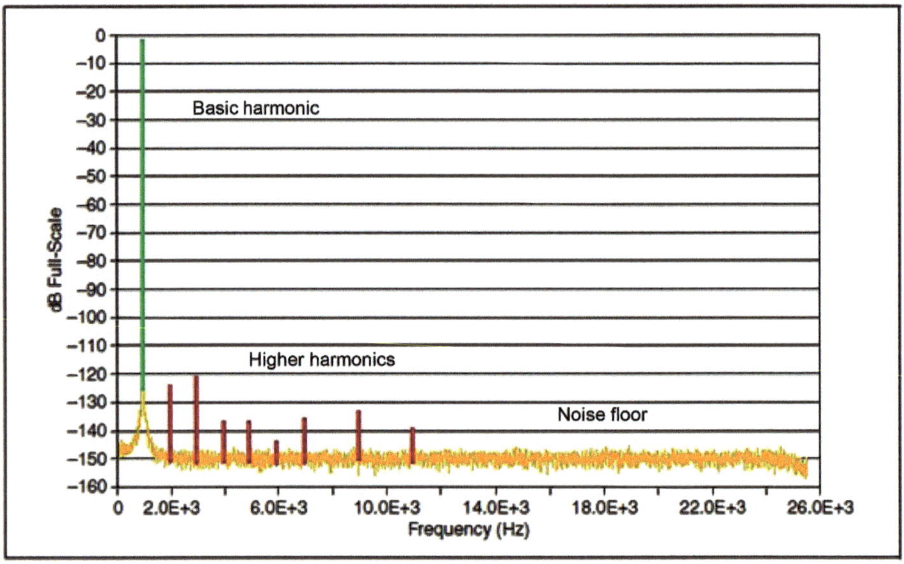

Figure 10.20 Amplitude spectrum of the NI 9234 24bit digitizer from National Instruments.

The amplitude frequency spectrum of the reconstructed excitation harmonic signal with a frequency of 1 kHz also contains higher harmonic components with frequencies of 2, 3, ..., 11 kHz. The *noise floor* of the frequency spectrum with a level of *SNHD* = 146 dB together with distortion by higher harmonic components *THD* = 115 dB determines the signal-to-noise ratio and distortion *SINAD* = 115 dB, which corresponds to the number of effective bits of 18.8 at a digitized signal frequency of 1 kHz.

10.2 Reconstruction of a Continuous Signal

In the reconstruction of continuous signals, interpolation methods and methods of direct digital synthesis including amplitude, frequency, and phase synthesis of these signals are used.

10.2.1 Stepwise reconstruction of signals

The most frequently used method of reconstruction of continuous signals is the *zero-order interpolation method*, in which the reconstructed signal is generated by a step-like function whose levels correspond to the values of the digital signal fed to the input of the D/A converter (Figure 10.21).

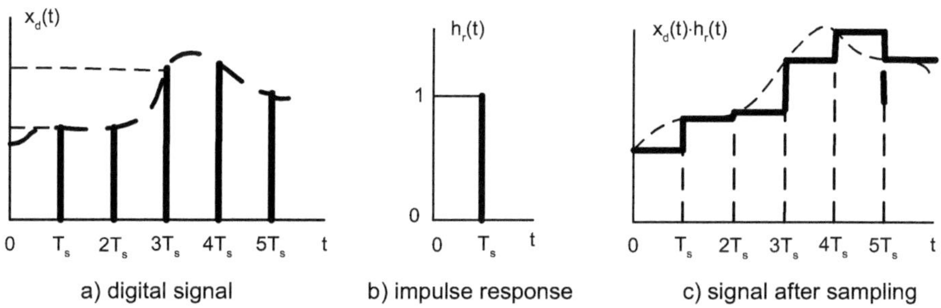

Figure 10.21 Time course of signals during zero-order interpolation.

The frequency spectrum of the reconstructed signal is determined by the convolutional product of the frequency spectrum of the impulse response $\boldsymbol{H}_{r0}(\omega)$ with the frequency spectrum $\boldsymbol{X}_{d0}(\omega)$ of the digital reconstructed signal

$$\boldsymbol{Y}_{d0}(\omega) = \boldsymbol{H}_{r0}(\omega) * \boldsymbol{X}_{d0}(\omega). \tag{10.33}$$

Because the frequency spectrum of the impulse response is based on the *discrete Fourier transform*

$$\boldsymbol{H}_{r0}(\omega) = \int_0^{T_s} e^{-j\omega t} dt = -T_s \frac{\sin\left(\frac{\omega}{2}\cdot T_s\right)}{\frac{\omega}{2}\cdot T_s} e^{-j\frac{\omega}{2}\cdot T_s}, \tag{10.34}$$

where $T_s = 1/\omega_s$ is the period of the sampling signal, the frequency spectrum of the reconstructed signal can be expressed in the form

$$\boldsymbol{Y}_{d0} = \frac{\sin\left(\frac{\omega}{2}\cdot T_s\right)}{\frac{\omega}{2}\cdot T_s} e^{-j\cdot\frac{\omega}{2}\cdot T_s} \cdot \sum_{k=-\infty}^{\infty} \boldsymbol{X}_{d0}(\omega - k\cdot\omega_s). \tag{10.35}$$

The frequency spectrum of this interpolation thus corresponds to the transmission of a lowpass filter in the frequency band $\pm 0.5 \cdot \omega_s$

$$\boldsymbol{Y}_{r0}(\omega) = \boldsymbol{H}_{r0}(\omega)\cdot \boldsymbol{X}_d(\omega) = \frac{\sin\left(\frac{\omega}{2}\cdot T_s\right)}{\frac{\omega}{2}\cdot T_s} \boldsymbol{X}_d(\omega) e^{-j\cdot\frac{\omega}{2}\cdot T_s}. \tag{10.36}$$

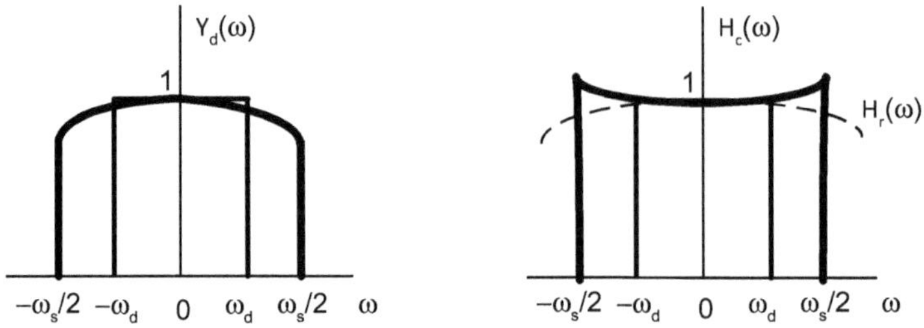

a) signal spectrum without correction b) signal spectrum with correction

Figure 10.22 Frequency compensation of the reconstructed signal.

To reconstruct the signal without distortion, this drop in the amplitude frequency response of the filter can be compensated by a *reciprocal frequency transfer filter*

$$\boldsymbol{H}_{c0}(\omega) = \frac{1}{\boldsymbol{H}_{r0}(\omega)} = \frac{\frac{\omega}{2} \cdot T_s}{\sin\left(\frac{\omega}{2} \cdot T_s\right)} \cdot e^{-j \cdot \frac{\omega}{2} \cdot T_s}. \tag{10.37}$$

For this correction of the frequency characteristic, digital filters with $\boldsymbol{H}_c(\omega)$ transmission are used, placed before the reconstruction D/A converter. It then applies to the frequency spectrum of the reconstructed signal corrected in this way

$$\boldsymbol{Y}_{c0}(\omega) = \boldsymbol{H}_{d0}(\omega) \cdot \boldsymbol{H}_{c0}(\omega) \cdot \boldsymbol{H}_{r0}(\omega). \tag{10.38}$$

The frequency spectrum compensation of the reconstructed signal is shown in Figure 10.22.

Reduction of the reconstruction error can be achieved either by increasing the sampling frequency and subsequently limiting the frequency band of the reconstructed signal, or by using higher-order interpolations. The most frequently used interpolations of higher orders are interpolations by *Lagrange polynomials.*

10.2.2 Direct digital synthesis

The method of direct digital synthesis (DDS) is used to reconstruct harmonic and non-harmonic signals. This method makes it possible to synthesize the time courses of signals, their frequency, and phase shift with respect to another signal generated in this way.

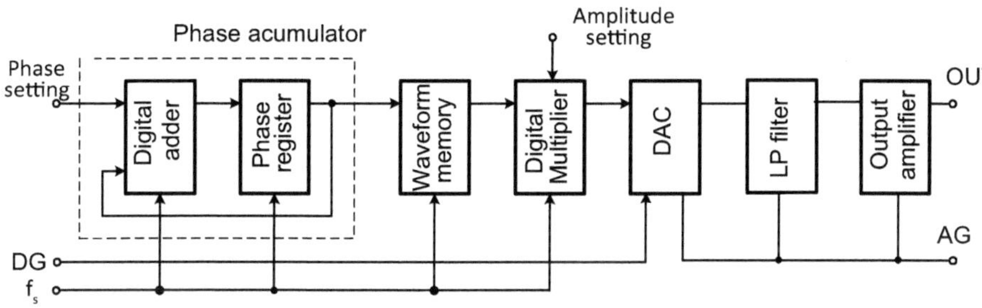

Figure 10.23 Block diagram of a direct digital synthesis synthesizer.

The synthesizer consists of a phase accumulator, a data memory, a digital multiplier, a D/A converter, a filter, and an output amplifier (Figure 10.23).

In the *phase accumulator*, the digital signal, which determines the *phase shift* of the generated signal with respect to the sampling signal CLK, is always shifted by the duration of its period after each arrival of the sampling pulse. The data memory is addressed with the signal shifted in this way, in which the waveforms of the generated signals are written, which are most often simple sine waves, or amplitude and frequency modulated signals.

If an M-bit digital signal is used to set the phase, then the phase resolution of the synthesizer is

$$\Delta_{\varphi} = 360 \cdot 2^{-M}. \tag{10.39}$$

For example, when using a 16-bit phase accumulator, the phase resolution is 0.0055°.

The *digital multiplier* allows you to set the amplitudes of the signals generated in this way, which are converted to corresponding analog waveforms by 12- to 16-bit D/A converters. Signals stored in the data memory can also be amplitude modulated using the multiplier. A lowpass filter is designed to suppress higher harmonic components of signals that arise during stepwise signal reconstruction. Active RC filters up to the fourth order are used to filter signals in the frequency range of hundreds of kHz. For higher frequencies, passive LC filters are used. The signal filtered in this way is subsequently amplified by the output amplifier.

In this *stepwise reconstruction*, the sinusoidal signal is reconstructed in equidistant time segments M and its frequency spectrum consists of a fundamental harmonic component with $f_1 = f_s/M$, where f_s is the sampling frequency and other higher odd harmonic components with frequencies $f_i = (i \cdot M) \pm 1$, $i = 1, 2, \ldots$, whose amplitudes decrease with the inverse value of the order of the corresponding higher harmonic component (Figure 10.24).

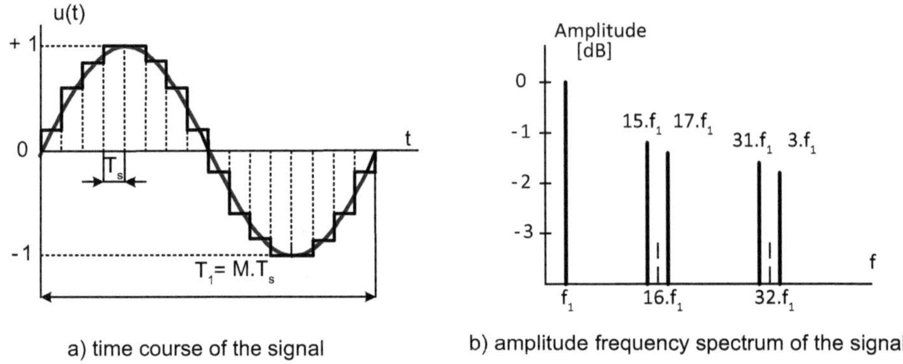

Figure 10.24 Stepwise reconstruction of a sine signal.

The signal reconstructed in this way is distorted by higher harmonic components *THD* (*total harmonic distortion*), which is determined by the ratio of the effective value of all higher harmonic components U_i, i = 2, 3, ... to the effective value of the basic harmonic component U_1

$$\text{THD} = \frac{\sqrt{\sum_{i=2}^{\infty} U_{ki}^2}}{U_i}. \tag{10.40}$$

For example, when reconstructing a sine signal in 64 sections, THD = 54.2 dB can be achieved without filtering, after filtering with a first-order filter with a cutoff frequency equal to the frequency of the generated sine signal, THD = 65.1 dB can be achieved, and when using a second-order filter, THD = 65.1 dB.

Table 10.3 shows the parameters of the frequency spectrum of the reconstructed sine signal by zero-order interpolation and subsequent digital filtering with first- and second-order filters.

Active filters of the second to eighth order of the *lowpass filter* type are used to filter the step-generated signal, the structure of which is usually based on the structure of second-order filters of the *Sallen Key* type (Figure 10.25).

Assuming that the resistors and capacitors in this filter are the same, i.e., that $R_1 = R_2 = R$ and $C_1 = C_2 = C$, then by DC transfer of the filter

$$k = \frac{R_3 + R_4}{R_4} \tag{10.41}$$

determine the type of approximation of its frequency, i.e., amplitude and phase transfer frequency characteristics

Table 10.3 Properties of the frequency spectrum of the reconstructed sinusoidal signal.

Number of samples in period *M*	The first pair of higher harmonics	THD [dB] without filtration	THD [dB] first-order filter	THD [dB] second-order filter
4	3.5	6.8	15.4	25.4
8	7.9	13.5	29.2	46.8
16	15.17	21.2	41.7	67.4
32	31.33	50.4	53.6	77.3
64	63.65	54.2	65.1	90.4
128	127.129	56.3	78.2	95.1
256	255.257	75.4	91.1	99.2

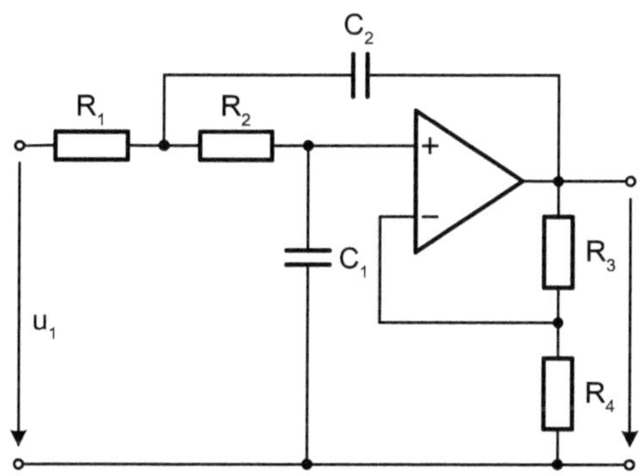

Figure 10.25 Low pass filter of the second order *Sallen Key* type.

$$F_{\mathrm{LP2}}(\omega) = -\frac{A_0}{1 + j \cdot \dfrac{\omega}{\omega_0} \cdot (3 - k) - \left(\dfrac{\omega}{\omega_0}\right)^2}. \tag{10.42}$$

Table 10.4 lists the types of approximations of the frequency characteristics of transmissions of second-order filters of the *Sallen Key* type.

Considering that in this type of filter, in addition to the negative voltage feedback, determining its direct transfer, an alternating voltage positive feedback is introduced by the capacitor C_2, determining the stability limit of the filter, which is given by the condition $R_3 < 2 \cdot R_4$.

From the course of these characteristics, it follows that filters with a *Bessel approximation* of frequency characteristics do not overshoot their transition characteristics and are therefore suitable for filtering a step-like reconstructed signal with high accuracy and low distortion. A suitable compromise between the quality of filtering and the time response of the filter to

Table 10.4 DC transfers of second-order filters with identical resistors and capacitors.

Approximation	Critical damping	Bessel	Butterworth	Chebyshev –3 dB	Limit of stability
DC transfer k	1000	1268	1586	2234	3000

step changes in its input voltage is made possible by the *Butterworth approximation* of its frequency characteristic. If the priority of choosing an approximation is a short reaction time of the filter, then the *Chebyshev approximation* is a suitable choice, which, however, exhibits undulations of its amplitude frequency characteristic in the passband of the filter, which can be a defect in some applications.

Higher-order filters can be created by connecting second-order filters in series. In this way, filters of up to the eighth order can be created. In some more sophisticated synthesizers, in addition to the type of approximation, the limit frequencies of the filters can be set with regard to the frequency range of the generated signals.

For a higher frequency range, exceeding hundreds of kHz, *passive LC filters* of up to fourth order with coils with ferrite cores, which are optimized in terms of their input and output impedances, are used instead of active filters.

To digitally adjust the frequency of the sampling signal f_s, a PLL (*phase locked loop*) phase lock is used, consisting of a crystal oscillator KO with a frequency of f_r and a frequency divider with a division ratio D_1 with a division ratio of $1/N_1$, whose output digital signal is in phase lock PD (*phase detector*) compared with the output signal of the frequency divider D_2 with a division ratio of $1/N_2$ (Figure 10.26).

The difference in the frequency of the output signals of these frequency dividers is converted into a voltage in the FD (frequency detector) and subsequent filtering, which is controlled by a voltage-controlled oscillator (VCO), whose frequency of the output digital signal in the steady state has the value

$$f_2 = \frac{N_2}{N_1} f_r. \tag{10.43}$$

These can generate impulse signals ranging from hundreds of Hz to hundreds of MHz with a temperature stability of up to 10^{-6}/K and a frequency stability of up to 10^{-3}/year.

A typical representative of a two-channel synthesizer is, for example, the AD 9854 circuit from *Analog Devices* with a 12-bit D/A converter, a maximum sampling frequency of 300 MS/s, 48-bit frequency resolution,

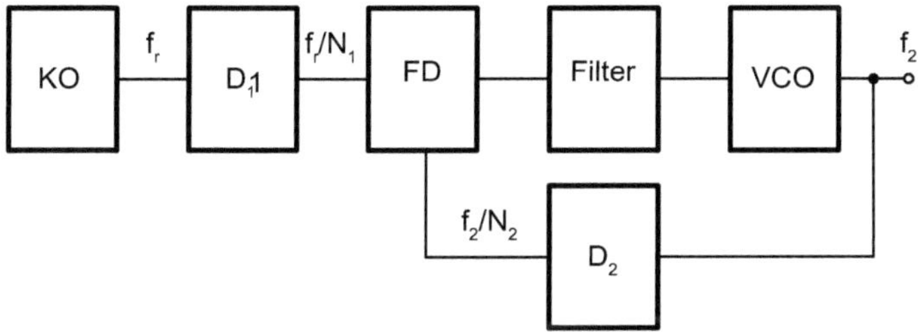

Figure 10.26 Frequency synthesizer.

14-bit phase resolution, and distortion by higher harmonic components THD < 80 dB at a frequency of 1 MHz of the generated harmonic signal. This synthesizer makes it possible to generate amplitude-, frequency-, and phase-modulated signals, including modulations of FSK, BPSK, and PSK signals. A typical representative of a four-channel synthesizer is, for example, the AD 9959 circuit with a 10-bit D/A converter with a maximum sampling frequency of 500 MS/s, enabling the generation of harmonic signals in the frequency range of 0.1 Hz to 150 MHz with 32-bit frequency resolution, 12-bit phase resolution, and distortion higher harmonic components THD < 60 dB at a frequency of 10 MHz.

10.2.3 Digital-to-analog converters

To reconstruct time-fast signals, D/A converters with identical or weight-graded switched current sources are used.

The block diagram of the D/A converter with *identical switched current sources* Z_i, $i = 1, 2,..., n$ with identical currents I is shown in Figure 10.27.

When current source Z_1 is switched on, current I flows to the inverting input of the operational amplifier, which is at ground potential. When the current source Z_2 is switched on, the current I is divided so that one half of it flows to the inverting input of the amplifier and the other half flows to the rest of the network, the resulting resistance of which is R.

Based on this binary distribution of currents, the output voltage of the converter is

$$u_{\mathrm{OUT}} = -R_z \cdot I \cdot \sum_{i=1}^{n} a_i \cdot 2^{-i}, \qquad (10.44)$$

where a_i, $i = 1, 2, \ldots, n$ are the bits of the converted digital signal, starting with bit a_1 with the MSB (*most significance bit*) and ending with the LSB (*least significance bit*).

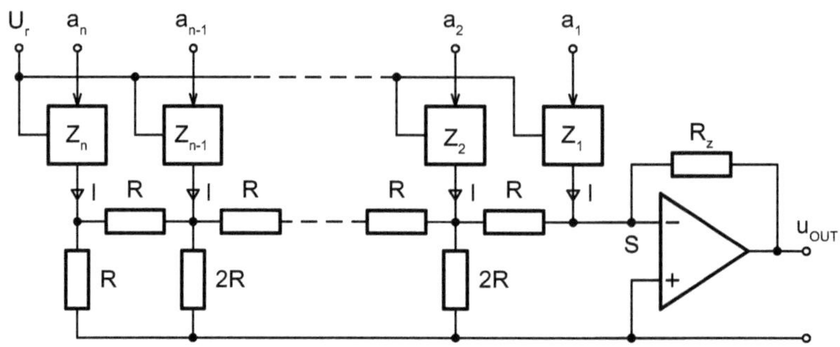

Figure 10.27 D/A converter with switched current sources.

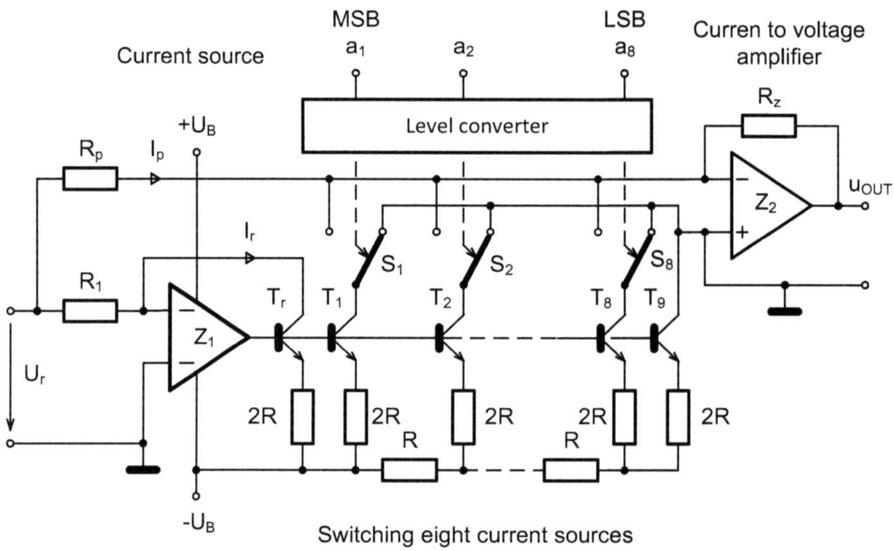

Figure 10.28 D/A converter with switched weight currents (DAC08).

The block diagram of the D/A converter with *switched weight current sources* is in Figure 10.28.

In this type of converter, a current $I_r = U_r/R_1$ is generated in the current source, consisting of an operational amplifier and a resistor R_1, which drives transistors T_1–T_8, whose emitter currents are scaled by weight in binary scaled values of $2^{-n} \cdot U_r/R$, $n = 1, 2, \ldots$. These currents are switched by switches S_1–S_n into complementary current buses, which are connected to the inputs of the differential operational amplifier Z_2, whose feedback resistor R_z determines the voltage range of the converter.

An increase in the accuracy of D/A converters is achieved by replacing simple current sources with combined switched current sources, and the surfaces of the BE transitions are dimensioned in such a way that they achieve the same current density. In this way, approximately the same transition temperature of the BE transistors can be achieved, which makes it possible to achieve higher accuracy and temperature stability of the generated voltage.

The conversion of the digital signal to the output voltage of both polarities is achieved by connecting the polarity resistor $R_p = 2 \cdot R_1$, connected between the reference voltage source and the inverting input of the operational amplifier. The output voltage range of the converter is

$$u_{\mathrm{OUT}} = \pm \frac{R_z}{2 \cdot R_i} \cdot U_r. \quad (10.45)$$

In this case, the value of the output voltage of the D/A converter is expressed as Offset Binary.

10.2.4 Sources of reference voltages

Reference voltage sources use either temperature-compensated Zener diodes or temperature-compensated junctions of BE bipolar transistors.

The *buried Zener diode* is governed by two phenomena. At a voltage less than 6 V, the *Zener phenomenon* with a negative temperature coefficient prevails, and at a higher voltage, the avalanche phenomenon with a positive temperature coefficient prevails. Around the Zener voltage, which is approximately 6 V, the temperature coefficient α_{ZD} of the *Zener diode* is approximately zero and its differential resistance is minimal.

For this reason, sources with Zener diodes use a connection consisting of a series connection of a Zener diode with a positive voltage temperature coefficient α_{UZD} and several diodes connected in the forward direction with a negative voltage temperature coefficient a_{UD} (Figure 10.29).

Because the voltage drop on the diodes connected in this way is determined by the sum of the voltages on them

$$U_{\mathrm{TCZD}} = U_{\mathrm{ZD}} + N \cdot U_D. \quad (10.46)$$

is the change of this voltage with temperature

$$\Delta U_{\mathrm{TCZD}} = \alpha_{\mathrm{TCZD}} \cdot U_{\mathrm{TCZD}} = \alpha_{\mathrm{ZD}} \cdot U_{\mathrm{ZD}} + N \cdot \alpha_{\mathrm{UD}} \cdot U_D. \quad (10.47)$$

The resulting temperature coefficient of this connection is then determined by the ratio

$$\alpha_{\mathrm{TCZD}} = \frac{\alpha_{\mathrm{ZD}} \cdot U_{\mathrm{ZD}} + N\alpha_{\mathrm{UD}} \cdot U_D}{U_{\mathrm{ZD}} + N \cdot U_D}. \quad (10.48)$$

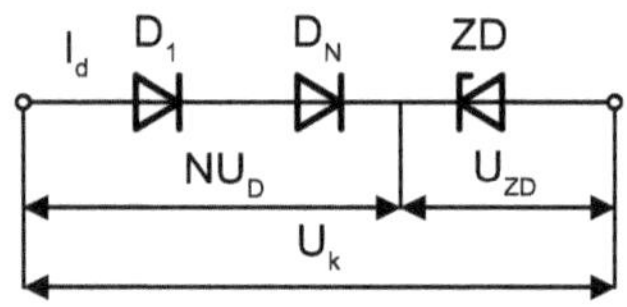

Figure 10.29 Temperature controlled Zener diode.

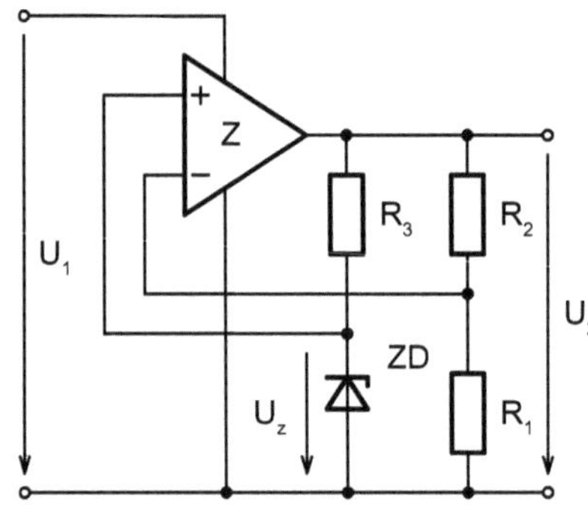

Figure 10.30 Voltage terferene source wirh Zener diode.

If, for example, a Zener diode with a voltage U_{ZD} = 8 V and a temperature coefficient α_{UZD} = 5.10^{-4}/K is connected in series with three diodes with a forward voltage U_D = 0.65 V with a voltage temperature coefficient α_{UD} = –2.10^{-3}/K, then the total voltage across this arrangement is U_{TCZD} = 9.95 V and has a temperature coefficient α_{TKZD} = 10^{-5}/K. If we place such a temperature-compensated diode in a microthermostat, maintaining the diode temperature at approximately 35 °C, a voltage-temperature coefficient of up to 10^{-6}/K can be achieved.

Such temperature-compensated *Zener diodes* are connected to the feedback circuit of the operational amplifier, which suppresses the effect of changes in the supply voltage of the source and the current load on the change in its output voltage (Figure 10.30).

Figure 10.31 shows voltage source with Zener diode.

$$U_2 = \left(1 + \frac{R_2}{R_1}\right) \cdot U_{ZD}. \tag{10.49}$$

Reference voltage sources with *temperature-compensated transitions of bipolar transistors* (*band gap*) are formed by a pair of bipolar transistors whose collector currents are in the ratio 1/*N* (Figure 10.31).

Assuming that the transistors T_1 and T_2 are identical, the voltage on the resistor R_1

$$\Delta U_{BE} = U_{BE2} - U_{BE1} = U_T \cdot \ln\left(\frac{I_{C1}}{I_{C2}}\right). \tag{10.50}$$

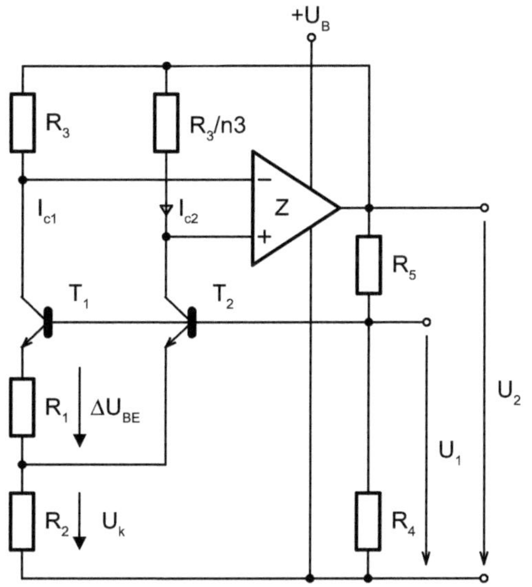

Figure 10.31 Voltage source with temperature compensated BE transitions.

At the ratio of collector currents $N = I_{C2}/I_{C1}$, the voltage of the resistor R_2 is

$$U_k = R_2 \cdot (I_{C1} + I_{C2}) = \frac{R_2}{R_1} \cdot (1 + N) \cdot \Delta U_{\text{BE}} = \frac{R_2}{R_1} \cdot (1 + N) \cdot U_T \cdot \ln N, \tag{10.51}$$

where $U_T = k \cdot \Theta/q_e$ is the thermal stress, Q is the temperature in K, $k = 1.38 \cdot 10^{-23}$ [J/K] is the *Boltzmann constant*, and $q_e = 1.59 \cdot 10^{-19}$ [C] is the charge electron.

The condition of temperature independence of voltage U_1 is fulfilled when the temperature change in voltage ΔU_k is equal to the temperature change in voltage ΔU_{BE}

$$\frac{\Delta U_k}{\Delta \theta} = \frac{R_2}{R_1} \cdot (1 + N) \cdot \frac{k}{q_e} \cdot \ln N = 2[\text{mV} / \text{K}]. \tag{10.52}$$

This condition holds for equality

$$\frac{R_2}{R_1} \cdot (1 + N) \cdot \ln N = 23, \tag{10.53}$$

upon fulfillment of which there is a voltage on resistor R_4

$$U_1 = 23 \cdot U_T + U_{\text{BE2}} = 1.205[\text{V}]. \tag{10.54}$$

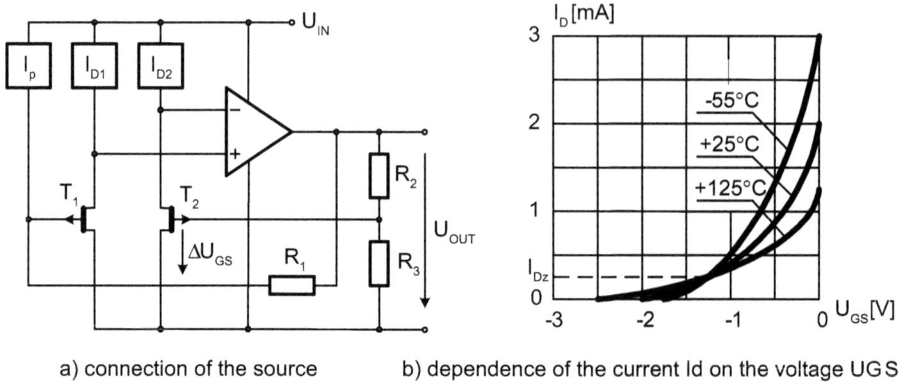

Figure 10.32 Temperature compensated voltage source with unipolar transistors.

This voltage, corresponding to the *band gap potential* of silicon at a temperature of 300 K, i.e., 27 °C, is then amplified to

$$U_2 = \left(1 + \frac{R_5}{R_4}\right) \cdot U_1. \tag{10.55}$$

The nominal output voltages of the reference sources range from 1.25 to 10 V at nominal output currents up to 10 mA.

Reference voltage sources with unipolar transistors are formed by a pair of JFET transistors, whose differential voltages U_{GS} are set to 1.205 V by the current source I_p and resistor R_1 (Figure 10.32).

At this voltage, corresponding to the bandwidth of the Si semiconductor at a temperature of 27 °C, the temperature dependence of the voltage difference ΔU_{GS} is zero, since the dependences of the currents I_d on the voltage U_{GS} intersect at one point for both JFET transistors at different temperatures of their GS transitions.

It then applies to the output voltage of the source

$$U_2 = \left(\Delta U_{GS} - R_1\right) \cdot I_P \cdot \left(1 + \frac{R_3}{R_4}\right). \tag{10.56}$$

These voltage sources have significantly less consumption compared to sources with bipolar transistors. That is why they are mainly used in devices with battery power.

Table 10.5 shows the parameters of typical reference voltage sources used in A/D and D/A measuring systems.

Each A/D and D/A converter also contains a reference voltage source whose temperature stability and noise should correspond to the resolution

Table 10.5 Parameters of typical reference voltage sources.

Principle of operation	Type	Nominal voltage	Temperature coefficient	RMS noise
Temperature compensated Zener diode	LTC 1021-10	10.00 V	50 µV/ K	3.5 µV
	LM 199	6.95 V	5 µV/K	3 µV
BE transition of bipolar transistors	REF 195 (6)	2.048 V (4.096 V)	5 µV/K	5 µV
GS transition of unipolar transistors	ADR 291	2.5 V	12 µV/K	8 µV
	ADR35xx	2.5 V, 5 V	7 µV/K	16 µV
	ADR45xx	2.5 V, 5 V	7 µV/K	10 µV

of these converters. If, for example, a voltage signal is to be digitized or reconstructed by n-bit A/D and D/A converters with a resolution of 2^{-n}, then the temperature stability and noise of their reference sources should not be greater by just this when their digital signals are allowed to change by 1 LSB value. For example, for 16-bit resolution, the temperature change of the reference voltage in the given temperature range should not be greater than $1.5 \cdot 10^{-5}$, and also the effective noise value should not exceed this limit.

In connection with the effect of noise of reference voltage sources and other noise sources in the entire measurement and reconstruction chain of A/D and D/A measuring systems, their *ER* is defined. For A/D systems, this is determined by the ratio of their voltage ranges to the effective value of the noise component of their output digital signals

$$\mathrm{ER} = \log_2\left(\frac{U_{\mathrm{FS}}}{U_n}\right)[\mathrm{bit}]. \tag{10.57}$$

Similarly, the effective discriminability of D/A systems can be defined.

The effective resolution of A/D and D/A systems is always less than their nominal resolution and, conversely, greater than their effective number of *ENOB* bits, which is also frequency dependent. For example, a 16-bit A/D system can achieve an effective resolution of up to 15 bits and an effective number of bits in the range of 12.5–13.5 bits at a given frequency of the digitized signal.

References

[1] Kester, W. The Data Conversion Handbook, Norwood 2005, ISBN 0-7506-7841-0.

[2] Tietze, U., Schenk, CH. *Electronic Circuits*. Handbook for Design and Application. Springer_Verlag Berlin Heidelberg, 2008. ISBN 978-3-540-78655-9.

[3] Horowitz, P., a Hill, W. *Art of Electronics*. Third Edition. Cambridge University Press, 2015. ISBN 9-7805-21809-269.

[4] Crecraft, D., Gergely, S. *Analog Electronics: Circuits, Systems and Signal Processing*. Butterworth-Heinemann, 2002. ISBN 13-978-07506-5095-3.

[5] Openheimer, A., a Schafer, R. *Discrete Time Signal Processing*. Prentice Hall, 1989. ISBN 3-978-013198842

[6] Mitra, S. K., Kaiser, J.F. *Handbook for Digital Signal Processing*. New York: John Wiley & Sons, 1993. ISBN 0-471-61995

[7] Norsworthy, S. R., R. Schreier, G. C. Temes. *Delta Sigma Converters, Theory and Simulation*. IEEE Press Marketing, 1997. ISBN 0-7803-1054-4

[8] Philips, K., Roermund, A.H. *Delta Sigma ADC Conversion for signal conditioning*. Springer Verlag 2010. ISBN:9789048171699.

[9] Vedral, J., Svatoš, K. Analog *Signal Processing and Digitization in Measurement*. Česká technika 2010. ISBN 978-80-0-06717-8.

11

Signal Processing in Measurement Systems

The processing of measured data is an integral part of all measuring systems. For this purpose, deterministic and statistical methods of data processing in the time and frequency domain are used. Typical methods of data processing in the time domain are determination of mean and RMS (*root mean square values*) of measured data, their interpolation, approximation, correlation, and autocorrelation. Data processing in the frequency domain includes their frequency analysis using the *discrete Fourier transform*.

11.1 Data Processing during Repeated Measurements

In the case of repeated measurements of one or more quantities, based on the recommendations of the *International Organization for Standardization* (ISO), the most probable value of the measured quantity is considered to be their *mean value*

$$X_s = \sqrt{\frac{1}{N}\sum_{i=1}^{N} X_i}, \tag{11.1}$$

which is determined from N consecutive measured values of one measured quantity.

The *uncertainty* of the determination of the measured quantity is determined from the mean value of the squared deviations of individual measures from their mcan values and is called the *mean squared deviation*

$$\sigma = \sqrt{\frac{1}{N}\sum_{i=1}^{N}\left(X_i - X_s\right)^2}. \tag{11.2}$$

It follows from eqn (11.2) that with N independent measurements of one quantity, its mean square deviation can theoretically be reduced in the ratio $1/\sqrt{N}$. In reality, this ratio may be smaller due to partial correlation between individual measurement results.

Assuming that we perform a larger number of independent measurements, e.g., several hundreds to thousands, the *frequency of occurrence*

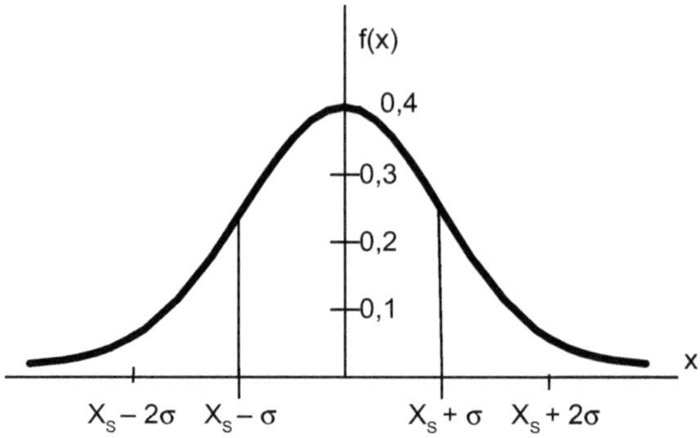

Figure 11.1 Probability density of data with normal distribution.

(*probability density*) of their values can be defined by a *normal* (*Gaussian*) *distribution*

$$f(x) = \frac{1}{\mu \cdot 2 \cdot \pi} \exp\left(-\frac{(x-\mu)^2}{2 \cdot \sigma^2} \right), \tag{11.3}$$

where $\mu \approx X_s$ is the mean value and σ is the standard deviation (Figure 11.1).

The relationship between the probability density $f(x)$ and the probability that the random variable x is in the interval (x_1, x_2)

$$p(x_1 \langle X \rangle x_2) = \int_{x_1}^{x_2} f(x) \cdot dx. \tag{11.4}$$

In technical practice, we usually require that the probability of occurrence of measured values, i.e., their level of significance varied within the specified interval. The most used intervals are the intervals $\alpha = \pm 2 \cdot \sigma$, or $\pm 3 \cdot \sigma$, at which the probability that the measured values are located in these intervals with 95% and 99.8%.

Table 11.1 shows the probabilities of occurrence of measured values for the most used intervals.

If we have a smaller number of measurements N available, e.g., only a few dozen, it is more appropriate to use the *Student's distribution* to determine the mean square deviation, the author of which is the English statistician *William Sealy Gosset*, publishing under the pseudonym *Student*.

The probability density of the measured data occurrence is defined by this distribution by the equation

$$f(x) = \frac{\Gamma\left(\frac{N+1}{2}\right)}{\Gamma\left(\frac{N}{2}\right) \cdot \sqrt{\pi \cdot N}} \left(1 + \frac{x^2}{N}\right)^{-\frac{N+1}{2}}, \tag{11.5}$$

Table 11.1 Probability of occurrence of measured values depending on the size of the interval α.

α	$\mu \pm \alpha \cdot \sigma$
0.674	0.500
1.000	0.683
1.645	0.900
2.000	0.950
3.010	0.998

where $\Gamma(N)$ is the *gamma function* given by the factorial of the variable N

$$\Gamma(N) = N! = \prod_{i=1}^{N} N_i. \tag{11.6}$$

This distribution, having a mean square deviation

$$\sigma = \sqrt{\frac{N}{N-2}} \tag{11.7}$$

with its properties for $N > 20$ practically no different from the normal *Gaussian distribution.*

The mean square deviation of the functional dependencies of multiple variables can be determined by the sum of the partial derivatives of these functional dependencies.

If they are, e.g., mean squared deviations $\sigma(A)$ and $\sigma(B)$ of two measured quantities A and B, then the mean squared deviation of their *sum* or *difference* is given by the squared sum of their deviations

$$\sigma(A \pm B) = \sqrt{\sigma^2(A) + \sigma^2(B)}. \tag{11.8}$$

If the functional dependencies are determined by the product or *quotient* of the measured quantities, then the resulting root mean square deviations of these dependencies are determined by the equations

$$\sigma(A \cdot B) = \sqrt{\mu^2(B) \cdot \sigma^2(A) + \mu^2(A) \cdot \sigma^2(B)}, \tag{11.9}$$

$$\sigma(A / B) = \sqrt{\sigma^2(A) / \mu^2(B) + \mu^2(A) \cdot \sigma^2(B) / \mu^4(B)}. \tag{11.10}$$

11.2 Processing Signals in the Time Domain

The processing of data and signals in the time domain includes the determination of their mean and effective values, periods of alternating signals, correlation, and autocorrelation of signals.

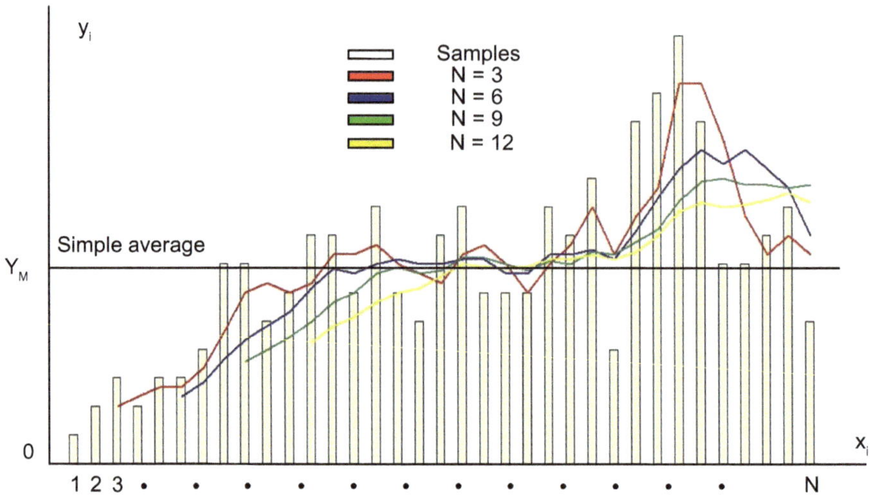

Figure 11.2 Time courses of averaged signals.

11.2.1 Determination of mean and effective values of signals

The *mean value* (MV), or its *simple average*, is determined by the simple sum of the values of signal samples y_i, $i = 1, 2, \ldots, N$, where N is the number of samples in the set in which the mean value is determined

$$Y_M = \frac{1}{N}\sum_{i=1}^{N} y_i, \qquad i = 1, 2, \ldots, N. \tag{11.11}$$

In addition to this simple average, *moving averages* (MA) are also used, which are defined by the average of the signal samples in the given files, in which the previous signal samples are gradually replaced by its following samples:

$$Y_{\mathrm{MA}} = \frac{1}{N}\sum_{i=j}^{j+N} w_i \cdot y_i, \qquad j = 1, 2, \ldots, N, \tag{11.12}$$

where w_i, $i = 1, 2, \ldots, N$ are the weights belonging to the ith sample in the given dataset.

From the point of view of the weights of individual signal samples, we distinguish *simple moving averages* (SMA) with the same weights $w_i = 1$, *binary stepped moving averages* (BSMA) with weights $w_i = 2^n$ and *exponential moving averages* (EMA) with weights $w_i = e^{-n}$.

Figure 11.2 shows the time courses of moving averaged signals, obtained by unweighted averaging from 3–12 consecutive samples.

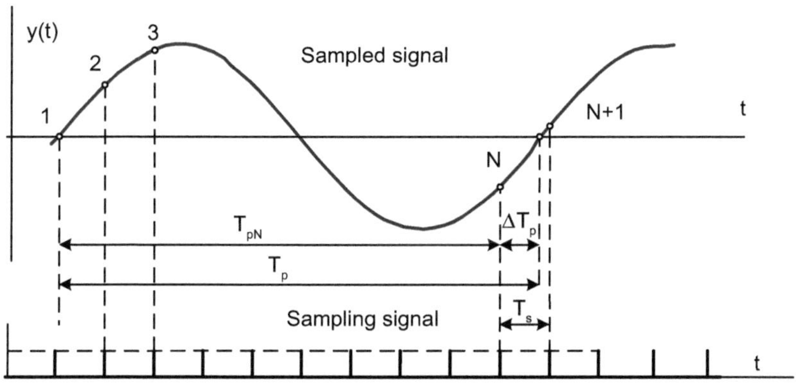

Figure 11.3 Time course of the signal during incoherent sampling.

From the courses of averaged periodic signals, it is clear that significant changes in the values of successive signal samples can be eliminated by moving averaging, while the degree of their elimination is proportional to the number of samples N from which the averaged signal samples are determined.

The RMS (*root mean square*) signal value is defined by the root mean square of the samples in a given dataset

$$Y_{\mathrm{RMS}} = \sqrt{\frac{1}{N}\sum_{i=1}^{N} y_i^2}, \quad i = 1, 2, \ldots, N. \tag{11.13}$$

This value expresses the normalized signal power in this file.

When determining the mean and effective values of periodic signals whose period time T_p is not an integral multiple of the period times of the sampling signal T_s, this so-called *incoherent sampling* and the low number of samples N taken from one period of the signal may result in errors that can be approximately determined from the equations (Figure 11.3)

$$\Delta X_{\mathrm{MA}} \approx \frac{y_{N+1} - y_N}{2}, \qquad \Delta X_{\mathrm{RMS}} \approx \sqrt{y_{N+1}^2 - y_N^2}. \tag{11.14 a, b}$$

A similar problem arises when determining the period time of a *periodic signal* from one of its periods. In this case, the uncertainty of determining the period time is given by the period of the sampling signal

$$\Delta T_p = \pm\frac{T_s}{2}. \tag{11.15}$$

Table 11.2 Parameters of periodic signals.

Signal	U_M	U_{RMS}	CF	FF
Sinuses	$2U_m/\pi$	$U_m/\sqrt{2}$	$1/\sqrt{2}$	$\pi/2\sqrt{2}$
Triangular	$U_m/2$	$U_m/\sqrt{3}$	$\sqrt{3}$	$2/\sqrt{3}$
Impulsive	U_m	U_m	1	1

To more accurately determine the period time of this signal, you can use, for example, *linear interpolation* between its neighboring samples N and $N + 1$, and the period time can be determined by the equation

$$T_p = T_{pn} + \Delta T_p = T_{pn} + \frac{y_N(t)}{y_{N+1}(t) - y_N(t)}, \tag{11.16}$$

where $y_N(t)$ and $y_{N+1}(t)$ are the signal values corresponding to N and $N + 1$ signal samples.

If the signal is *strictly periodic*, i.e., its period and shape do not change, the mean and rms values including its period time can be determined from one period. If we perform sampling of this signal in M periods, then it is possible, assuming that the variances of the values of the samples taken are uncorrelated, i.e., statistically independent, the variance of the determination of these parameters is approximately reduced in the ratio $1/\sqrt{M}$ (see Section 10.1 in Chapter 10).

In connection with the determination of the *mean* and RMS values of periodic signals, their *crest factors* (*CF*) and fluctuation *form factor* (*FF*) can be determined

$$\mathrm{CF} = \frac{Y_m}{Y_{RMS}}, \qquad \mathrm{FF} = \frac{Y_{RMS}}{Y_M}, \tag{11.17 a, b}$$

where Y_m is the amplitude of the signals.

Table 11.2 shows the mean and effective values of periodic signals, including their shape and fluctuation factors

11.2.2 Interpolation

By *interpolation*, we mean the insertion of additional values between the measured values with the aim of achieving greater amplitude and, in the case of time-varying signals, also temporal resolution. *Interpolation* differs from *approximation* in that the sought curve passes through all measured points. The opposite of interpolation is extrapolation, which determines approximate values of a function *outside* the interval of known values.

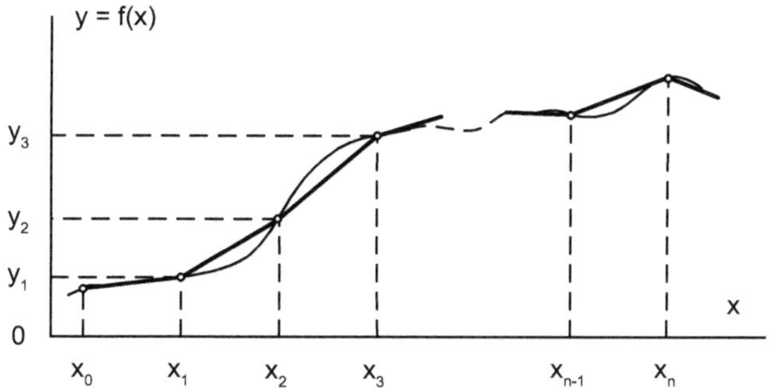

Figure 11.4 Linear interpolation.

The *polynomial Newton interpolation* is most often used to interpolate waveforms

$$N_n(x) = a_0 + a_1 \cdot (x - x_0) + + \ldots + a_n \cdot (x - a_0) \cdot (x - x_1) \ldots (x - x_{n-1}), \tag{11.18}$$

where a_i, $i = 1, 2, \ldots, N$ are the relative differences between adjacent signal samples.

In *linear interpolation*, where adjacent points are connected by a straight line, the difference is proportional

$$a_1 = y[x_i, x_{i+1}] = \frac{y(x_{i+1}) - y(x_i)}{x_{i+1} - x_i}. \tag{11.19}$$

In *quadratic interpolation*, in which adjacent points are connected by a parabola, the difference is proportional

$$a_2 = y[x_i, x_{i+1}, x_{i+2}] = \frac{y[x_{i+1}, x_{i+2}] - y[x_i, x_{i+1}]}{x_{i+2} - x_i}. \tag{11.20}$$

Figure 11.4 shows the course of the signal during linear interpolation.

In addition to these polynomial interpolations, *Lagrangian interpolation* polynomials are also used to interpolate waveforms

$$L_n(x) = \sum_{i=0}^{n} y(x_i) \cdot L_i^n(x), \tag{11.21}$$

where

$$L_i^n(x) = \prod_{\substack{k=0,\\ j \neq i}}^{n} \frac{x - x_j}{x_i - x_j} \tag{11.22}$$

are the *Lagrange coefficients*.

Lagrange polynomials of the second to fourth order are usually used, as polynomials of higher orders are already difficult to interpret, and moreover do not yield significantly better results. A typical area of use of interpolation of functional waveforms by *Lagrange polynomials* is the generation of signals by their direct digital synthesis DDS (*direct digital synthesis*). Compared to Newtonian interpolation polynomials, *Lagrange polynomials* interpolate functional dependencies with identical left and right derivatives at the points between which the function is interpolated. As a result, this significantly reduces the occurrence of higher harmonic components of the generated waveforms.

11.2.3 Approximation

The task of the *approximation* is to fit the functional dependence into such an approximation function that the differences between the individual values of the approximated and approximate functional dependence are minimal. The least squares (MSE (*minimum square error*)) method is used for this calculation algorithm.

In general, to approximate any functional dependence, given by the values y_i, $i = 1, 2, \ldots, k$ at the points x_i, $i = 1, 2, \ldots, k$, a polynomial $k + 1$ of the 10th order can be used

$$P_k(x) = \sum_{i=0}^{k} p_i \cdot x_i^i, \tag{11.23}$$

where its coefficients $p_0, \ldots, p_k$ are chosen so that the sum of the squares of the deviations

$$e_i = y_i - P_k(x_i), \qquad i = 1, 2, \ldots, k \tag{11.24}$$

was minimal, i.e. to apply

$$\boldsymbol{F}(p_0,\ldots,p_k) \equiv \sum_{i=1}^{n} e_i^2(p_0,\ldots,p_k) = \sum_{i=1}^{n}\left(y_i - \sum_{j=0}^{k} p_j \cdot x_i^j\right)^2 \tag{11.25}$$

where $\boldsymbol{F}(p_0,\ldots, p_k)$ is the functional of the sum of squared deviations.

Since this functional is always non-negative and therefore has one global minimum, its zero partial derivatives can

$$\frac{\partial \boldsymbol{F}(p_0,\ldots,p_k)}{\partial p_{kj}} = 0, \qquad j = 0,1,\ldots,k \tag{11.26}$$

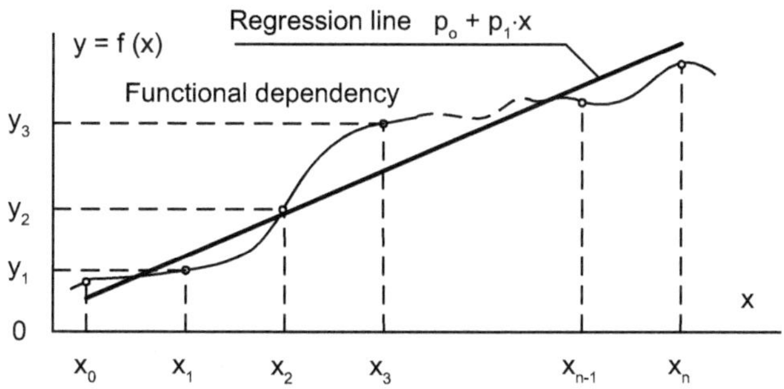

Figure 11.5 Linear regression.

coefficients p_j, $j = 0, 1, 2, \ldots, k$ to the approximation polynomial P_k can be determined.

The most frequently used regression function is a straight line. In this *linear regression*, a straight line is fitted through the approximate functional dependence (Figure 11.5)

$$P(x) = p_0 + p_1 \cdot x. \tag{11.27}$$

In this case, eqn (11.27) reduces to the form

$$\boldsymbol{F}(p_0, p_1) = \sum_{i=1}^{n} \left(p_0 + p_1 \cdot x_i - y_i\right)^2. \tag{11.28}$$

To find the minimum of the sum of squared deviations, we perform partial derivatives of this function, and after adjustment, we obtain a system of two equations

$$p_1 \sum_{i=1}^{n} x_i^2 + p_0 \sum_{i=1}^{n} x_i = \sum_{i=1}^{n} x_i \cdot y_i, \tag{11.29}$$

$$p_1 \sum_{i=1}^{n} x_i + p_0 \cdot n = \sum_{i=1}^{n} y_i, \tag{11.30}$$

from which the parameters of the regression line can be determined

$$p_1 = \frac{n\sum_{i=1}^{n}(x_i \cdot y_i) - \sum_{i=1}^{n} x_i \cdot \sum_{i=1}^{n} y_i}{n\sum_{i=1}^{n} x_i^2 - \left(\sum_{i=1}^{n} x_i\right)^2}, \tag{11.31}$$

$$p_0 = \frac{\sum_{i=1}^{n} x_i^2 \cdot \sum_{i=1}^{n} y_i - \sum_{i=1}^{n} x_i \cdot \sum_{i=1}^{n} y_i}{n\sum_{i=1}^{n} x_i^2 - \left(\sum_{i=1}^{n} x_i\right)^2} \tag{11.32}$$

Based on linear regression, functional dependencies can be approximated by a power or exponential regression function. If the regression function is a *power function*

$$y = a \cdot x^b, \tag{11.33}$$

then after its logarithm

$$\ln(y) = \ln(a) + b \cdot \ln(x) \tag{11.34}$$

the linear regression method can be applied, with the difference that the approximated function, defined by the data $\ln(x_i)$, $\ln(y_i)$, can be approximated by a linear function

$$y = c + b \cdot x, \tag{11.35}$$

while knowing the coefficient c and the coefficient a can be determined

$$a = e^c. \tag{11.36}$$

If the regression function is an exponential function

$$y = a \cdot b^x, \tag{11.37}$$

can be obtained after logarithmizing this function

$$\ln(y) = \ln(a) + x \cdot \ln(b) \tag{11.38}$$

determine the linear regression function

$$\ln(y) = \mathrm{c} + x \cdot \mathrm{d}, \tag{11.39}$$

and from the coefficients c and d determine the coefficients

$$a = e^c, \qquad b = e^d. \tag{11.40 a, b}$$

The quality of the approximation is determined by the *root mean square deviation* (*RMSD*), defined by the square of the differences between the y_i values of the approximated function and the y_{ia} values of the approximation function

$$\sigma = \frac{\sum_{i=1}^{n} \left(y_i - y_{is}\right)^2}{n}. \tag{11.41}$$

The quality of the approximation is also determined by the *regression coefficient*, defined by the equation

$$R = \sqrt{1 - \sigma}. \tag{11.42}$$

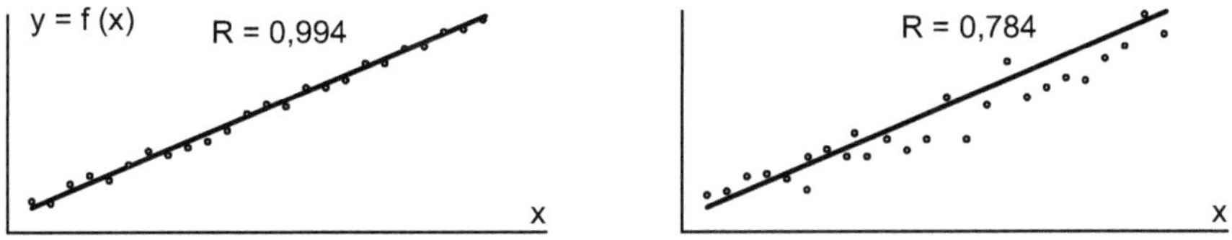

Figure 11.6 Linear regression of two different data sets.

Figure 11.6 shows the linear regressions of the two datasets, indicating the *regression coefficient R*.

11.2.4 Correlation and autocorrelation

Correlation makes it possible to compare the time course of signals and from their similarity, it is possible to determine their periodicity, the delay between signals, or even recognize the wanted signal in a heavily disturbed signal.

The *correlation* of two functional waveforms of signals, determined by their samples (x_i), $g(x_i)$, $i = 1, 2, \ldots, N$, is defined by the equation

$$R\left[f(x) * g(x)\right]_k = \frac{1}{N}\sum_{i=0}^{N-k-1}\left[f(x_i)\cdot g(x_{i+k})\right]. \tag{11.43}$$

The *correlation* is thus given by the sum of the products of the samples of one *reference signal* with the samples of the second *correlated signal*, which are delayed from the reference signal by k samples.

Figure 11.7 shows the time courses of *two binary signals* whose samples are shifted by k samples and the course of their correlation function, which has a linear course in the range of overlapping signal samples. The correlation function takes on its maximum value of unity when the two correlated signals are most similar. The relative displacement of the signal samples can then be determined from the position of this maximum. Assuming that both signals are periodic with the same period time, their period time can be determined from the time differences of the maximum values of the correlation function.

Autocorrelation of a signal allows determining the similarity of its waveforms in several different sections, determined by its samples $f(x_i)$, $i = 1, 2, \ldots, N$

$$R\left[f(x)\right]_k = \frac{1}{N}\sum_{i=0}^{N-k-1}\left[f(x_i)\cdot f(x_{i+k})\right], \tag{11.44}$$

where k determines the size of the correlated signal section.

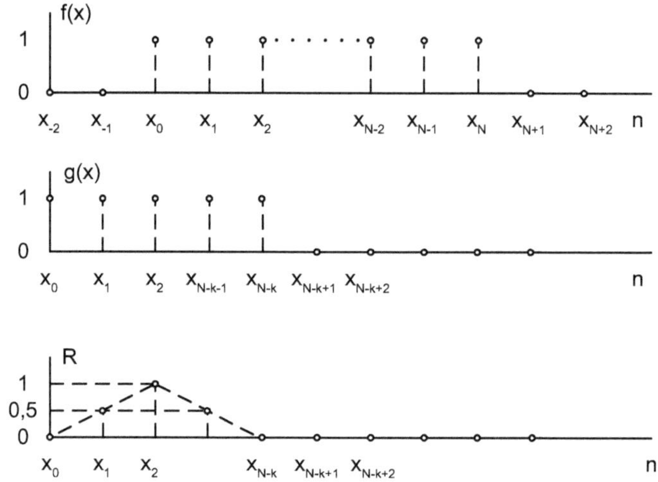

Figure 11.7 Waveforms of correlated signals.

While correlation determines the similarity of the waveforms of two signals, autocorrelation determines the similarity of the waveforms of several sections of one signal. In the case of autocorrelation, its maximum value always corresponds to a zero coefficient *k*, i.e., a zero delay between sections of the correlated waveform.

To more accurately determine the period time T_p between two signals, it is necessary to use a significantly higher number of samples than that would correspond to the minimum number of samples whose period time is T_s

$$N_{\min} >> \frac{T_p}{T_s}. \tag{11.45}$$

In this case, there are multiple maxima in the correlation function, which are separated by the period of the analyzed signal. At the same time, possible false detection of the period of the signal, which may not be periodic, can be avoided.

To assess the degree of correlation, the *Pearson correlation coefficient* is used for autocorrelation of signals

$$\mathrm{CF}\left[f(x),g(x)\right] = \frac{\sum_{i=1}^{N}\sum_{j=k}^{N+k}\left[f(x_i) - g(x_i)\right]^2 \quad F_s(x)\cdot G_s(x)}{\sqrt{\sigma(f(x)\cdot \sigma(g(x)}}, \tag{11.46}$$

where

$$F_s(x) = \sqrt{\frac{1}{N}\sum_{i=1}^{N} f(x_i)}, \qquad G_s(x) = \sqrt{\frac{1}{N}\sum_{i=k}^{N+k} g(x_i)} \qquad (11.47\text{ a, b})$$

are their mean values and

$$\sigma[f(x)] = \sqrt{\frac{1}{N}\sum_{i=1}^{N}\left[(f(x_i) - F_s(x)\right]^2},$$
$$\sigma[g(x)] = \sqrt{\frac{1}{N}\sum_{i=k}^{N+k}\left[(g(x_i) - G_s(x)\right]^2}, \qquad (11.48\text{ a, b})$$

are their RMS values in the range 1 to N, or k to $N + k$.

11.3 Spectral Analysis of Signals

Each *periodic signal* $x(t)$ with period T_p can be expressed as the sum of an infinite, in practice finite, number of harmonic waveforms with frequencies that are determined by an integral multiple of the frequency of the original signal using the *Fourier series*

$$x(t) = \frac{a_0}{2} + \sum_{k=1}^{n}\left[(b_k \cdot \sin(k \cdot \omega \cdot t + a_k \cdot \cos(k \cdot \omega \cdot t)\right]), \qquad (11.49)$$

where $\omega = 2 \cdot \pi/T_p$ is the circular frequency corresponding to the period T_p of the signal and n is theoretically infinite, in practice always a finite, integer.

Coefficients a_k and b_k are determined by the following equations:

$$a_k = \frac{2}{T}\int_0^T x(t) \cdot \cos(k \cdot \omega \cdot t)dt, \qquad b_k = \frac{2}{T}\int_0^T x(t) \cdot \sin(k \cdot \omega \cdot t)dt, \qquad (11.50\text{a, b})$$

where $k = 0, 1, 2, \ldots$ is an integer.

The analyzed course can also be expressed by the amplitudes and phases of the individual spectral components of the signal

$$x(t) = \sum_{k=1}^{n} X_k \cdot \sin(k \cdot \omega \cdot t + \varphi_k), \qquad (11.51)$$

which applies

$$X_k = \sqrt{a_k^2 + b_k^2}, \qquad \varphi_k = \text{actg}\left(\frac{b_k}{a_k}\right), \qquad k = 1,2,\ldots,n. \qquad (11.52\text{ a, b})$$

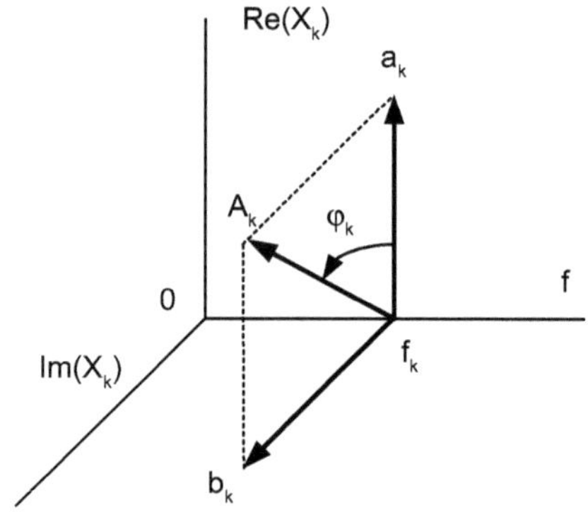

Figure 11.8 Display of the X_k component in the complex plane.

The three-dimensional representation of the amplitude $X(f_k)$ and the phase φ_k of the kth harmonic is shown in Figure 11.8.

We can also express the spectral components of the signal in a complex form

$$\boldsymbol{X}(f_k) = \frac{1}{T} \int_{-T_p/2}^{T_p/2} x(t) \cdot \exp(-j \cdot 2\pi \cdot f_k \cdot t \cdot dt), \tag{11.53}$$

where T is the period of the signal $x(t)$ and $f_k = kf_1 = k/T_p$ is the frequency of the kth harmonic.

The time course of the signal can be reconstructed by the *inverse Fourier transformation*

$$x(t) = \sum_{k=-n}^{n} \boldsymbol{X}(f_k) \cdot \exp(j \cdot 2\pi \cdot f_k \cdot t), \tag{11.54}$$

where n is an integer.

11.3.1 Spectral analysis of sampled signals

A *non-periodic sampled signal* with equidistant time periods $T_s = 1/f_s$ has a frequency spectrum determined by the DFT (*discrete Fourier transform*)

$$\boldsymbol{X}(f) = \sum_{n=-\infty}^{\infty} x(t_n) \cdot \exp(-j \cdot 2\pi \cdot f \cdot t_n), \tag{11.55}$$

where $t_n = n \cdot T_s = n/f_s$ are the time points of taking individual samples.

The frequency spectrum of a *non-periodic sampled signal* is therefore periodic with a period of $1/f_s$. Copies of the spectrum of the original signal occur symmetrically around integer multiples of the sampling frequency f_s (Figure 11.9).

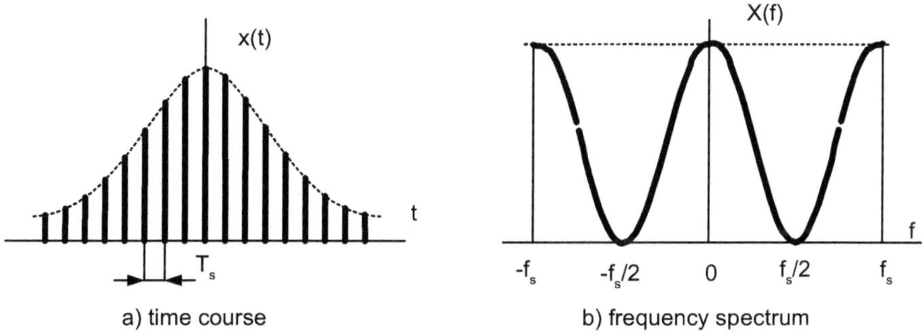

Figure 11.9 Non-periodic sampled signal.

Periodically sampled signals have a frequency spectrum

$$\boldsymbol{X}_k = \frac{1}{N}\sum_{n=0}^{N-1} x\left(n \cdot T_s\right) \cdot \exp\left(-\frac{j \cdot 2\pi \cdot n \cdot k}{N}\right), \tag{11.56}$$

where k is an integer (harmonic order) in the relation $f_k = k \cdot f_1 = k/T_p$, n is an integer in the relation $t_n = \mathrm{n} \cdot T_s$, and N is the number of samples taken during the period T_p of the transformed periodic course $x(t)$ in time equidistant intervals T_s.

The reconstructed time course of the signal then has the form

$$x\left(t_n\right) = \sum_{k=0}^{N-1} \boldsymbol{X}_k \cdot \exp\left(\frac{j \cdot 2\pi \cdot n \cdot k}{N}\right). \tag{11.57}$$

The frequency spectrum of a *periodically sampled signal* is therefore also *discontinuous* and *periodic*. The number of samples N must be chosen in accordance with the *sampling theorem*, i.e., even the highest significant harmonic component of the signal $x(t)$ must be sampled more than twice in its period (Figure 11.10).

The following quantities are used to assess the properties of the frequency spectrum of the analyzed signal:

Amplitude spectrum $||\mathbf{X}_k|$ is defined by the absolute value of the kth harmonic component of the signal. For a voltage signal, it is given in dB, with the reference voltage level being 1 V (dBV), 1 mV (dBm), or 1 μV (dBμV).

The *power spectrum* $|\mathrm{X}_k|^2$ is defined by the square of the amplitudes of the harmonic components of the signal. In the case of a voltage signal, the power spectrum represents the mean power of the corresponding harmonic component on the 1 Ω resistor. Power spectra are given in dBW, dBmW, and dBμW at reference powers of 1 W, 1 mW, and 1 μW.

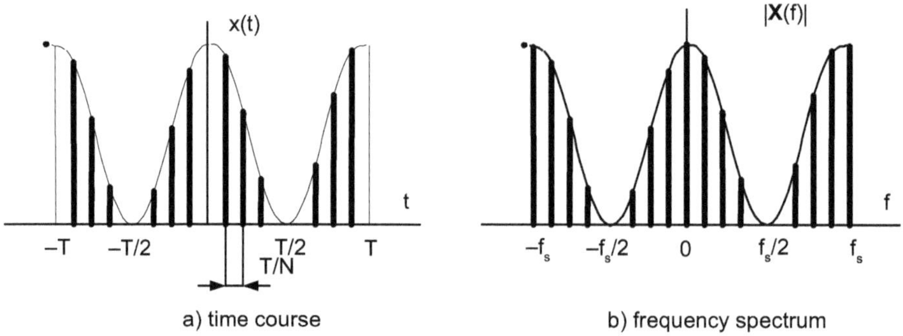

Figure 11.10 Periodic sampled signal.

The *energy spectral density* $|X_k|^2 \cdot t/\Delta t$ expresses the energy of part of the signal spectrum in the frequency range Δf. This quantity is mainly used in the analysis of non-stationary impulse signals.

The *power spectral density* $||X_k|^2/\Delta t$ expresses the spectrum power of a part of the signal spectrum in the frequency range Δf. This parameter is used when analyzing the characteristics of stationary signals.

The *power cepstrum* (a portmanteau of the word spectrum) is defined by the square of the absolute value of the Fourier transform of the logarithm of the power spectrum or spectral power density

$$C(\tau) = \left|\boldsymbol{F}\left\{\log\left|\boldsymbol{X}(f)^2\right|\right\}\right|^2 \ [\mathrm{s}], \tag{11.58}$$

where $\boldsymbol{F}\{.\}$ is the *Fourier transform symbol*. The variable τ has the dimension of time and is called kveference (a portmanteau of the word frequency).

Cepstral analysis is especially suitable for the analysis of phenomena whose power spectrum has the character of a product or share of several partial power spectra. After logarithming the cepstrum, the sum or difference of the partial spectra is thus obtained. This makes it possible to distinguish the frequency spectrum of the useful signal from the frequency spectrum of interfering signals.

11.3.2 Time windowing

In the case of *incoherent sampling*, i.e., when the number of samples N is not an integral multiple of the periods of the sampled signal, spectral components that do not correspond to the components of the coherently sampled signal occur in the frequency analysis of the signal sampled in this way. This so-called *spectrum leakage* can be partially suppressed by using *time*

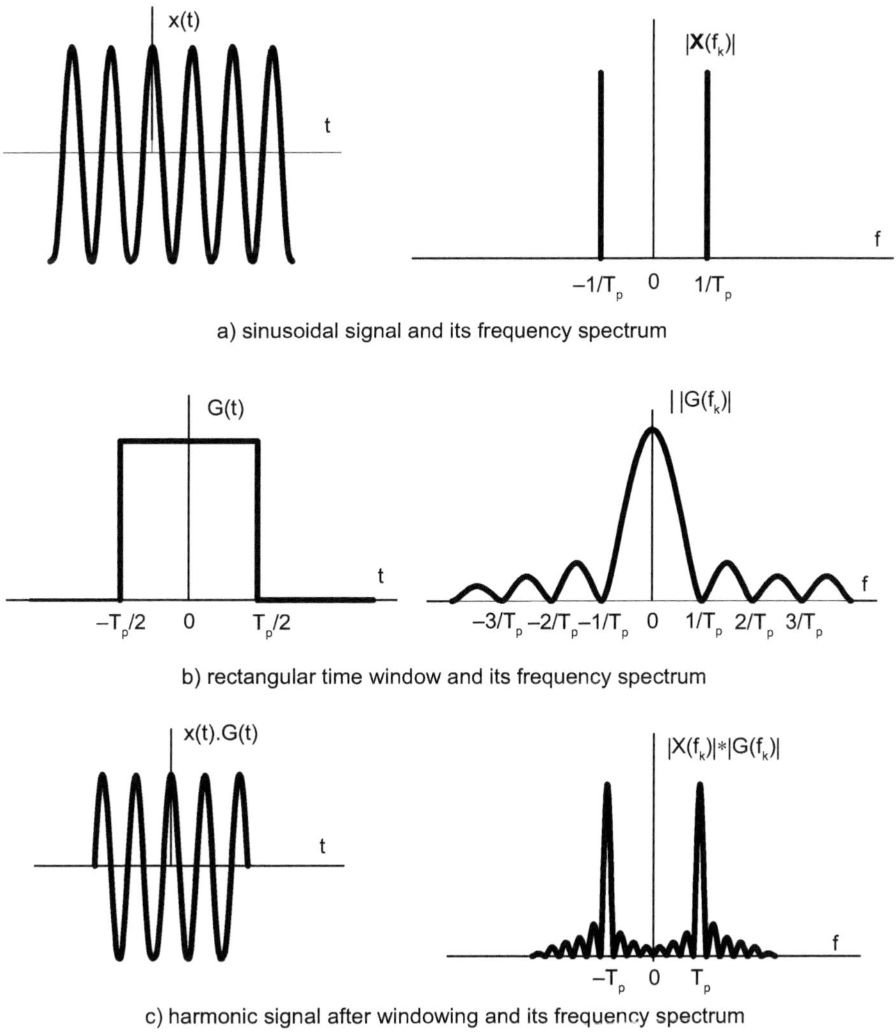

Figure 11.11 Passage of a harmonic signal through a rectangular window.

windows. The use of a *rectangular time window* of width T_p, applied to the analyzed harmonic signal, which has a line frequency spectrum, is shown in Figure 11.11.

The amplitude frequency characteristic of the *rectangular window* has, in addition to the basic lobes, symmetrically distributed around the zero frequency axis, additional lateral lobes, the largest two of which have a level of 13 dB compared to the amplitude of the basic lobes. By passing the signal $x(t)$ through this window, we obtain a time-trimmed time harmonic course, whose frequency spectrum is determined by *convolution* of the frequency spectrum of the original signal $X(f)$ and the frequency spectrum of the windowed signal

$$G(f) = \frac{1}{T_p} \cdot \frac{\sin(\pi \cdot f \cdot T_p)}{\pi \cdot f \cdot T_p} \tag{11.59}$$

with zero points for frequencies $1/T_p$, $2/T_p$, etc., and a maximum value of $1/T_p$ at zero frequency.

If the course of the analyzed signal $x(t)$ is periodic within a window of width T_p, i.e., there is an integer multiple of the signal periods in the window, the effect of the window on the frequency characteristic of the signal is not manifested, since the components of the signal spectrum $X(f)$ are multiplied by the frequency spectrum $G(f)$ of a rectangular time window. The effect of the window is not manifested even if there are other components of the signal spectrum at frequencies n/T_p, as they are multiplied by the zero values of the frequency spectrum of the rectangular time window $G(f)$.

If the T/T_p ratio is not a whole number, a continuous spectrum with side lobes is produced even during harmonic signal processing due to the effect of the window. These lobes represent interfering components of the spectrum that are not contained in the original signal. Time windows without step changes at their ends are therefore used to limit this phenomenon. The most common time windows are, for example, *cosine windows* of the *Hanning*, *Hamming*, and *Blackman-Harris* types.

The width of the time window T_p determines the frequency resolution of the spectrum analyzer. It is not possible to distinguish the frequencies of waveforms whose period T is smaller than T_p. From this point of view, the time window behaves like a bandpass with a *frequency bandwidth* BW, which for a rectangular window is equal to

$$\mathrm{BW} = 1 / T_p. \tag{11.60}$$

For these time windows, the *equivalent noise bandwidth* (ENBW) is defined by the equation

$$\mathrm{ENBW} = \frac{1}{2 \cdot \pi} \int_0^{\infty} \left| \frac{G(\omega)}{G_m} \right|^2 d\omega, \tag{11.61}$$

where $G(\omega)$ are their transfer functions and G_m their maximum values at circular frequency $\omega = 2 \cdot \pi/T_p$ In

Table 11.3 shows the suppression of the main lobes of the transmission frequency characteristics of the most commonly used time windows, including their equivalent noise bandwidths.

Table 11.3 Parameters of the most used window functions.

Window type	Main lobe suppression	ENBW
Rectangular (Boxcar)	13 dB	1.00
Triangular (Barlettovo)	25 dB	1.27
Hanning	32 dB	1.51
Hamming	43 dB	1.37
Blackman-Harris	58 dB	2.01

If the signal is not coherently sampled, i.e., the input and sampling frequencies are *unequal*, the dynamic range of the digitizer is determined by the equation

$$\mathrm{DR} = 6{,}02 \cdot n + 1.76 + 10\log\frac{3 \cdot N}{\pi \cdot \mathrm{ENBW}}[\mathrm{dB}], \qquad (11.62)$$

where N is the number of samples in the period of the analyzed signal.

For example, coherent sampling of the signal by a 16-bit A/D converter with the number of samples taken $N = 32$ kB makes it possible to achieve a *dynamic range* of spectral analysis DR = 134 dB. We achieve almost the same result when using a *rectangular window* in the spectral analysis of an incoherently sampled signal. If we use *a Blackman-Harris window* with an equivalent bandwidth of ENBW = 2.01, then with the same number of samples taken, the dynamic range of the analysis will be reduced to 129 dB, but the effect of leakage on the resulting frequency spectrum will be significantly reduced.

11.4 Digital Filtering of Signals

Signal filtering is used to suppress or, on the contrary, to highlight part of their frequency spectrum. Analog or digital filters are used to filter signals. *Analog filters*, made up of resistors, capacitors, and operational amplifiers, use negative and positive linear feedback to realize their transmissions. *Digital filters* use algorithms based on the properties of *discrete Fourier transform* DFT and *Z transform* to generate their transmissions.

11.4.1 Properties of filters

According to the course of the transmission frequency characteristics of the filters, we distinguish filters of the *lowpass filter*, *highpass filter*, *bandpass filter*, and *band-stop filter*. The normalized courses of the amplitude-frequency characteristics of these types of filters are shown in Figure 11.12, where n denotes the order of the filter, which determines the steepness of

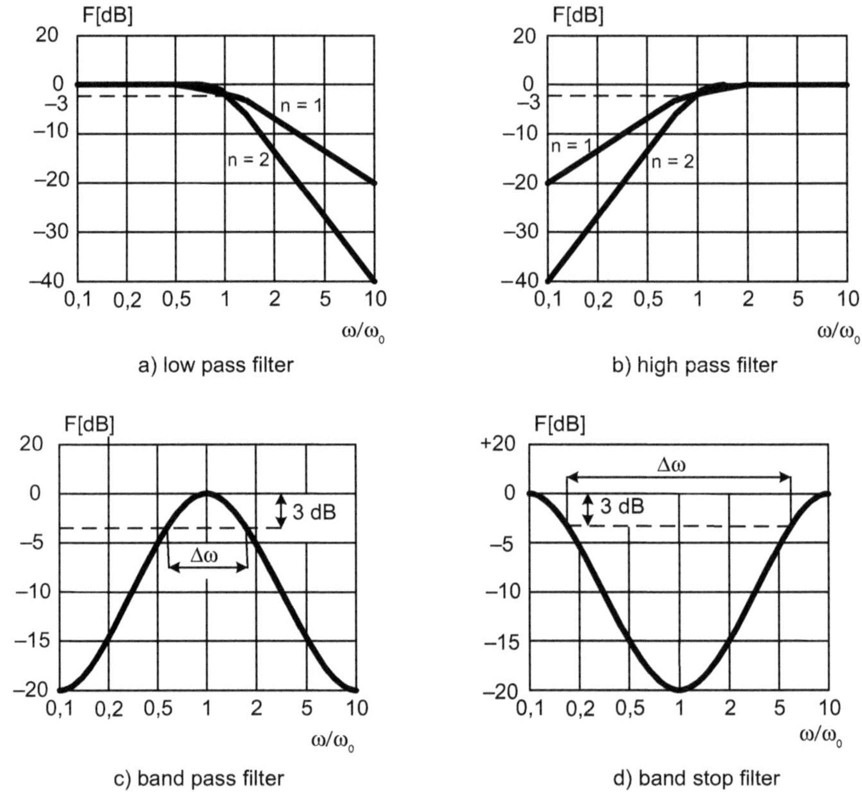

Figure 11.12 Amplitude frequency characteristics of filters.

the transition of the amplitude-frequency characteristic between the filter's permeability and non-permeability bands.

The *passbands* of the filters are determined by their lower and upper cutoff ring frequencies at which the filters' transmissions differ by 3 dB from their reference transmissions. In addition, *bandpass filters* define the *bandwidth* $\Delta\omega$, given by the difference of the *cutoff circular frequencies*, and the *mean circular frequency*, given by the square mean of these cutoff circular frequencies

$$\Delta\omega = \omega_h - \omega_d, \qquad \omega_o = \sqrt{\omega_d \cdot \omega_h}. \qquad (11.63\ \text{a, b})$$

The *quality factor* of bandpass filters is then defined for these filters by the reciprocal of their *normalized bandwidth*

$$Q = \frac{1}{B} = \frac{\omega_0}{\omega_h - \omega_d}. \qquad (11.64)$$

11.4.2 Approximation of frequency characteristics of filters

Bessel, Butterworth, and *Chebyshev polynomials* are most often used to approximate the frequency characteristics of filters.

Lowpass filters with *n*th *Bessel approximations* express the frequency characteristics of their transmissions by the equations

$$F_{\mathrm{LP1}}(\omega) = \frac{1}{1 + j \cdot \dfrac{\omega}{\omega_d}}, \qquad F_{\mathrm{LP2}}(\omega) = \frac{1}{1 + j \cdot \dfrac{\omega}{\omega_0} - \dfrac{1}{3}\left(\dfrac{\omega}{\omega_0}\right)^2}, \tag{11.65 a, b}$$

$$F_{\mathrm{LP3}}(\omega) = \frac{1}{1 + j \cdot \dfrac{\omega}{\omega_0} - \dfrac{2}{5} \cdot \left(\dfrac{\omega}{\omega_0}\right)^2 - j \cdot \dfrac{1}{15} \cdot \left(\dfrac{\omega}{\omega_0}\right)^3}, \tag{11.66}$$

$$F_{\mathrm{LP4}}(\omega) = \frac{1}{1 + j \cdot \dfrac{\omega}{\omega_0} - \dfrac{3}{7} \cdot \left(\dfrac{\omega}{\omega_0}\right)^2 - j \cdot \dfrac{2}{21} \cdot \left(\dfrac{\omega}{\omega_0}\right)^3 + \dfrac{1}{105} \cdot \left(\dfrac{\omega}{\omega_0}\right)^4}. \tag{11.67}$$

The frequency dependence of the phase shift of the output signal relative to its input signal is approximately *directly proportional* to the order of the filter *n* and the frequency

$$\varphi_n(\omega) \approx \operatorname{arctg}\left(n \cdot \frac{\omega}{\omega_0}\right) \approx n \frac{\omega}{\omega_0}. \tag{11.68}$$

It follows that the *group delay* of the filter, defined by the ratio of the change in phase to the change in angular frequency, is approximately directly proportional to the order of the filter for this type of filter

$$T_{\mathrm{LP}n}(\omega) = \frac{d\varphi}{d\omega} = -\frac{n}{\omega_0}. \tag{11.69}$$

Lowpass filters with the *Butterworth n*th approximation express the frequency characteristics of their transmissions by the equation

$$F_{\mathrm{LP}n}(\omega) = \frac{1}{\sqrt{1 + \left(\dfrac{\omega}{\omega_0}\right)^{2n}}}. \tag{11.70}$$

The drop in the amplitude-frequency characteristic is steeper compared to *Bessel filters*, and their transition characteristic has a slight overshoot, increasing with the order of the filter.

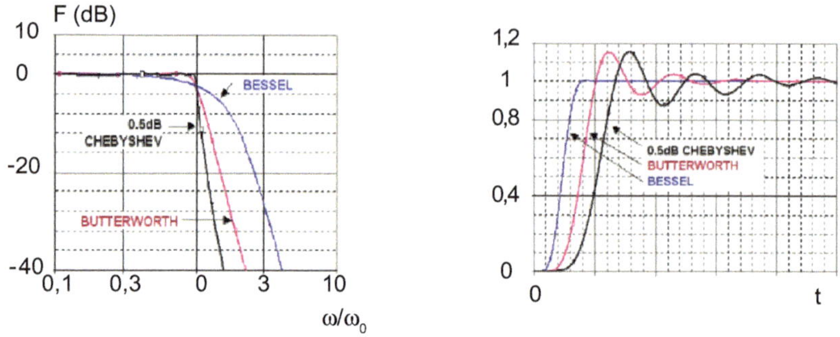

a) amplitude frequency characteristics b) transient characteristics

Figure 11.13 Frequency and transition characteristics of low-pass filters of the 2nd order.

Lowpass Chebyshev filters have the steepest drop-off of the amplitude frequency characteristic of all filters, which, however, is wavy even at frequencies lower than their cutoff frequency. Their frequency transmissions are approximated by

$$F_{\mathrm{LP}n}(\omega) = \frac{A_0}{\sqrt{1 + \varepsilon^n \cdot C_n^2 \frac{\omega}{\cdot \omega_0}}}, \tag{11.71}$$

where c_n are *Chebyshev polynomials* of order n and ε determines the proportional *ripple* of their amplitude frequency characteristics in the passband

$$C_n = \cos\left[n \cdot arg\left(\cosh \frac{\omega}{\omega_0}\right)\right]. \tag{11.72}$$

The *ripple amplitude* of the amplitude frequency characteristic of the filters is for $C_n = 1$

$$A_1 = \frac{1}{\sqrt{1 + \varepsilon^2}}. \tag{11.73}$$

The number of pulsations of the amplitude frequency characteristic is determined by the order of the filter n.

Figure 11.13 shows the amplitude-frequency characteristics and the corresponding transition characteristics of second-order lowpass filters with *Bessel, Butterworth*, and *Chebyshev approximations* with 0.5 dB ripple in the passband.

11.4.3 Relationship between analog and digital filters

The transmission of an *analog filter* is defined in the frequency domain by its *transfer function*

$$\boldsymbol{H}(\omega) = \frac{\boldsymbol{Y}(\omega)}{\boldsymbol{X}(\omega)}, \qquad (11.74)$$

where $\boldsymbol{X}(\omega)$ and $\boldsymbol{Y}(\omega)$ are the frequency spectra of the input and output signal of the filter.

The time course of the filter output voltage is given by the product of its *transition characteristic* $h(t)$ with the time course of its input signal $x(t)$

$$y(t) = h(t) \cdot x(t). \qquad (11.75)$$

The transfer of a digital filter is defined by its *transfer function* based on the *Z transform*

$$\boldsymbol{H}(z) = \frac{\boldsymbol{Y}(z)}{\boldsymbol{X}(z)}, \qquad (11.76)$$

where $\boldsymbol{X}(z)$ and $\boldsymbol{Y}(z)$ are the frequency spectra of the input and output signal of the filter.

The time sequence of samples of the output signal is determined by the product of the transition characteristic of the filter $h(n)$ with the time sequence of samples of its input signal

$$y(n) = h(n) \cdot x(n) = \sum_{k=0}^{N} b_k \cdot x(n-k), \quad n = 1,2,\ldots. \qquad (11.77)$$

If, for example, the transfer function $\boldsymbol{H}(z)$ is expressed by a polynomial

$$h(z) = b_0 + b_1 \cdot z^{-1} + b_2 \cdot z^{-2} + \ldots + b_N \cdot z^{-N}, \qquad (11.78)$$

then the time course of the output signal of the filter is determined by the sequence of samples

$$y(n) = \sum_{k=0}^{N} b_0 \cdot x(n) + b_1 \cdot x(n-1) + \ldots + b_N \cdot x(n-N), \qquad (11.79)$$

where $x(n)$ to $x(n-N)$ are samples of the filter input signal.

The transfer function of each digital filter can be expressed as a quotient of polynomials

$$\boldsymbol{H}(z) = \frac{B(z)}{A(z)} = \frac{b_0 + b_1 \cdot z^{-1} + b_2 \cdot z^{-2} + \ldots + b_N \cdot z^{-N}}{1 - \left(a_1 \cdot z^{-1} + a_2 \cdot z^{-2} + \ldots + a_M \cdot z^{-M}\right)}. \qquad (11.80)$$

If it is true that for a polynomial $A(z) = 1$, ($a_i = 0$), then this transfer function expresses the transfer of a filter with a FIR (*finite impulse response*), which is always stable, because the poles of its transfer lie always in the left half-plane

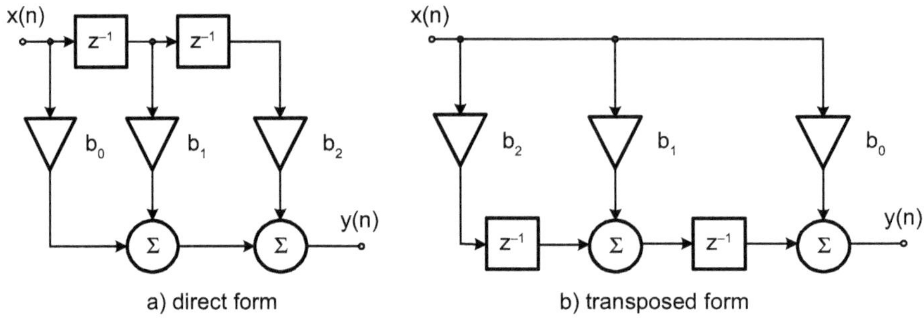

Figure 11.14 Basic structures of FIR filters.

of the complex Gaussian surface. Their disadvantage is that in order to achieve greater steepnesses between their permeable and impervious bands, it is necessary to choose a higher order of the polynomial B. As a result, this lengthens the response time of the filter to the time change of the input signal.

If it is true that for a polynomial $A(z) \neq 1$, $(ai \neq 0)$, then this transfer function expresses the transfer of a filter with an IIR (*infinite impulse response*), which generally does not have to be stable, because the poles of its transfer may not always lie in the left half-plane of the complex Gaussian surface. The advantage of these filters is that the same steepness of their frequency characteristics can be achieved with a lower order of polynomials and thus shorter settling times of their output signals.

11.4.4 Digital filter structures

The basic structures of FIR *digital filters* include their direct and transposed forms.

The *direct form* of the FIR filter is formed by connecting delay elements with transmission z^{-1}, weight elements with weights b_n and adders Σ (Figure 11.14(a)). The *transposed structure*, given by the dual connection of the direct form, has its inputs and outputs interchanged, and its connection nodes are interchanged with adders (Figure 11.14(b)).

It applies to the transfer function and the output signal of both structures

$$\boldsymbol{H}(z) = b_0 + b_1 \cdot z^{-1} + b_2 \cdot z^{-2}, \tag{11.81}$$

$$y(n) = b_0 \cdot x(n) + b_1 \cdot x(n-1) + b_2 \cdot x(n-2). \tag{11.82}$$

By *serially arranging* FIR *filters* according to Figure 11.15, transmissions of higher-order filters can be carried out, since the resulting transmission of these filters is given by the product of transmissions of individual basic filter structures

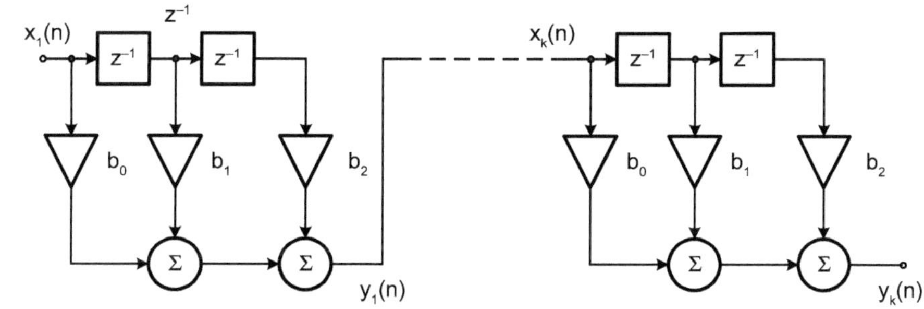

Figure 11.15 Series ordering of FIR filters.

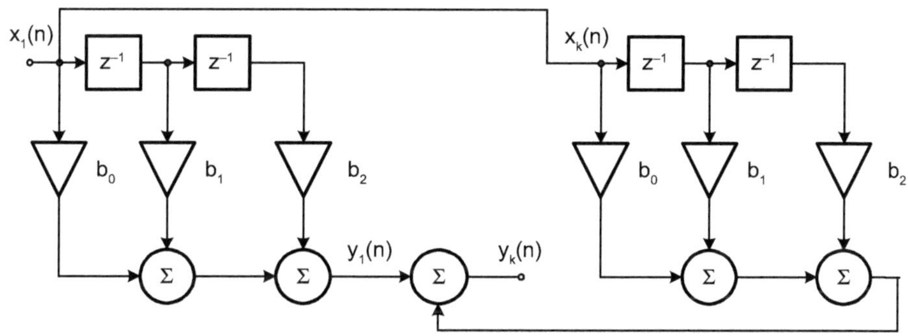

Figure 11.16 Parallel ordering of FIR filters.

$$\boldsymbol{H}_s(z) = \boldsymbol{H}_0(z) \cdot \boldsymbol{H}_1(z) \ldots \boldsymbol{H}_k(z), \qquad k = 1,2,\ldots,k. \tag{11.82}$$

By arranging FIR filters in parallel according to Figure 11.16, transmissions of higher-order filters can also be carried out, since the resulting transmission of these filters is given by the sum of the transmissions of the individual basic filter structures

$$\boldsymbol{H}_p(z) = \boldsymbol{H}_0(z) + \boldsymbol{H}_1(z) \ldots + \boldsymbol{H}_k(z), \qquad k = 1,2,\ldots,k. \tag{11.83}$$

The basic structures of IIR digital filters are shown in Figure 11.17. Again, as with digital FIR filters, we also distinguish with IIR filters their direct and transposed form of connection.

It applies to the transfer function of both structures

$$\boldsymbol{H}(z) = \frac{b_0 + b_1 \cdot z^{-1} + b_2 \cdot z^{-2}}{1 - a_1 \cdot z^{-1} + a_2 \cdot z^{-2}}. \tag{11.84}$$

Similar to FIR filters, higher-order filters can also be realized with the basic structures of IIR filters.

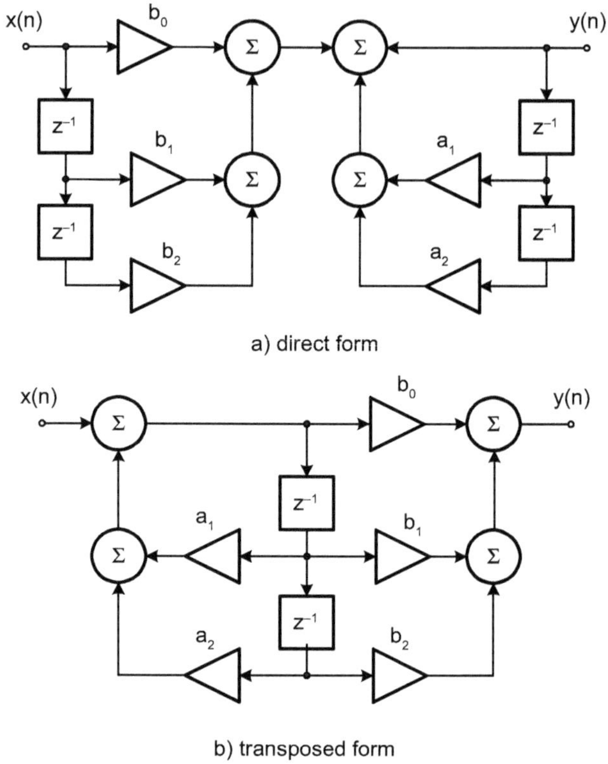

Figure 11.17 Basic structures of IIR filters.

Compared to finite impulse response FIR filters, IIR filters are always simpler, have a faster response, and also require less computing power. However, their disadvantages are generally *lower frequency stability* and non-linear frequency dependence of their phase delay.

11.4.5 Programs for the design of digital filters

To apply digital filtering, programs are used in measurement systems that enable the generation of transfer functions of digital filters with specified types of frequency characteristic approximations. Typical representatives of these programs are, e.g., *MATLAB*, *LabWindows*, and *LabVIEW*. A special program intended for the design and use of digital filters for signal processing is, for example, the *Scope FIR program*, which enables the design and implementation of FIR and IIR filters up to the 64th order with various types of frequency characteristic approximations.

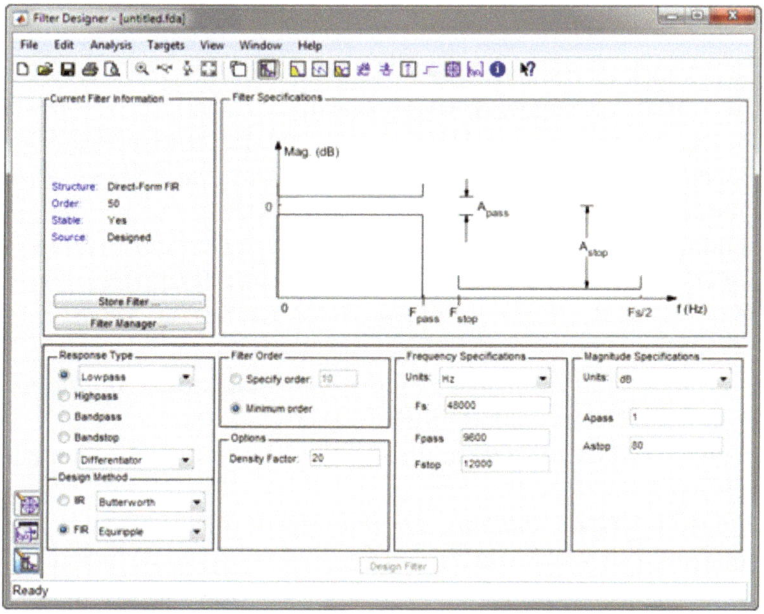

Figure 11.18 Default MATLAB filter design panel.

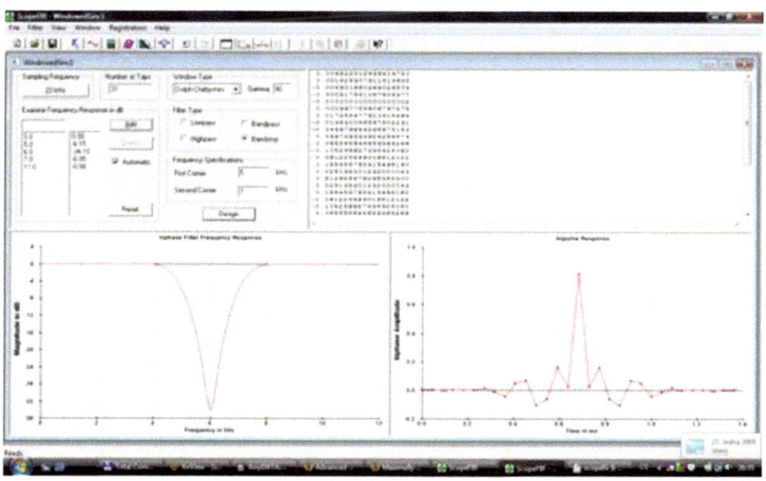

Figure 11.19 Amplitude and transition characteristic of band-stop.

Figure 11.18 shows the default panel in the *MATLAB program*, intended for the design of digital filters.

Figure 11.19 shows the amplitude and transient characteristics of a band-stop with a center frequency of 6 kHz and a bandwidth of 2 kHz in the *Scope FIR program.*

References

[1] Ewiadomski, S. *Filter handbook.* Heinemann, 1989. ISBN 0-434-91378-2.

[2] Crecraft, D.A., Gergely, S. *Analog Electronics: Circuits, Systems and Signal Processing.* Butterworth-Heinemann, 2002. ISBN 13-978-0750-650-953.

[3] Oppenheim, A.V., Schafer, R.W. *Discrete Time Signal Processing.* Prentice Hall. 1989. ISBN 978-0131-9884-22.

[4] Schrufer, E. *Signalverarbeitung. Numerische Verarbeitung digitaler Signale.* Carl Hauser Verlag. Munchen - Wien. 1990. ISBN 3-446-15944-4.

[5] Mitra, S. K. A., Kaiser, J.F. *Handbook for Digital Signal Processing.* John Wiley & Sons, 1993. ISBN 0-471-61995.

[6] Dorf, R. C. *The Electrical Engineering Handbook – Analog Frequency Filters.* CRC Press, 2000. ISBN: 0-8493-2177-8.

[7] Sophocles, J.O. *Introduction to Signal processing.* Prentice Hall 2009. ISBN 0-13-209172-0.

[8] Lyons, R. *Understanding Digital Signal Processing.* E text book, Amazon. ISBN 978-0137027415.

[9] Proakis, J. K., Manolakis, D. G. *Digital Signal Processing. Principles, Algorithms, and Applications.* Prentice Hall 1996. ISBN 0-13-3TM33-9.

[10] Vedral, J., Svatoš, K. *Analog Signal Processing and Digitization in Measurement.* Czech Technical University in Prague, ISBN 978-80-0-06717-8.

List of Symbols

Abbreviation	Importance
AC	Alternating Current
ADC	Analog Digital Convertor
AG	Analog Ground
AH	Acceptor Handshake
AM	Amplitude Modulation
APK	Amplitude Phase Keying
ARG	Automatic Repeat Request
ASK	Amplitude Shift Keying
AUI	Attachment Unit Interface
AXIe	Advanced Extension for Instrumentation
b/s	bit per second
B/s	Byte per second
BG	Band Gap
BiFET	Bipolar Field Effect Technology
BIOS	Basic Input-Output
BISDN	Broadband-Integrated Services Digital Network
BP	Band Pass Filter
BPSK	Binary Phase Shift Modulation
BR	Bit Rate
BS	Band Stop Filter
BSF	Band Stop Filter
BSMA	Binary Step Moving Average
CAMP	Cyclic Adenosin Monophosphate
CAU	Central Acquisition Unit
CCD	Charge Capacitor Devices
CDC	Cellular Data Card
CF	Crest Factor
CLK	Clock
CLL	Current Locked Loop
CMRR	Common Mode Rejection Ratio

CPU	Central Processing Units
CRC	Cyclic Redundancy Check
CSD	Compression Switched Data
CSS	Chirp Spread Spectrum
CSSM	Chirp Spread Spectrum Modulation
CTE	Charge Transport Effectivity
DAC	Digital Analog Convertor
DAQ	Data Acquisition Systems
DAV	Data Valid
DC	Direct Current
DCE	Data Communication Equipment
DDRM	Double Data Rate Memory
DDS	Direct Digital Synthesis
DFT	Discrete Fourier Transformation
DG	Digital Ground
DMA	Direct Memory Access
DPM	Double Pole Multiplexing
DR	Dynamic Range
DS ADC	Delta Sigma ADC
DSM	Delta Sigma Modulation
DT	Duplex Transfer
DTE	Data Terminal Equipment
DVD	Digital Versatile Disc
EBR	Enable Byte Register
ECL	Emitter Coupled Logic
ECSD	Enhanced Circuit Switched Data
ECU	Electronic Control Unit
EER	Enable Event Register
EMA	Exponential Moving Average
ENBW	Equivalent Noise Bandwidth
ENOB	Effective Number of Bits
EOU	End of Identity
ER	Effective Resolution
EXOR	Exclusive Or
FBW	Frequency Bandwidth
FF	Form Factor
FG	Flux Gate
FHSS	Frequency Hopping Spread Spectrum
FIFO	First In, First Out
FIR	Finite Impulse Response

FM	Frequency Modulation
FMC	Full Modem Connection
FSB	Fast System Bus
FSK	Frequency Shift Keying
FT	Fourier Transformation
GET	Group Executive Trigger
GPRS	General Packet Radio Service
GSM	Global System Mobile
GTL	Go to Local
HDD	Hard Disk Drive
HIPRLAN	High Performance Radio Local Area Network
HPF	High-Pass Filter
HPPM	High Pulse Position Modulation
HUB	Host Bus Universal
CHEMFET	Chemically Sensitive Field Effect Transistors
ICP	Integrated Circuit Piezoelectric
IIR	Infinite Impulse Response
IME	International Mobile Equipment Identity
IMRR	Isolation Mode Rejection Ratio
IP	Internet Protocol
IrDA	Infrared Data Connection
IRTP	Isochronous Real-Time Protocol
ISA	Industrial Standard Architecture
JFET	Junction Field Effect Transistor
LAN	Local Area Network
LDF	Length Data Field
LE	Listener Extended
LED	Light Emitting Diode
LFM	Linear Variable Frequency Modulation
LIP	Link Protocol
LLO	Local Lock Out
LMS	Location Monitoring Service
LPF	Low Pass Filter
LPC	Low Pin Count
LPF	Low Pass Filter
LSB	Least Significant Bit
LTE	Long-Term Evolution
LVDS	Low Voltage Differential Signaling
LVDT	Linear Variable Differential Transformator
MAU	Medium Attachment Unit

MBLT	Multi Byte Letter Transfer
MCP	Multi Core Processor
MDAC	Multiplying Digit to Analog Convertor
MEMS	Micro Electro Mechanical Systems
MLA	My Listen Address
MM	Matrix Multiplexers
MMS	Multimedia Message
MRP	Media Redundancy Protocol
MSB	Most Significant Bit
MSE	Minimum Square Error
MSS	Master Summary Bit
MTA	My Talk Address
NDAC	Not Data Accepted
NRFD	Not Ready for Data
NS	Network Slicing
NTC	Negative Temperature Coefficient
OCD	Open Collector Driver
OTRA	Open Tree Ring Area
PA	Priority Algorithm
PCI	Personal Computer Interface
PCM	Pulse Code Modulation
PCS	Personal Communications Service
PD	Phase Detector
PH	Power of Hydrogen
PID	Photo Ionized Detector
PLL	Phase Locked Loop
PRBFM	Pseudo Random Binary Frequency Modulation
PSD	Position Sensor Detector
PSK	Phase Shift Keying
PTC	Positive Temperature Coefficient
PTN	Public Telephone Network
PWM	Pulse Width Modulation
QAM	Quadrature amplitude modulation
QPM	Quad Pole Multiplexing
QPSK	Quadrature Phase-Shift Keying
RAM	Random Access Memory
RC	Regression Coefficient
RL	Remote Local
RMS	Root Mean Square
ROM	Read Only Memory

RRS	Round Robin Select
RTD	Resistive Temperature Detector
RTP	Real-Time Protocol
RTS	Real Time Sampling
RVDT	Rotary Variable Differential Transformator
SAR	Successive Approximation
SAV	Surface Acoustic Wave
SBR	Standard Byte Register
SCP	Single Core Processor
SCS	Satellite Communication System
SCXI	Signal Conditioning eXtension for Instrumentation
SDC	Synchro to Digital Convertor
SDRAM	Synchronous Dynamic Random Access Memory
SER	Standard Event Register
SFM	Step Frequency Modulation
SG	Signal Ground
SH	Sample Hold Unit
SINAD	Signal to Noise and Distortion
SM	Smart Sensors
SMA	Simple Moving Average
SMRR	Series Mode Rejection Ratio
SMS	Short Message System
SN	Scatternet Net
SNHD	Signal Non Harmonic Distortion
SNR	Signal-to-Noise Ratio
SPD	Serial Poll Disable
SPE	Serial Poll Enable
SPI	Serial Peripheral Interface
SPM	Single Pole Multiplexing
SR	Service Request
SRAM	Static Random Access Memory
SSD	Solid State Drive
SSG	Solid State Switch
SSHD	Solid State Hybrid Drive
TA	Terminal Adapter
TCT	Talk Control
TDMA	Time Division Multiple Access
TE	Talker
THD	Total Harmonic Distortion
TO	Time Out

TPA	Twisted Pair
UART	Universal Asynchronous Receiver Transmitter
UMTS	Universal Mobile Telecommunication System
USB	Universal Serial Bus
VME	Versa Modul Europe Bus
VXI	Extension for Instrumentation
WAP	Wireless Application Protocol
WCDMA	Wideband Code Division Multiple Access
WiFi	Wireless Fidelity
WLAN	Wireless Local Area Network

List of Mathematical and Physical Constants

Mathematical Constants

Euler's number	2.718
Ludolf's number	3.1415926535

Physical constants

Avogadro's constant	$6.0221 \cdot 10^{23}$ [mol^{-1}]
Planck's constant	$6.626472 \cdot 10^{-34}$ [J.s]
Boltzmann's constant	$1.38 \cdot 10^{-23}$ [$J \cdot K^{-1}$]
Speed of light	$3 \cdot 108$ [$m \cdot s^{-1}$]
Electron charge	$1.60234 \cdot 10^{-19}$ [C]
Electron mass	$9.1 \cdot 10^{-31}$ [kg]
Earth acceleration	9.81 [$m \cdot s^{-2}$]
Vacuum permittivity	$8.854 \cdot 10^{-12}$ [$F \cdot m^{-1}$]
Vacuum permeability	$1.2563 \cdot 10^{-6}$ [$H \cdot m^{-1}$]

List of Physical Quantities

Quantities	Symbol	Si units	Unit dimensions
Absolute temperature	Θ	Kelvin	K
Acceleration	a	meter per quadrat second	m/s^{-2}
Active power	P	Watt	W
Admittance	Y	Siemens	S
Angle	ϕ	radian	rad
Apparent power	S	Volt ampere	VA
Capacity	C	Farad	F
Concentration	M	mol	Mol/m^3
Conductivity	γ	Siemens per meter	S/m^1
Current density	J	Ampere per quadrat meter	A/m^2
Damping	b	Newton per quadrat meter	N/m^2
Distance	x	meter	m
Electrical current	I	Ampere	A
Electrical induction	D	Coulomb per	C/m^2
Electrical voltage	U	Volt	V
Energy	E	Joule	$J = W \cdot s$
Force	F	Newton	$kg \cdot m/s^2$
Frequency	f	Hertz	Hz
Humidity	Φ	kilogram per cubic meter	kg/m^3
Charge	Q	Coulomb	C
Impedance	Z	Ohm	Ω
Inductance	L	Henry	H
Intensity of magnetic field	H	Ampere per meter	$A/ \cdot m^1$
Intensity of electrical field	E	Volt per meter	V/m^1
Light flux	F	lumen	lm
Luminous intensity	I	candela	$W \cdot m^{-2}$
Magnetic flux	Φ	Weber	Wb
Magnetic induction	B	Tesla	T

Quantities	**Symbol**	**Si units**	**Unit dimensions**
Mass	m	kilogram	kg
modulus of elasticity	G	Newton per quadrat meter	N/m^2
Period		second	s
Permeability	μ	Henry per meter	H/m
Permittivity	ε	Farad per meter	F/m
Power factor	$\cos\phi$	-	-
Pressure	p	Pascal	Pa
Radiation area activity	I	Berquell per quadrat meter	Bq/m^2
Reactive power	Q	Volt ampere reactive	VAR
Resistance	R	Ohm	Ω
Resistivity	ρ	Ohm meter	W · m
Rigidity	k	Newton per meter	N/m
Square	S	quadrat meter	m^2
Temperature in Celsius		Celsius	^oC
Time	t	second	s
Torque moment	M	Newton meter	N · m
Velocity	v	meter per second	m/s^{-1}
Viscosity	ν	quadrat meter per second	$m^{2/}s$
Volume	V	cubic meter	m^3

List of Names of International Institutions

Abbreviation	Name of the institution
NPSTC	National Public Safety Telecommunications Council
ETSI	European Telecommunications Standards Institute
OFCOM	Independent Regulator and Competition Authority for the UK Communications Industries
EIA	Electronic Industries Association
PTB	Physikalisch-Technische Bundesanstalt
ISO	International Standard Organisation

Word Special Protocol Commands

The name of the command	Meaning of the command
Abort Normal Operation	The device immediately stops all activities and goes into a rest state. It is ready to accept commands. The command can be used during system configuration as well as during normal operation.
Assign Handler Line	Assigns the handling of a given IRQ line to the appropriate interrupt manager. The command can only be used during system configuration.
Assign Interrupter Line	Commands the given IEQ line to the appropriate interrupt source. The command can only be used during system configuration.
Asynchronous Mode control	Determines how information about asynchronous events and changes is sent to the response register. The command can be used during system configuration as well as during normal operation.
Begin Normal Operation	Returns the device to normal state. The command can be used during system configuration as well as during its normal operation.
Byte Available	The command sends 8-bit data to the *servant* through the *master*. The command can only be used during normal system operation.
Byte Request	The command receives 8-bit data from the *servant* through the master. The command can only be used during normal system operation.
Clear	The command resets the VXI interface of the *servant*. The command can be used during system configuration as well as during normal operation.
Clear Lock	The command ends the locked state of the *Servant*, i.e., the *servant* can only respond to the commands of its *master*. The command can only be used during normal system operation.

The name of the command	Meaning of the command
Control Event	The command enables or disables the generation of specific events by the Servant. The command can be used during system configuration as well as during normal operation.
Control Response	The command enables or disables the generation of responses to changes to the specified bits of the response register. The command can be used during system configuration as well as during its operation.
End Normal Operation	The command is used to properly terminate the normal activity of the *manager* and all his *servants*. The command can be used during system configuration as well as during normal operation.
Grant Device	The command commands the *servant* to the *manager*. The command can only be used during system configuration
Identify Commander	By command, the *manager* gives his address to the *servant*. The command can only be used during system configuration.
Read Handler Line	The IRQ *Master* command specifies the line whose service is assigned to the specified interrupt *manager*. The command can be used during system configuration as well as during normal operation.
Read Handlers	The command finds the number of Interrupt *managers* in the specified *slave*. The command can be used during system configuration as well as during normal operation.
Read Interrupt Line	The command specifies the IRQ line that is assigned to the specified interrupt source. The command can be used during system configuration as well as during its normal operation.
Read Interrupters	The command finds the number of *interrupt sources* in the specified *servant*. The command can be used during system configuration as well as during normal operation.
Read MODID	The command detects the status of the MODID lines controlled by the device in slot 0. The command can be used during system configuration as well as during its normal operation.

The name of the command	Meaning of the command
Read Protocol	The command finds out which protocols, in addition to the *word serial protocol*, are supported by the *servant*. The command can be used during system configuration as well as during normal operation.
Read Protocol Error	The command instructs the *servant* to pass the error code. The command can be used during system configuration as well as during normal operation.
Read Servant Area	With this command, the *resource manager* reads the size of the fields of the *servants* of each lord. The command can only be used during normal system operation.
Release Device	Upon receiving the command, the *master* will remove the specified device from the list of his *servants*. The command can only be used when configuring the system.
Set Lock	The command puts the *servant* in a locked state. The command can only be used during normal system operation.
Set Lower MODID	The command controls the state of the MODID0 to MODID6 lines. The command can be used during system configuration as well as during normal operation.
Set Upper MODID	The command controls the state of the MODID7–MODID12 lines. The command can be used during system configuration as well as during normal operation.
Trigger	The command triggers a prearranged *servant* activity. The command can only be used during normal system operation.
User Defined	User-defined command. All commands defined by this VXI standard have the most significant bit at the log level. 1. User defined commands have this bit at the log level. 0. The command can only be used during normal system operation.

Note: The VXI standard does not define *Long Word Serial and Extended Longword Serial* commands.

Advantech Plug-in Cards

Module functions	Designation	Module parameters
Multiplex cards	PCI-1711	Multiplexed 2 × 16 channel isolated PCI card, 12 bit ±625 V to ±10 V, 1 GΩ, 100 kS/s, 4 KB FIFO
	PCI-1713	Multiplexed 2 × 16 channel isolated PCI card, 12 bit ±625 mV to ±10 V, 1 GΩ, 100 kS/s, 4 KB FIFO
	PCI-1715	Multiplexed 2 × 16 channel isolated PCI card, 12 bit ±625 mV to ±10 ±0.625 V to ±10 ±0.625 V to ±10 V, 1 GΩ, 0.5 MS/s, FIFO 1 KB
	PCI-1718	Multiplexed 2 × 8-channel isolated PCI card, 12 bit ±625 mV to ±10 V, 1 GΩ, 100 kS/s, 4 KB FIFO
	PCI-1720	Multiplexed 2 × 8 channel isolated PCI card, 12 bit ±625 mV to ±10 V, 100MGΩ, 100 kS/s, FIFO 1 kB, 16 TTL outputs, 1× counter/timer 10 MHz
	PCI-1723	Multiplexed 2 × 8-channel isolated PCI card, output 16 bit ±10 V, 4–20 mA, 20 mA, 16 TTL outputs
		Multiplexed 2 × 8-channel isolated PCI card, 14 bit ±10 V, 20 mA, 16 TTL inputs, 16 TTL outputs
	PCI-1727	Multiplexed 2 × 8-channel isolated PCI card, 14 bit ±10 V, 20 mA, 16 TTL inputs, 16 TTL outputs
Digital cards	PCI-1730	Digital universal isolated digital PCI card, 16 TTL inputs, 16 TTL outputs
	PCI-1733	Digital isolated digital PCI card, 32 TTL inputs
	PCI-1750	Digital universal isolated digital PCIE card, 16 TTL inputs, 16 TTL outputs, 6 bit, 10 MHz counter/timer
	PCI-1751	Digital universal PCI card, 48 TTL IO/O inputs, 3 × 16 bit counters/timers, 10 MHz
	PCI-1753	Digital universal PCI card, 96 I/O TTL inputs, TTL outputs, 3 × 16 bit counters/timers, 10 MHz
	PCI-1756	Digital isolated PCIE card, 8 TTL IN channels, 2 × 16 bit counter/timer, 10 MHz
	PCIE-1760	Digital universal isolated digital PCIE card, 8 iz. of TTL inputs, 2 × 16 bit counter/timer, 10 MHz, 8 relays 125 V/0.5 A (AC), 30 V/0.2 A (DC)
	PCI-1761	PCI digital card, 8 TTL inputs, 8 relays 250 V/2 A (AC), 30 V/2 A (DC)
	PCI-1762	PCI digital card, 16 TTL inputs, 16 relays 250 V/2 A (AC), 30 V/2 A (DC)

CAMAC Modules from Wiener Power Electronic

Module function	Designation	Module parameters
Controller	CC-USB	Programmable controller, 24 bit, 3 (12) Mb/s, LAN
PGAmodul	CMDGG8	8 I/O FPGA XC35400, FIFO 32 kB,
Connection module	CFB BRIDGE	Interconnection module between Le Croy, Ortec, Silena and CAMAC FERA systems, FiFo 28 kB 1 MS/s
Charge ADC	CQ16	16-channel, 12-bit charge A/D converter, 2 nC – 250 pC, 5.5 μs, DNL ≤ 1%, FIFO 16 (32) Mb/s
Power supply	4 MUC	Power supply ±6 V, ±12 V, ±24 V
Frames	CAMAC300	25 slot frame CAMAC, ±6 V, ±12 V, ±24 V, 300 W
	CAMAC600	2× 25 slots frame CAMAC, ±6 V, ±12 V, ±24 V, 600 W

VME Modules from Wiener Power Electronic

Module function	Designation	Module parameters
Controller	VME-USB	FPGA controller Spartan 3, 4 kB ROM, 28 kB RAM, USB. 2
PGAmodul	XLM727	72 channel I/O FPGA module, 64 BB/s,
Diagnostic modul	VM-DBA	Bus diagnostic module, 2 kB, LED, USB. 2
Multifunctional FPGA modules	MDGG-8	Multifunctional FPGA module, 8 LVD (ECL I/O, 36 kB RAM
	MDGG-16	Multifunctional FPGA module, 16 LVD (ECL I/O, 36 kB RAM
Power supply modules	VML-64	Power supply ±5 V, ±12 V
	VML-64	Power supply ±3.3, ±5 V, ±12 V
	VML-5V	Power supply ±5 V
	VML-V430	Power supply +2 V, +5 V, +5 V, ±12 V

VME Modules from Mestyc

Module function	Designation	Module parameters
Charge amplifiers	MPR-1	Simple charge amplifier
	MDS-8	8 channel charge amplifier
	MPR-16	16 channel charge amplifier
	MPR-32	32 channel charge amplifier
	MPR-64	64 channel charge amplifier
Multi-channel shapers, filters and discriminators	MCF-16-FV	16-channel fast spectroscopic amplifier
	MCF -16-LN	Low noise amplifier for germanium radiation detectors
	MCF -16-FC	Fast current spectroscopic amplifier
	MCFD-16	16-channel discriminator with result processing
Multichannel preamplifiers	MUX-16	16-channel preamplifier with variable, leading edge discriminators
	MUX-32	32-channel preamplifier with variable, leading edge discriminators
	NTM-16	16-channel preamplifier with variable leading edge discriminators for scintillation radiation detectors
Digitizers and impulse processors	MADC-32	32-channel 12-bit digitizer, 1.2 GS/s, 8 KB
	MQDC-312	32-channel 12-bit digitizer, 4 GS/s, 4 KB
	MTDC-32	32-channel 12-bit digitizer, 1.6 GS/s
	MDPP-16	16-channel low-noise pulse processor
	MDPP-32	32-channel low-noise pulse processor
Neutron detectors	MPD-4	4-channel semiconductor radiation detector amplifier
	MHV-4	4-channel power supply for semiconductor detectors
Power supplies	MVNV-4	Unit of power supplies of preamplifiers
Controller VME	MRC-1	Remote controller for the MESYTEC control bus.
Frame VME	MHL-32	32-channel cable converter 31 pin - LEMO 00

VXI Modules from National Instruments

Designation	Module function	Module parameters
VXI-1000	System framework	Designed for 8 positions
VXI-1001	System framework	Designed for 12 positions
VXI-1100	Amplifier for thermocouples	3 channel amplifier, 50 mV to 10 V, frequency limiting 4 Hz
VXI-1102	Amplifier for thermocouples and RTD sensors	32-channel amplifier, 50 mV to 5 V, current range 0 to 20 mA, bandwidth 2 Hz to 333 kHz
VXI-1102B	Differential voltage amplifier	32 differential channels, 0.1 V to 10 V, gain 1 - 100
VXI-1102C	Amplifier for thermocouples and RTD sensors	32 differential channels, 0.1 V to 10 V, gain to 100, 0 to 20 mA
VXI-1104	Voltage amplifier	32 channels, 30 V AC, frequency range 2 Hz
VX-1112	Amplifier for thermocouples	8-channel amplifier, with LP filter
VX-1120	Voltage amplifier	8-channel amplifier for thermocouples, RTDs and strain gauges, gain 1 to 2000
VXI-1124	Voltage amplifier	6-channel voltage amplifier with 12-bit ADC sample rate 555 S/s for one channel and 125 S/s for all channels
VXI-1125	Voltage amplifier	8-channel isolated amplifier for thermocouples with internal EEPROM memory for storing calibration constants

VXI-1126	Frequency - voltage converter	8-channel isolated frequency to voltage converter with adjustable frequency ranges 250 Hz to 128 kHz, trigger levels, hysteresis and filtering
VXI-1127	Multiplex (matrix) switch	Relay multiplexer (1 × 32, 2 × 16, 4 × 8), matrix switch (1 × 8 × 8, 2 × 4 × 8) for switching up to 250 V/0.2 A and 30 V/1 A
VXI-1128	Multiplex (matrix) switch	Configurable relay multiplexer (1 × 32, 2 × 16, 4 × 8), matrix switch (1 × 8 × 8, 2 × 4 × 8) for switching up to 250 V/0.03 A, RON < 1.2 kΩ
VXI-1129	Matrix switch	Matrix relay thermocouple switch (2 × 4 ×16, 2 × 4 × 32, 2 × 8 ×16), closed contact voltage < 1 mV
VCXI-1140	Differential sampling amplifier	Low-noise 8-channel differential sampling signal amplifier with a time variation of 15 V/μs and a spectral voltage density of 25 nV/√Hz
VXI-1141	Switched filter	8-channel programmable switched LP filter with a sample rate of up to 333 kS/s and a drop of 135 dB/octave
VX-1540	LVDT amplifier	8-channel signal amplifier of LVDT sensors with sampling frequency up to 333 kS/s with internal EEPROM memory of calibration constants
VXI-1531	Sound and vibration amplifier	Two-channel audio and vibration signal amplifier with a sampling rate of 50 kHz and adjustable channel baud rates of 2.5 kHz, 5 kHz and 5 kHz
VXI-1600	Multiplex data collection system	256-channel multiplexed data acquisition system with 16-bit resolution and sampling rate up to 200 kS/s

PXI Modules from National Instruments

Module function	Designation	Module parameters
Frames	PXIe-1073	4 slots, 3 Gb/s, 38 W
	PXIe-1084	18 slots, 4 Gb/s, 58 W (with 2.4 kW cooling)
	PXIe-1085	18 slots, 24 Gb/s, 38 W (with 1.6 kW cooling)
	PXIe-1095	18 slots, 24 Gb/s, 82 W (with 2.4 kW cooling)
Controllers	PXIe 884x	Intel Core I5, 4 Gb/s, HD 512 GB, W10, LabView
	PXIe 886x	Xeon 4 Core, 16 Gb/s, HD 512 GB, W10, LabView, Linux
	PXIe 888x	Xeon 8 Core, 24 Gb/s, HD 512 GB, W10, LabView, Linux
PC connection modules	PXIe-8360	1 channel MXI-Express x1, 250 MS/s
	PXie-8381	1 channel MXI-Express Gen 2 × 8, 4 GS/s
	PXie-8398	4 channels MXI-Express Gen 3 × 16, 16 GS/s
	PXie-8399	8 channels MXI-Express Gen 3 × 16, 16 GS/s
	PXI-4357	4-terminal 100BASE-T1 Ethernet connection
Digital multimeters	PXI-4065	6,5 decade multimeter, ±300 V, ±3 A, 3 kS/s
	PXI-4070	6,5 decade multimeter, ±300 V, ±1 A, 1,8 MS/s
	PXI-4071	7,5 decade. multimeter, ±300 V, ±3 A, 1,8 MS/
	PXI-4072	6,5 decade. multimeter, ±300 V, ±1 A, 1,8 MS/s
Impedance meter	PXI-4190	Impedance meter Z, Y, R, C, L, 40 Hz – 2 MHz
Multifunction modules	PXI-6396	8 diff. channel, 18 bit, ADC 14 MS/s, 2 × 18 bit DAC
	PXI-6229	8 diff. channels, 16 bit ADC, 4 × 16 bit DAC
	PXI-6365	16 diff. channel 16 bit ADC, 1,2 MS/s, 2 × 16 bit DAC
	PXI-6396	8 diff. channel 18 bit, ADC 14 MS/s, 2 × 18 bit DAC, 3.3 MS/s
	PXI-4309	32 channel 24 bit ADC, ±0,1 V to ±10 V, 2 MS/s

Multifunctional FPGA module	PXI-7856	8 I/O analog channels, 48 I/O digital channels, 1 MS/s
	PXI-7862	8 I/O analog. channels, 32 I/O digital channels, 1 MS/s
	PXI-7865	8 I/O analog. channels, 32 I/O digital channels, 1 MS/s
	PXI-7868	6I/O analog channels, 48 I/O digital channels 1 MS/s
Scope modules	PXI-5140	2 (4) channels, 10 bit ADC, 100 MHz, FIFO 1 GS/s, 2 GB,
	PXI-5160	2 (4) channels 10 bit ADC, 500 MHz, 2,5 GS/s, 2 GB
	PXI-5162	2 (4) channels, 10 bit, 1,5 GHz, 5 GS/s, FIFO 2 GB
	PXI-5185	2 channel, 8 bit ADC, 3 GHz, 12,5 GS/s, FIFO 1 GB
	PXI-5186	2 channel, 8 bit ADC, 5 GHz, 12,5 GS/s, FIFO 1 GB
Counters, timers	PXI-6601	4 channels, 32 bit, Up/Down, 80 MHz
	PXI-6608	8 channel, 32 bit, Up/Down, 80 MHz
Function generators	PXI-5433	1 (2) channel, 80 MHz, 320 MS/s, 16 bit, 512 MB, ±12 V
	PXI-5441	1 (2) channel, 40 MHz, 100 MS/s, 16 bit, 512 MB, ±12 V
	PXI-5450	1 (2) channel, 145 MHz, 400 MS/s, 16 bit, 512 MB, ±1 V
	PXI-5451	1 (2) channel, 145 MHz, 400 MS/s, 16 bit, 2 GB, ±1 V
Spectrum analyzers	PXI-5650	500 kHz – 1,3 GHz, 85 dB, retuning time 1,5 ms
	PXI-5651	500 kHz – 3,3 GHz, 82 dB, retuning time 1,5 ms
	PXI-5652	500 kHz – 6,6 GHz, 80 dB, retuning time 1,5 ms
	PXI-5654	500 kHz – 10 (20,40) GHz, 76 dB, retuning time 1,5 ms
Tensometric modules sensors	PXI-4330	8 channel, 25,6 kS/s, ±25 mV – ±100 mV, 120 Ω, 1 kΩ
	PXI-4331	8 channel, 102 kS/s, ±25 mV – ±100 mV, 120 Ω, 1 kΩ
	PXI-4339	8 channel, 25,6 kS/s, ±10 mV, ±50 mV, 120 Ω, 350 Ω, 1 kΩ
	PXI-4499	16 channel, 204,8 kS/s, DR = 144 dB
Modules of vibrometric sensors	PXI-4464	4 channel, HP 3,4 Hz, 204 kS/s, DR = 119 dB
	PXI-4468	4 channel, HP 0,8 Hz, 250 kS/s, DR = 121 dB
LVDT modules	PXI-4340	4 channel LVDT, 400 Hz – 10 kHz, 24 bit ADC, 25,6 kS/s
Temperature sensor modules	PXI-4553	32 channel isolated thermocouple amplifier J, K, E, T
	PXI-4537	20 channel. amplifier Pt100, 100 S/s

CompactQ Modules from National Instruments

Module function	Designation	Module parameters
Frames	cDAQ-9178	8 slots, USB 2.0
	cDAQ-9179	14 slots, USB 2.0
	cDAQ-9185	4 slots, Bus Ethernet
	cDAQ-9810	8 slots, Bus Ethernet,
Frame controllers	cDAQ-9133	4 core processor Intel Atom, 1,3 GHz, 16 GB, Windows 7
	cDAQ-9134	4 core processor Intel Atom, 1,3 GHz, 32 GB, Windows 7
	cDAQ-9137	4 core processor Intel Atom, 1,9 GHz, 32 GB, Windows 7
Digital I/O modules	NI-9375	16 I/O channels, 12 V, 24 V, 7 μs
	NI-9425	32 I/O channels, 12 V, 24 V, 48 V, 8 μs
	NI-9477	32 OUT channels 12 V, 24 V, 48 V, 8 μs
Multifunction modules	NI-9218	2 diff. channels, 24 bit ADC, ±100 mV, 512 kS/s
	NI-9219	2 diff. channel, 24 bit ADCt, ±100 mV, 512 kS/s
ADC modules	NI-9205	16 diff. channel, 16 bit ADC, ±0,2 V to ±10 V, 250 kS/s
	NI-9215	16 diff. channel, 16 bit ADC, ±10 V, 100 kS/s
	NI-9216	16 diff. channel, 16 bit ADC, ±10 V, 500 kS/s
	NI-9223	4 diff. channel, 16 bit ADC, ±10 V, 1 MS/s
	NI-9239	4 diff. channel, 24 bit ADC, ±10 V, 50 kS/s
	NI-9251	2 diff. channel, 24 bit ADC, ±3 V, 100 kS/s
	NI-9252	4 diff. channel, 24 bit ADC, ±10 V, 50 kS/s

Tensometric modules sensors	NI-9235	8 diff. channel, 24 bit ADC, ±30 mV, 10 kS/s
	NI-9236	8 diff. channel, 24 bit ADC, ±30 mV, 10 kS/s
	NI-9236	8 diff. channel, 24 bit ADC, ±25 mV, 51 kS/s
Modules of vibrometric sensors	NI-9231	8 diff. channel, 24 bit ADC, ±5 V, 51 kS/s
	NI-9232	8 diff. channel, 24 bit ADC, ±3 V, 102 kS/s
	NI-9234	4 diff. channel, 24 bit ADC, ±5 V, 102 kS/s
	NI-9250	4 diff. channel, 24 bit ADC, ±5 V, 102 kS/s
Temperature sensor modules	NI-9210	4 diff. channel, 24 bit isolated ADC, 14 S/s
	NI-9212	8 diff. channel, 24 bit isolated ADC, 96 S/s
	NI-9213	16 diff. channel, 24 bit isolated ADC, 95 S/s
	NI-9216	8 diff. channel, 24 bit ADC, Pt100, 400 S/s
	NI-9217	4 diff. channel, 24 bit ADC, Pt100. 400 S/s
	NI-9226	8 diff. channel, 24 bit ADC, Pt1000, 400 S/s

External Modules from Advantech

Group	Designation	Module parameters
ADAM-30x DIN bar	ADAM-3011	Thermocouple amplifier B, E, J, R, S, T, output 10 V
	ADAM-3013	Amplifier of RTD temperature sensors, output 5 V, 10 V, 0 to 20 mA,
	ADAM-3014	Voltage amplifier ±10 mV to ±10 V, currents ±20 mA, output 10 V
	ADAM-3016	Tensometric amplifier, ±10 mV to ±100 mV, output 10 V
	ADAM-3017	Tensometric amplifier ±10 mV to ±100 mV, output 10 V
ADAM-36x DIN bar	ADAM-3600	Sensor and wireless connection module terminal
	ADAM-3613	Module of 4 of RTD temperature sensors, OUT 5 V, 10 V, 20 mA
	ADAM-3617	Module of 4 diff. channel, 16 bit, 10 S/s ±2,5 V, ±10 V, 20 mA, 2 kV
	ADAM-3618	Thermocouple amplifier, J, K, R, S, T, ±10(100) mV
	ADAM-3624	Voltage amplifier, 4 channel, 12 bit, 10 S/s, ±10 V, 20 mA, 4 – 20 mA
	ADAM-3651	Digital 8-channel input module, expander to 3650
	ADAM-3656	Digital module, 8 OUT channels, 1 OC, DIN Rail
ADAM-40x Independently Modules RS485	ADAM-4011	I/O module, isolation 3 kV, 1x AI (mV, V, mA), 1× DI, 2× DO, built-in watchdog timer, open wire detection RS-485
	ADAM-4013	I/O module isolated 3 kV, 1× AI (RTD Pt, Ni), built-in watchdog timer, RS-485
	ADAM-4015	I/O modul, 6× AI (RTD Pt, Ni, Balco), RS485 ModBus
	ADAM 4017	I modul, 8× AI (mV, V, mA), ModBus
	ADAM-4018	I/O module isolated 3 kV, 8× AI (mA, TC), independent configuration of all channels, interrupt detection, ModBus
	ADAM- 4021	I/O module isolated 3 kV, 1× AO (V, mA), built-in watchdog timer, communication via RS-485, power supply 10 V to 30 V
	ADAM-4024	I/O module isolated 3 kV, 4× AO (V, mA), independent configuration for all channels, built-in watchdog timer, RS-485,
	ADAM-4050	I/O module, 7× DI, 8× DO, timer, RS-485
	ADAM-4051	16 channel digits. output module, 12 V, 24 V, 48 V, RS-485

	ADAM-4052	I/O module isolated 5 kV, 8× isolated DI (6× fully isolated independent, RS-485, power supply 10 V to 30 V
	ADAM-4053	I/O module, 16× DI, built-in watchdog timer, communication via RS-485 ASCII ADAM protocol, power supply 10 V to 30 V
	ADAM-4055	I/O module, 8× IDI, 8× IDO, RS485 Modbus
	ADAM-4056	I/O module, 12× IDO, prod from output, RS-485 Modbus
	ADAM-4060	I/O module, 4× relays, RS-485 Modbus, power supply 10 V to 30 V
	ADAM-4068	I/O module, 8× relays, RS-485 Modbus, power supply 10 V to 30 V
	ADAM-4080	Module two 32 bit counters-timers, ModBus, RS-485
ADAM-45× Independently Modules RS-485	ADAM-4520	Isol. RS-232 to RS-485 converter, 1.2 kbps to 115.2 kb/s
	ADAM-4550	Radio modem, 2.442 GHz, BW 322 MHz, DSSS modulation, 1 Mb/s, 300 kb/s – 115.2 kb/s, RS-232, RS-485
	ADAM-4656	12 channel digital output module, incl. 5 kV, RS-485
	ADAM-4561	Isolated USB to RS-232/422/485 converter, 3 kV, 115.2 kb/s
	ADAM-4570	2 channel RS-232 to RS-485 converter, 1.2 kbps to 115.2 kb/s
	ADAM-4561	Isolated USB to RS-232/422/485 converter, 3 kV, 115.2 kb/s
	ADAM-4570	2 channel RS-232 to RS-485 converter, 1.2 kbps to 115.2 kb/s
ADAM-60× Independently Modules Ethernet 10/100	ADAM-6015	I/O module, 7× AI diff. (RTD), input type Pt, Ni RTD, iz.2 kV, input impedance 10 kΩ, Ethernet10/100, power supply 10 V to 30 V
	ADAM-6017	I/O mod, 8× AI diff. (mV, V, mA), isolation 2 kV, input impedance 20 MΩ, digital outputs OC, Ethernet10/100
	ADAM-6018	8 channel Thermocouple Isolation amplifier J, K, T, E, R, S, B, 10 S/s, Ethernet 10/100, Power Supply 10 - 30 V
	ADAM-6022	I/O module, 2× PID controller 2× (3 ×AI, 1×AO, 1×DI, 1×DO), iz.2 kV, input. Imp. 20 MΩ, Ethernet10/100, 10 V to 30 V supply
	ADAM-6024	I/O isolated 6 IN diff. Channels modul, 10 VDC, 0 ~ 20 mA, 4 ~ 20 mA, Real Time Clock, Modbus TCP Ethernet 10/100
	ADAM-6050	I/O module, 12× DI, 6× DO, Ethernet 10/100, 2 kV isolation, digital outputs OC, Ethernet10/100
	ADAM-6051	I/O module, 12× DI, 6× DO, isolation 2 kV, digital outputs OC, Ethernet 10/100
	ADAM-6052	I/O module 8× DI, 8× DO, 2 kV, digital outputs 35 V/1A, digital inputs Ethernet 10/100
	ADAM-6060	I/O module 6× relay (switching), 6× DI, 2 kV isolation, digital inputs Ethernet 10/100
	ADAM-6066	I/O module 6× relay (switching), AC: 240 V, 5 A, DC: 30 V, 5 A, 100 Hz, Ethernet 10/100

Group	Designation	Module parameters
ADAM-61× Independently Modules Ethernet, ProfiNet	ADAM-6117EI	I/O module, 8× differential AI, 8× diff. channels, 16 bit, 10 S/s, ±150 mV to ±10 V; ±20 mA 0 to 20 mA 4 to 20 mA, Ethernet 10/100
	ADAM-6117P	I/O module, 8× differential AI, 8× diff. channels, 16 bit, 10 S/s, ±150 mV to ±10 V; ±20mA 0~20mA 4 to 20mA, Ethernet 10/100
	ADAM-6150P	I/O module, 8× isolated DI, 7× isolated DO, Ethernet 10/100
	ADAM-6150N	I/O module, 8× differential AI, 8× diff. channels, 16 bit, 10 S/s, ±150 mV to ±10 V; ±20 mA 0 to 20 mA 4 to 20 mA, Ethernet 10/100
ADAM-62× Independently Modules Ethernet 10/100	ADAM-6217	I/O module, 8 differential inputs, ±150 mV to ±10 V; ±20 mA, Ethernet 10/100
	ADAM-6224	I/O module, 4 analog inputs, 4 digital outputs, Ethernet 10/100
	ADAM-6226	I/O module, 4× relays, 4× DI output, Ethernet 10/100
	ADAM-6250	I/O module, 16 iz. of channels, counter/timer, 3 kHz, Ethernet 10/100
	ADAM-6251	I module, 16 iz. of channels, counter/timer, 3 kHz, Ethernet 10/100
	ADAM-6256	O module, 16 iz. of channels, counter/timer, 3 kHz, Ethernet 10/100
	ADAM-6260	6× relays, 250 V/0.5 A AC, 30 V/5 A DC, Ethernet 10/100
ADAM-63× Modules Ethernet 10/100	ADAM-6350	I/O modul, 18× DI, 18× DO, OPC UA, Ethernet 10/100
	ADAM-6360	I/O modul, 8× relay, 14× DI, 6× DO, Ethernet 10/100
	ADAM-6366	I/O modul, 6× relay, 18× DI, 6× DO, Ethernet 10/100
	ADAM-6371	I/O modul, 8× AI, 11× DI, 10× DO, OPC UA Ethernet 10/100
ADAM-6520	ADAM-6520	Ethernet switch, half, full duplex, 3 kV protection, Ethernet 10/100
ADAM-67× Modules Ethernet	ADAM-6717	Compact intelligent EDGE gateway, Ethernet 10/100
	ADAM-6750	Compact multifunctional EDGE gateway, Ethernet 10/100
	ADAM-6760	Compact I/O gateway with 8DI/8SSR Relay, Ethernet 10/100

WISE-40××	WISE-4010	I/O module, 4 IN analog. 20 mA, 4 mA to 20 mA, 2 DO, Ethernet 100
Independently	WISE-4012	I/O modul, 4 IN analog., 2 DO, WiFi 2,4 GHz, Ethernet 10/1000
Modules LAN	WISE-4021	I/O modul WiFi, 4× AI, 2× DO, Ethernet 10/1000
	WISE-4050	I/O modul, 4× DI + 4× DO WiFi 2,4 GHz, Ethernet 10/1000
	WISE-4051	I/O modul, 8× digitál input, RS-485 WiFi 2,4 GHz,
	WISE-4060	I/O Module, 4× DI + 4× relay, Ethernet 10/1000
WISE-42××	WISE-4210	LPWAN Wireless access to Ethernet 10/1000
Wireless modules	WISE-4220	Wireless I/O module with temperature and humidity measurement, 2.4G WiFi
WISE-46×	WISE-4610	Wireless I/O module, LoRa, LoraWAN, accumulator
Radio modules	WISE-4614	4 diff. channel. 16 bits. A/No, 1 S/s/channel, ±150 mV to ±10 V, 20 mA, ±20 mA, 4 mA to 20 mA, 4 no. output channels OC, 200 kHz
	WISE-4615	4 diff. channel. RTD amplifier, Pt100, Pt1000, 15 bit A/No, 1 S/s
	WISE-4617	2 channel 16 bit A/C module, 1 S/s, ±1 V to ±10 V, 20 mA,
	WISE-4671	Radio module, 1.2 kbs to 115.2 kbs, 3GPP standard, GPS, Glonass, Galileo, 15 s repeat, 2.5 m accuracy, 31 s response
	WISE-4672	Radio module, 6 channel. digital output OC, 1.2 kbps to 115.2 kb/s

Index

W

Y

Z

About the Author

Josef Vedral is a graduate of the Faculty of Electrical Engineering CTU in Prague. From 1971 to 1974 he worked at the Research Institute of Aviation Technology on the development of measuring systems for data collection. In 1976, he defended his doctoral thesis and in 1990 he obtained his habilitation in the field of measuring technology at the CTU in Prague. Between 1975 and 1980, he participated in the development of liquid and gas chromatography measuring systems at the company Laboratory Instruments. Between 1981 and 1990 he participated in the development of CAMAC modular measuring systems at the Research Institute of Nuclear Technology Instruments. In the following years, he participated in the research of methods of measuring the dynamic properties of digitizers in the time and frequency domain.

Between 2012 and 2020 he led two projects of the Technology Agency of the Czech Republic in the field of non-disassembly diagnostics of HV electrical machines.

He is the author of 82 scientific articles in foreign journals, 12 CZ patents, 12 university textbooks in CZ and 3 textbooks in EN.